# SIX SIGMA FOR TRANSACTIONS AND SERVICE

# SIX SIGMA FOR TRANSACTIONS AND SERVICE

**PARVEEN S. GOEL**

**PRAVEEN GUPTA**

**RAJEEV JAIN**

**RAJESH K. TYAGI**

**McGraw-Hill**
New York   Chicago   San Francisco   Lisbon
London   Madrid   Mexico City   Milan   New Delhi
San Juan   Seoul   Singapore   Sydney   Toronto

The *McGraw·Hill* Companies

Library of Congress Cataloging-in-Publication Data

Transactional Six Sigma for service / Parveen Goel ... [et al.].
    p. cm.
    ISBN 0-07-144330-4
    1. Process control.   2. Six sigma (Quality control starndard.   I. Goel,
Parveen.

    TS156.8.T73   2004
    658.4'013—dc22

                                                            2004058796

1 2 3 4 5 6 7 8 9 0   DOC/DOC   0 10 9 8 7 6 5 4

ISBN 0-07-144330-4

*The sponsoring editor for this book was Kenneth McCombs and the production supervisor was Sherri Souffrance. It was set in Fairfield Medium by Patricia Wallenburg.*

*Printed and bound by RR Donnelley.*

  This book is printed on recycled, acid-free paper containing a minimum of 50 percent recycled, de-inked fiber.

# CONTENTS

# FOREWORD

Six Sigma has become a proven methodology across industries, be it manufacturing, pharmaceutical, plastics, software, or service. During the last 5 to 7 years, companies like GE, Honeywell, Caterpillar, Bank of America, and Citibank have announced their gains with implementation of Six Sigma in respective corporations. Savings of the order of a billion dollars by some of these corporations have been reported. Based on the amount of published material, it appears that that Six Sigma has become a matured methodology in the manufacturing area, however, there is a lack of literature regarding implementation of Six Sigma in the service sector. I believe *Six Sigma for Transactions and Service* will serve as a handy reference for service professionals. Most Six Sigma books are aimed at a manufacturing audience, and this book does a nice job in providing a roadmap for services desiring to apply the same principles.

*Six Sigma for Transactions and Service* fills the gap by providing readers a comprehensive information about Six Sigma in the service sector. I am impressed with the multidisciplinary experience of the authors of the book, and the organization of the information. The authors have adapted the existing framework of Six Sigma to the service environment. Readers would appreciate this customization that lacks in the Six Sigma domain. Experts try to use Six Sigma as a 'silver bullet.' Contrary to the convention, this one book contains both the improvement (DMAIE) as well as design (DMADO) aspects, which are adaptation of the well known DMAIC and DMADV aspects of Six Sigma.

I have worked with Rajesh Tyagi and Praveen Gupta at the college, and recognize their experience in the Service and Six Sigma areas. This collaboration has resulted in a very timely book. I am sure readers worldwide will find this book a good read and valuable in their Six Sigma endeavors, as I have.

Scott Young
Chair, Department of Management
College of Commerce, DePaul Univesity

# PREFACE

After successfully implementing Six Sigma in manufacturing, Motorola applied Six Sigma in support functions and realized significant savings. Banks, insurance, hospitals, schools, and many service organizations have implemented Six Sigma successfully. Interestingly, many books lack adaptation of Six Sigma to the service (transaction or non-transactional) environment. However, questions arise about how to go about adapting and implementing Six Sigma in service organizations. Clearly, there is a need to provide integrative perspective with solid academic background combined with practical applications and case studies. Various new tools and techniques need to be applied to the transaction and service world.

*Six Sigma for Transactions and Service* offers an evolving set of principles and methods. This book presents a strategic, operational, and design perspective of Six Sigma for the service environment. Having co-authors and contributors with engineering, business, academic, quality, customer service, and consulting backgrounds brings out a diversity of ideas, and leads to an innovative application of Six Sigma in service organizations.

The book is organized in three parts. Part One focuses on the role of services in the economy, service benchmarks, the service component in a corporation's value chain, introduction to quality in transaction and services, and recent trends and challenges faced by service sector.

Part Two expounds Six Sigma methodology fo transaction and services. An adaptation of DMAIC (define, measure, analyze, improve and control) to DMAIE (define, measure, analyze, innovate, embed), which is more suitable to the service environment. Here the focus is to emphasize the creative and cultural aspects of service environment. Several tools including service blueprinting, regression analysis, mind mapping, data envelopment analysis, and TRIZ (Russian acronym for the theory of solving inventive problems) have been presented besides standard DMAIC tools. A set of templates is provided to ease the practice of DMAIE in capitalizing opportunities in the service operations.

Part Three makes this book unique in the sense that the authors have developed two basic axioms of service operations. These axioms are used to comprehend service operations and utilize necessary Six Sigma tools to design service processes with improved performance. Also, Design for Six Sigma is expounded in this section to make this handbook a reference book for service applications. The conventional Design for Six Sigma methodology DMADV (define, measure, analyze, design, and validate) has been retooled as DMADO (design, measure, analyze, design, optimize) to ensure effective implementation in the service environment. Several tools from Pugh's concept selection, quality function deployment (voice of customers), and robust design have been expanded for practical implementation.

Part Three also covers the implementation of Six Sigma methodology in service firms. A road map is presented to assist in the implementation of Six Sigma. Chapter 14 looks into sample service functions for Six Sigma implementation. In addition, a chapter has been devoted to human capital, a critical component of service functions. Finally, today's hot topic, outsourcing, is investigated and looked into for application of Six Sigma.

The authors believe this book will present Six Sigma for the service environment in an innovative way to enhance its value proposition, and guide corporations in practicing Six Sigma in an effective way at the corporate, project, and design of service level.

# ACKNOWLEDGMENTS

Six Sigma has been in practice for many years now. There are more than 200 books that have been published on this topic. To write a Six Sigma book for the service arena meant adapting current practices to the service industry in an innovative and effective way. Besides, creating an integrated book that helps understand the peculiarities of the service environment an includes Six Sigma concepts, as well as Design for Six Sigma methods, one must present Six Sigma in a way that will allow the reader to go to one source for all information. Such a book could only be developed through collaboration. Our team that collaborated on *Six Sigma for Transactions and Service*, consists of individuals who have enormous experience in the area of design, business, performance improvement, and service. We all are privileged to come together for this project and be able to complete the book. Working on a project with experienced team members is a learning experience and fun. It has been a very enjoyable and intellectually rewarding experience to explore new frontiers in the service sector. We could discuss topics through teleconferences, late night meetings at home, or through emails. Having four co-authors work together, means many families suffered the fate of an author's family. We would like to acknowledge their contributions as follows:

**Rajesh Tyagi**
I am grateful for the love and encouragement my parents, brothers, and sister have always provided during every endeavor in my life. And, most important, I would like to thank my wife, Anjali, who makes my life wonderful and whom I love with all of my heart.

**Rajeev Jain**
Thanks are due to family, friends, publisher, and editors. Without the support of the family, completing such an undertaking would have been impossible. I would like to thank my wife Rashmi, and daughters Divya and Namrita, who allowed

me to devote many long hours to this project and sacrificed their right on my time. I am also grateful to my parents for the values they embedded in me. A special credit is due to Praveen who initiated the project and gave some of our conceptual discussions a shape.

**Praveen Gupta**
I would like to thank my family (Archana, Avanti, and Krishna) and friends for allowing me to continue my journey of experiencing life with them every where. I am sure sometimes I have talked too much in discussing my ideas, and not listened well when I should have. Well, I thank all of you for your patience, advice, and support. I also would like to acknowledge the privilege of working with Rajesh, Rajeev, and Parveen. It has been a wonderful experience.

**Parveen Goel**
I want to acknowledge my spiritual guide *Jagadguru Shri Kripalu Ji Maharaj* for his guidance and inspiration in my life. My extended family that includes my wife Rajni, son Dhruv, parents, sister, brother-in-law, and nieces Kanika and Surbhi, provided much needed support and encouragement for this project. My friends B.K. Mishra and Pradeep Kumar helped me by discussing some of the ideas and concepts included in the book.

We are especially indebted to Manu Vora for Human Capital, Arvin Srivastava for Six Sigma in Services, and Om Yadav for Customer Driven Transactional Processes chapters.

Finally, we would like to acknowledge Ken McCombs, Senior Acquisition Editor, who worked closely with us and guided us to the completion of this project. Without his continual support, this book could not become a reality. Our sincere appreciation is also extended to Ms. Patricia Wallenburg for providing excellent support in editing this book.

Rajesh Tyagi
Rajeev Jain
Praveen Gupta
Parveen Goel

# PART ONE

# SERVICE QUALITY BENCHMARKS

# ROLES OF SERVICES AND TRANSACTIONS IN GLOBAL AND U.S. ECONOMY

## SERVICES—THE KEY TO FUTURE COMPETITIVE ADVANTAGE

Today, there is significant concern among the working population in the United States regarding shifting of jobs from the United States to other economies of the world. We have seen manufacturing jobs go in the past, and now we are seeing service jobs move away. Is this something we should all be concerned about, or is this part of an evolutionary process? Why are these other economies able to offer us equivalent or better quality at lower cost? How can we minimize the loss of such jobs? What role can Six Sigma play in the process and what impact could it have? An understanding of the global and U.S. economic process would help in making the right decisions.

## THE U.S. ECONOMY

The twentieth century has been the most dynamic in terms of the dramatic shifts in focus we experienced globally. We saw

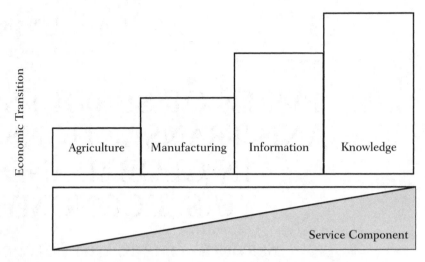

**FIGURE 1.1.** Changing World Economy and the Service Component

the world economies transition from an agricultural focus, to manufacturing, and then to information and knowledge (see Figure 1.1). Although the degree of this change has varied from one economy to another, no economy in the world has remained untouched by this change.

It would be wrong to say that we are moving from manufacturing to service. Services have been a component in all economics—it is only a matter of how significant a component. As we move to the information economy and knowledge-based economics, services become the dominant component of the value offered to a customer.

## U.S. ECONOMY CLASSIFIED BY INDUSTRY

Let us look at the composition of the U.S. economy. The U.S. economy is divided into 10 industries. Besides agriculture, the North American Industry Classification System (NAICS) categorizes the rest of the economy into nine industries as follows:

· Construction
· Finance, insurance, and real estate
· Government

- Manufacturing
- Mining
- Retail trade
- Services
- Transportation and public utilities
- Wholesale trade

In terms of employment, services is the largest of the industry divisions, and mining is the smallest. (NAICS is the newer version of the SIC classification system with which we may all be familiar. As a result of the North American Free Trade Agreement [NAFTA], this new classification system was developed to address additional needs to harmonize industry classifications with our other partners.)

In the preceding group of nine, we can consider government as a noncommercial enterprise (although it always helps to think of it in commercial terms). Of the remaining eight, manufacturing and mining are dominantly nonservice oriented, while the other six have very significant service components. As a minimum, six of ten industry classifications are service dominated. We will analyze this further to understand their contribution to the U.S. economy.

## INDUSTRY PERFORMANCE

In 2002, the total U.S. gross domestic product (GDP) was $10.4 trillion in current dollars ($9.4 trillion in 1996 dollars). The relative importance of various industry groups and their performance with respect to their real growth rates and their contributions to real GDP growth is presented in Figures 1.2 and 1.3.

In terms of GDP, agriculture represented a minimal 1.4% of GDP in 2002, and the government accounted for another 12.9% of the GDP. Hence, the rest of the private industries sectors comprise 85.7% of the GDP. The relative contribution to the U.S. economy by each of the nonagricultural industry sectors is shown in Figure 1.2.

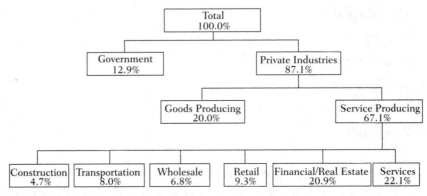

**FIGURE 1.2.** Composition of the Industries by Industry Groups

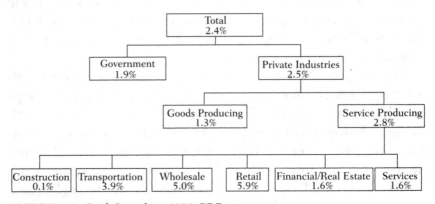

**FIGURE 1.3.** Real Growth in 2002 GDP

## SERVICE SECTORS

The dominance of the service sector is impressive. The two largest categories in the service sector, which represented 43% of our GDP in 2002, consist of:

· Finance, insurance, and real estate
· Services, the catchall category covering all other private services

They contributed over $1.5 trillion to GDP in 2002. Each of them alone exceeds the $1.4 trillion in output produced by the manufacturing sector of the economy.

The next major categories (in terms of size) are:

· Retail trade
· Transportation and public utilities
· Wholesale trade

Each contributes over a half trillion dollars to annual U.S. output.

Even construction (viewed as a service in the context of international trade), which is the smallest category, has higher economic output than either agriculture or mining.

Considering that two-thirds of our GDP is provided by service industries and that the rest of the industries also have service components, the importance of services in our economy is overwhelming.

## REAL GROWTH RATES AND CONTRIBUTIONS

In addition to the absolute contribution to the GDP, let us also review what is fueling the growth in the U.S. economy. The economy, which grew at an average of 4% annually between 1995 and 2000, slowed down to 0.3% in 2001. In 2002, growth in real GDP increased by 2.4%. Private industries, which represent five-sixth of the GDP in total, increased 2.5%, and government increased 1.9% in 2002. The economic growth was led by the private services sector, which increased 2.8%—more than twice the rate for the private goods-producing sector, which grew by 1.3%.

Real GDP in all five major industry groups in the service-producing sector increased (see Figure 1.3). Retail sector experienced the maximum growth with a 5.9% increase. This was followed by electric, gas, and sanitary services (5.6%), a component of the transportation and public utilities group, and wholesale trade (5.0%). The finance, insurance, and real estate (FIRE) sector and the services sector, both grew a modest 1.6%. The growth in services was restrained by weakness in the business and professional services industries, which includes consulting services and software development.

## REAL GROWTH IN 2002 GDP

Regarding GDP growth, the service industries grew by 2.8% in 2002 while the goods-producing industries grew at less than half that rate (1.3%) in 2002. The services sector has been growing at a higher rate in the past as well, although the differential was not as large.

As the growth rates in the service sectors continue to exceed those in the nonservice industries, services will become an even larger contributor to the U.S. economy and will continue to grow in importance.

## ROLE OF SERVICES

In the United States, the service sector has been the fastest growing for the last half century—and there is no stopping it. It is also the biggest sector. As per the Office of Service Industries, 50 years ago, the service sector accounted for about 60% of U.S. output and employment. Today, the service sector's (including government) share of the U.S. economy has risen to roughly 80%.

Even at the corporate level, looking at companies like IBM which were primarily manufacturing organizations, today almost half of their revenue is contributed by their IBM Global Services division, which focuses on providing services (as per the 2003 financial statements included in the 10K filed with the SEC, Global Services revenue was $43 billion out of a total revenue of $89 billion). Looking at profitability, a third of the gross profits ($11 billion) were contributed by this division, next to the software sales division, suggesting a high value addition by the services organization.

As per Robert Litan, Director, Economic Series, Brookings Institution, "Services trade is the unsung hero of our trade balance. World-wide, the U.S. trade deficit is set to surpass $450 billion. But while the gap in merchandise trade keeps growing, some economists see indications that the trend won't last the decade. The reason: the growing U.S. trade in services, which now tops $250 billion annually and runs a surplus of $80 billion. Both numbers, say economists, are poised to explode over the next 10 years."

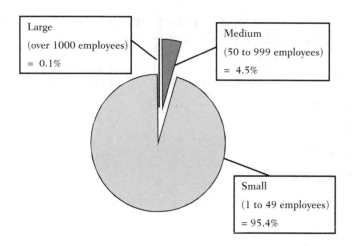

Total number of businesses: 6.7 million

**FIGURE 1.4.** Distribution of Business in the U.S Based on Number of Employees

## IMPORTANCE OF THE SERVICE SECTOR

In addition to its sheer size, the service sector is a critical component of everyday American economy. It is what keeps the entire economy integrated. Even the nonservice sector requires services to be useful to the rest of the economy. A unique characteristic of the service economy is the size of the firms. As shown in Figure 1.4, the U.S. economy is dominated by small firms, which represent over 95% of the businesses.

**Small firms.**    Today, nine out of every ten firms with less than 20 employees are in the services sector, and these small service companies account for 90% of jobs at small firms. On average, in each of the 50 states, about 350,000 jobs are created through the small firms in the services sector compared to about 27,000 jobs in the manufacturing sector. As a result, the small service firms are a significant component of the service economy.

**Large firms.**    A change in the Dow Jones Index of blue-chip stocks is one way to assess prominence of the service industries among large firms. In 1982, American Express became the first services firm to be added to the Dow. Today, eight of the thirty firms that now make up the Dow Jones Index are in the service sector. In addition to American Express, we have:

· AT&T
· Citigroup
· Disney
· McDonald's
· JP Morgan
· Sears
· Wal-Mart

In addition, many of the traditional manufacturing firms, such as GE and IBM, have transformed themselves to include a significant service component in their operations. Today, they compete globally on service operations. Further, by design, none of the firms listed among the Dow industrials is a transportation or a utilities concern. Each of these key service categories has its own market index.

**Investment.**    The global deployment of the Internet and rapidly emerging new information technologies are a key driving force enhancing the service economy. A recent survey by Gartner, identified the "changing business model" (caused by emerging technologies) as the leading cause of the investment in technology. If one ranked the U.S. industries according to the ratio of investment in information technology relative to their revenue, communication industry leads the group, followed by the various service industries (see Table 1.1). It is interesting to observe that all the top 10 rankings go to the service industries.

Therefore, the service economy is playing a significant role in attracting investment that is a key stimulus for economic growth. Why do we find this growth in the service economy? A key factor is that it is less of a commodity than the manufacturing sector. As manufacturing technology has become more widely available, that know-how no longer represents a competitive advantage. As a result, firms are adding services to complement products as a differentiator and competitive

**TABLE 1.1.** Service Industries

| RANK | INDUSTRY | IT SPEND AS % OF REVENUE | SERVICE |
|---|---|---|---|
| 1 | Communication | 13.19 | ✔ |
| 2 | Distribution—All—Wholesale and Retail | 4.06 | ✔ |
| 3 | Financial Services | 3.94 | ✔ |
| 4 | Public Administration | 3.03 | ✔ |
| 5 | Information Technology | 2.91 | ✔ |
| 6 | Services | 2.30 | ✔ |
| 7 | Transportation | 1.80 | ✔ |
| 8 | Health Services | 1.63 | ✔ |
| 9 | Education | 1.00 | ✔ |
| 10 | Utilities | 0.95 | ✔ |
| 11 | Manufacturing | 0.72 | |
| 12 | Agricultural Products and Services | 0.50 | |
| 13 | Petroleum | 0.47 | |
| 14 | Mining | 0.23 | |

advantage. Information is the fuel that keeps the services economy running. Service industries are also driving the investment in this economy.

**Industry employment.** As a result of this phenomenal growth in the services sector, today, 85% of all nonfarm workers are employed in the service sector. In the last two decades, almost 40 million more employees have joined the service sector. In addition, these new service sector jobs account for most of the net increase in employment since the 1970s.

As per the Department of Labor, over the 2000–2010 period, total employment is projected to increase by 15%, which is only slightly less than the 17% growth during the previous decade, 1990–2000 (see Figure 1.5 and Table 1.2). It is extremely revealing that the service industries will experience employment growth more than *three times* that of the nonservice sectors.

TABLE 1.2. Employment by Major Industry Division, 1990, 2000, and Projected 2010

| Industry | Thousands of Jobs | | | Change | | Percent Distribution | | | Average Annual Rate of Change | |
|---|---|---|---|---|---|---|---|---|---|---|
| | 1990 | 2000 | 2010 | 1990–00 | 2000–10 | 1990 | 2000 | 2010 | 1990–00 | 2000–10 |
| **Total (1)** | **124,324** | **145,594** | **167,754** | **21,269** | **22,160** | **98.2** | **98.8** | **99.0** | **1.6** | **1.4** |
| Nonfarm wage and salary | 108,760 | 130,639 | 152,447 | 21,879 | 21,807 | 87.5 | 89.7 | 90.9 | 1.8 | 1.6 |
| **Goods producing** | **24,906** | **25,709** | **27,057** | **803** | **1,347** | **20.0** | **17.7** | **16.1** | **0.3** | **0.5** |
| Mining | 709 | 543 | 488 | –167 | –55 | 0.6 | 0.4 | 0.3 | –2.6 | –1.1 |
| Construction | 5,120 | 6,698 | 7,522 | 1,578 | 825 | 4.1 | 4.6 | 4.5 | 2.7 | 1.2 |
| Manufacturing | 19,077 | 18,469 | 19,047 | –607 | 577 | 15.3 | 12.7 | 11.4 | –0.3 | 0.3 |
| Durable | 11,109 | 11,138 | 11,780 | 29 | 642 | 8.9 | 7.7 | 7.0 | 0.0 | 0.6 |
| Nondurable | 7,968 | 7,331 | 7,267 | –637 | –64 | 6.4 | 5.0 | 4.3 | –0.8 | –0.1 |
| **Service producing** | **83,854** | **104,930** | **125,390** | **21,076** | **20,461** | **67.4** | **72.1** | **74.7** | **2.3** | **1.8** |
| Transportation, communications, utilities | 5,776 | 7,019 | 8,274 | 1,243 | 1,255 | 4.6 | 4.8 | 4.9 | 2.0 | 1.7 |
| Wholesale trade | 6,173 | 7,024 | 7,800 | 851 | 776 | 5.0 | 4.8 | 4.6 | 1.3 | 1.1 |
| Retail trade | 19,601 | 23,307 | 26,400 | 3,706 | 3,093 | 15.8 | 16.0 | 15.7 | 1.7 | 1.3 |
| Finance, insurance, and real estate | 6,709 | 7,560 | 8,247 | 851 | 687 | 5.4 | 5.2 | 4.9 | 1.2 | 0.9 |
| Services | 27,291 | 39,340 | 52,233 | 12,049 | 12,893 | 22.0 | 27.0 | 31.1 | 3.7 | 2.9 |
| Government | 18,304 | 20,680 | 22,436 | 2,376 | 1,757 | 14.7 | 14.2 | 13.4 | 1.2 | 0.8 |
| Federal government | 3,085 | 2,777 | 2,622 | –308 | –154 | 2.5 | 1.9 | 1.6 | –1.0 | –0.6 |

TABLE 1.2.  Employment by Major Industry Division, 1990, 2000, and Projected 2010 (Continued)

| INDUSTRY | THOUSANDS OF JOBS | | | CHANGE | | PERCENT DISTRIBUTION | | | AVERAGE ANNUAL RATE OF CHANGE | |
|---|---|---|---|---|---|---|---|---|---|---|
| State and local government | 15,219 | 17,903 | 19,814 | 2,684 | 1,911 | 12.2 | 12.3 | 11.8 | 1.6 | 1.0 |
| Agriculture (2) | 3,340 | 3,526 | 3,849 | 186 | 323 | 2.7 | 2.4 | 2.3 | 0.5 | 0.9 |
| Private household wage and salary | 1,014 | 890 | 664 | −124 | −226 | 0.8 | 0.6 | 0.4 | −1.3 | −2.9 |
| Nonagriculture self-employed and unpaid family workers (3) | 8,921 | 8,731 | 9,062 | −190 | 331 | 7.2 | 6.0 | 5.4 | −0.2 | 0.4 |
| Secondary wage and salary jobs in agriculture (except agricultural services); forestry, fishing, and trapping; and private households (4) | 205 | 155 | 150 | −50 | −5 | 0.2 | 0.1 | 0.1 | −2.8 | −0.3 |
| Secondary jobs as a self-employed or unpaid family worker (5) | 2,084 | 1,652 | 1,582 | −432 | −70 | 1.7 | 1.1 | 0.9 | −2.3 | −0.4 |

Source: U.S. Department of Labor

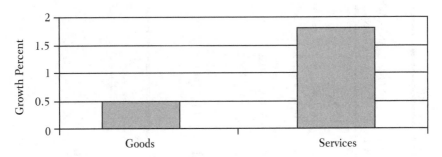

**FIGURE 1.5.** Employment Growth (2000–2010)

Over this decade, the service sector will continue to be the primary generator of employment in the economy. It is expected that 20.5 million new jobs will be created in the service sector by 2010. In comparison, the goods-producing sector will contribute relatively modest employment gains, primarily in construction and durable manufacturing industries.

The 20 million plus jobs represent a 19% increase in employment in the service sector, while the manufacturing sector employment is expected to increase by barely 3% over the 2000–2010 period. This will result in manufacturing returning to its 1990 employment level of 19.1 million. However, its share of total jobs will decline from 13% in 2000 to 11% in 2010.

More than half the employment growth opportunities will be in the health services. These sectors account for a significant share of the fastest-growing industries.

Before looking at the employment trends in each industry, the summary in Table 1.3 shows the concentration of the business establishments, and employment by industry.

## GROWTH OF SERVICE SECTOR

The above trends should be convincing enough to highlight the importance of the service sector. It is the dominant sector, as well as the provider of future growth. Let us also track the growth of the Dow Jones Industrial Average over the last decade. The growth of the service sector has been the most prominent in the last decade. Table 1.4 measures how long it took the Dow Jones to climb each successive 1,000-point milestone.

**TABLE 1.3.** Business and Employment Concentration by Industry

| INDUSTRY | # OF ESTABLISHMENTS (%) | EMPLOYMENT (%) | # OF EMPLOYEES (IN MILLIONS) |
|---|---|---|---|
| Construction | 10 | 5 | 6.6 |
| FIRE | 9 | 6 | 7.8 |
| Government | 3 | 16 | 21.3 |
| Retail | 20 | 18 | 23.3 |
| Services | 39 | 30 | 41.2 |
| Transportation | 4 | 5 | 6.8 |
| Wholesale | 9 | 5 | 6.7 |
| **Total—Service Industries** | **94** | **85** | |
| Manufacturing | 5 | 14 | 16.7 |
| Mining | 1 | 1 | 0.5 |
| **Total—All Industries** | **100** | **100** | |

**TABLE 1.4.** Dow Jones Growth Performance

| POINTS | DATE | TIME |
|---|---|---|
| 1,000 | November 14, 1972 | 76 years |
| 2,000 | January 8, 1987 | 14 years |
| 3,000 | April 17, 1991 | 4 years |
| 4,000 | February 23, 1995 | 4 years |
| 5,000 | November 21, 1995 | 9 months |
| 6,000 | October 14, 1996 | 11 months |
| 7,000 | February 13, 1997 | 4 months |
| 8,000 | July 16, 1997 | 5 months |
| 9,000 | April 6, 1998 | 9 months |
| 10,000 | March 29, 1999 | 12 months |
| 11,000 | May 3, 1999 | 1 month |

The phenomenal growth from 1995 through May 1999 causes us to disbelieve in the Dow. The Dow grew from 4,000 points to 11,000 points in these five years—more than it did in the whole century before that! Hence, it is the service sector that has fueled the growth in the last decade.

There is a significant concern today that service sector jobs are now moving out of the United States and being transferred offshore to nations where labor costs are lower. Will this bring a downturn in the service economy? Will this slow down the growth in the United States? The 15% annual growth predicted by the employment forecasts would definitely suggest that it will not slow down the economy. That is not to say that there may be a need for realignment of the type of services that are required—but services will definitely be required.

## EMPLOYMENT TRENDS

Let us look at the employment trends in each of these industry sectors in more detail, starting with the six industries in the service sector, the government, and then the two goods-producing sectors.

## SERVICE INDUSTRIES

**Construction.**   The construction industry ranks third of the nine industry divisions in number of establishments. Construction accounts for about 10% of establishments and about 5% of employment.

Looking at the employment data from 1991 to 2002, the annual average employment in the construction industry has increased from a bottom of about 4.5 million in 1992, to 6.7 million in 2001, and then declined slightly to 6.6 million in 2002. This would mirror employment trends in the economy as a whole, which increased every year until between 2001 and 2002.

The unemployment rate in construction reached 9.3% in 2002, compared to the overall unemployment rate of 5.8% across the United States (see Figure 1.6).

**Finance, insurance, and real estate.**   The finance, insurance, and real estate (FIRE) division is the median industry in terms of employment and fourth of nine in number of establishments. The division represents about 9% of all establishments and about 6% of all employment.

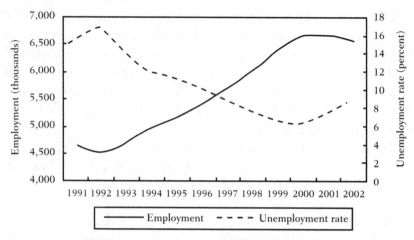

**FIGURE 1.6.** Employment and Unemployment in Construction

Again, the employment data for 1991 to 2002 show that the annual average employment in FIRE has almost grown each year from a low of 6.6 million in 1992 to 7.8 million in 2002. This represents an increase of 17.5% from the cyclical low and follows a pattern similar to the national economy.

The FIRE unemployment rate of 3.2% in 2002 was much lower than the overall unemployment rate of 5.8% (see Figure 1.7).

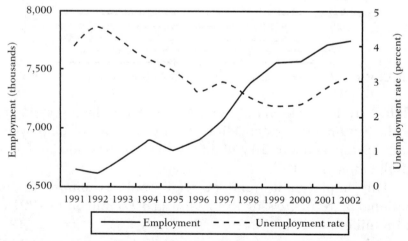

**FIGURE 1.7.** Employment and Unemployment in Finance, Insurance, and Real Estate

**Government.** The government sector is roughly the same size as manufacturing and fairly large in terms of number of employees. It represents about 16% of all employment and about 3% of all establishments. The government is comprised of federal, state, and local workers. The largest group of workers is in education and makes up about 45% of the total. Postal workers make up about 4% of the total.

The annual average employment in government has grown by 15.5% since 1991 to reach 21.3 million in 2001, primarily at the state and local level; federal employment has been trending downward since 1990.

The unemployment rate among government workers reached 2.5% in 2002, compared to the national unemployment rate of 5.8% (see Figure 1.8).

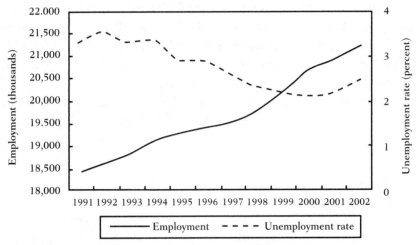

**FIGURE 1.8.** Employment and Unemployment in Government

**Retail trade.** The retail trade is the second-largest industry, both in number of establishments and number of employees, and represents about 20% of all establishments and about 18% of all employment.

The employment trend in retail trade is similar to the national pattern, growing from a low of 19.3 million in 1991 to 23.3 million in 2002.

The unemployment rate in retail trade has increased from a low of 5.5 percent in 2000, to 7.3% in 2002, com-

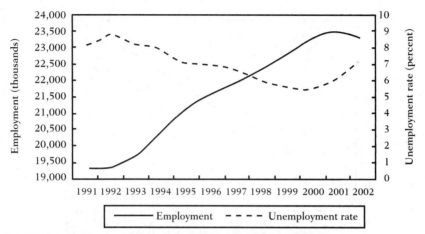

**FIGURE 1.9.** Employment and Unemployment in Retail Trade

pared to the overall unemployment rate of 5.8% (see Figure 1.9).

**Services.**    Services are the largest industry, both in number of establishments (about 39% of all establishments) and employment level (about 30% of all employment).

The average employment in services has grown every year to reach 41.2 million in 2002. This represents a 45.3% increase over 1991. Considering that the employment in the economy as a whole has increased at 22.1%, this represents more than double the growth rate.

The 2002 unemployment rate in services is 5.5% below the national level (see Figure 1.10).

**Transportation and public utilities.**    The transportation and public utilities industry includes establishments in transportation, communication, gas, electric, and sanitary services. It is still relatively small and represents about 4% of all establishments and about 5% of all employment covered by unemployment insurance.

The employment in transportation and public utilities has grown from about 5.7 million in 1992, by 19.3% to 6.8 million in 2002, a trend similar to the economy as a whole.

The 2002 unemployment rate in transportation and public utilities was similar to the services industry, at 5.5% (see Figure 1.11).

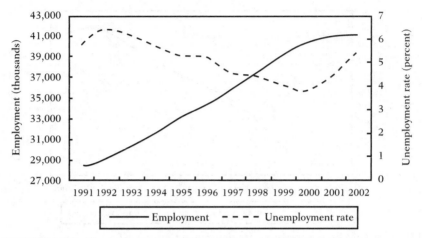

**FIGURE 1.10.** Unemployment and Unemployment in Services

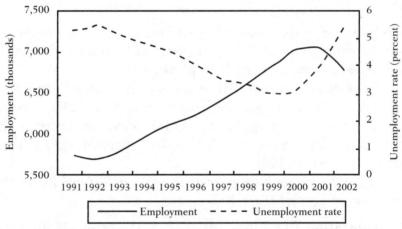

**FIGURE 1.11.** Employment and Unemployment in Transportation and Public Utilities

**Wholesale trade.**   The wholesale trade represents about 9% of all establishments and about 6% of all employment, with an annual average employment of 6.7 million in 2002.

The unemployment rate in wholesale trade in 2002 was 5.0% and also below the national level, falling to 2.8% in 2000. The overall unemployment rate in 2002 was 5.8% (see Figure 1.12).

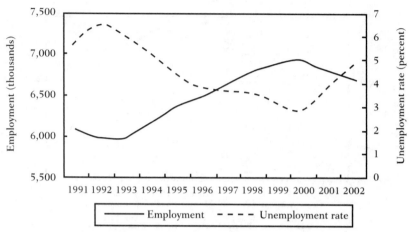

**FIGURE 1.12.** Employment and Unemployment in Wholesale Trade

## NONSERVICE INDUSTRIES

**Manufacturing.** The manufacturing industry is significantly larger than any of the other goods-producing industries in terms of employment. Factories account for approximately one-third of all goods-producing establishments. However, manufacturing represents about 75% of the employment in the non-service sector. In total, manufacturing represents about 5% of all establishments and about 14% of all employment.

The annual average employment in manufacturing has been just over 18 million in the last decade. However, it declined sharply in 2001 and 2002, and now stands at 16.7 million—a very different pattern from the national trend.

The unemployment rate in manufacturing rose to 6.7% in 2002, compared to the overall unemployment rate in 2002, which was 5.8%, although this is lower than the peak of 7.8% in 1992 (see Figure 1.13).

**Mining.** Mining is the smallest industry in number of employees and number of employers and represents less than 1.0% of the total establishments or workers.

There has been a general decline in mining employment between 1991 and 1999, with a slight increase between 1999 and 2002. This is represents a contrast to the national trend.

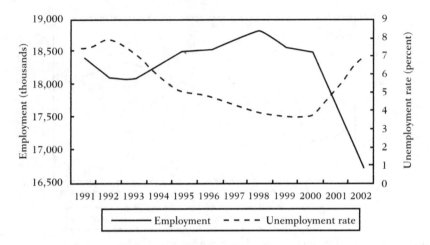

**FIGURE 1.13.** Employment and Unemployment in Manufacturing

The unemployment rate in mining for 2002 was 6.2%, slightly higher than the 5.8% for the economy as a whole (see Figure 1.14).

Looking at the employment trends and prospects for the future, the service industries clearly have an edge over the non-service sector.

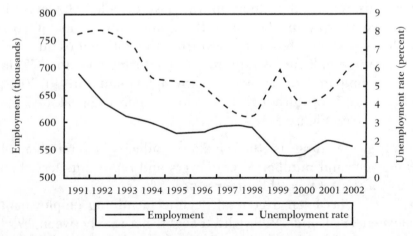

**FIGURE 1.14.** Employment and Unemployment in Mining

# TRENDS IN SERVICE INDUSTRIES

## Outsourcing

Given the phenomenal growth in service industries, a trend for "outsourcing" has been developing over the last decade even within the service industries. With the need for specialization, and to maintain focus on one's core competencies, "outsourcing" noncore activities appeals to many firms.

The concept of outsourcing is nothing new to the U.S. economy. Management gurus like Peter Drucker have advocated focus on core competencies of the firm for a long time. When Ross Perot formed EDS and embodied the concept of outsourcing the technology function to another specialist firm, U.S. corporations saw it as a means of transferring the responsibility for a complex operation, with spiraling costs and a "black box" image, to a company that was better equipped to provide a desired services level while controlling costs. Others saw this as a means of getting ready expertise where they themselves did not have a critical mass to hire and retain high-quality talent in the IT function.

Since that time, we have come a long way. Specialist firms have formed in a variety of areas. Hewitt Associates is a leading global firm providing outsourcing services for the human resources function—from benefits administration, to comprehensive workforce management. Today, you could outsource almost anything from mailing, printing, accounting, telemarketing, product design, or production. IBM is a great company that spawned a number of companies due to its decision to outsource. Microsoft, today's largest software company, found its roots when IBM decided that operating system software for personal computers was not its core competency, and it outsourced the requirement to Bill Gates. IBM also had a need to service its customers in remote locations—however, it had a policy that it would not sell products to its customers if it could not service them within 24 hours. FedEx saw a need in the market for providing fast and reliable delivery service, and today, most companies do not maintain a courier service in-house.

So, why the big concern with offshoring—which is the same as outsourcing, but across the border into another country? A *Business Week* article published early in 2003 predicted that over 3 million jobs will be lost from the United States to other countries around the world over the next five to ten years, which has caused an uproar in the political circles. There is a big movement to protect against such a drive. This emotional response comes because it primarily impacts you and me—the common person on the street. If jobs are lost in the U.S. economy, it is considered harmful to the U.S. economy.

The United States has been in the offshoring mode since the 1980s—subcontract manufacturing was the focus then. We had product components or parts manufactured in other parts of the world, and then assembled and packaged here. The '90s extended this trend to the manufacturing of complete products—China has been the primary beneficiary. In addition, IT related services—from software development to maintenance and implementation support—have also benefited from the offshoring trend. Since the turn of the century, we are starting to outsource administrative and transaction-based business processes related to human resources, finance, and call centers. By the end of this decade, we should see the United States transferring more knowledge-based services relating to design and analysis, utilizing more highly qualified professionals from around the world.

We outsourced a large number of manufacturing jobs over the last decade and reaped the benefit of being one of the most buoyant economies during that period (as evident from the growth of Dow Jones Industrial Index). That too was a play on moving work to lower cost areas around the world, whether the work moved to Mexico, Canada, or China. The U.S. economy has prospered. We have competed through higher productivity and also kept inflation under control.

Unless the countries around the world wish to move toward protectionism, there is no option for us but to participate in this process. The United States was late in recognizing the threat from Japan in the consumer electronics industry and in the small-car market segment. By not participating in the movement and by isolating ourselves, we lost all the jobs as well as the market to Japanese products. There is pioneering research by Michael Porter in his book, *Competitive Advantage of Nations*, which describes how different nations have developed different core competencies. That allows these nations to become world leaders in those product lines, as well as makes it very difficult for other nations to compete with them effectively if trade barriers are removed. As the United States propagates the removal of trade barriers through initiatives such as NAFTA and The World Trade Organization, we need to be prepared for their impact on our economy as well. It will change the dynamics in various industries. There will be some losses, but much more to gain as well.

## GAINS FOR THE U.S. ECONOMY

The United States has much to gain from offshoring. For every dollar of U.S. spending sent offshore (work performed on behalf of the United States), the U.S. economy derives a net benefit between $1.20 and $1.35 (see Figure 1.15).

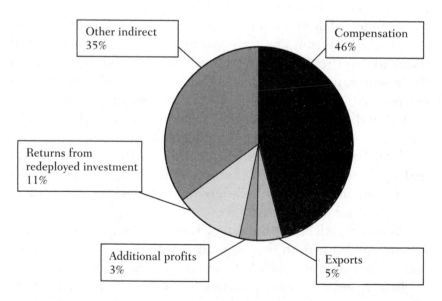

Other indirect
35%

Compensation
46%

Returns from
redeployed investment
11%

Additional profits
3%

Exports
5%

**FIGURE 1.15.** Relative Benefits to the U.S. of Sending Work Offshore ($1.20 to $1.35 per $1 or Spending Outsourced to a Foreign Country

The U.S. economy gains in various ways. The largest benefit to the U.S. corporations is the reduction in expense, largely compensation. This increases their profitability and makes them more competitive in the marketplace. Many countries provide other benefits as well that help reduce costs—from free real estate, to subsidized utilities, to tax holidays. These countries provide many incentives to get foreign companies to invest in their country.

Transfer of work offshore creates a demand for U.S. goods and services. The offshore companies that provide services to the United States, need goods and services from the United States, such as technology, computer equipment, software, telecommunication services, etc. This creates opportunities for U.S. corporations to sell to these other countries. For example, U.S. exports to India have increased by more than 60% since 1990.

Many times, the work performed offshore is by an operation owned by a U.S. conglomerate. Most large corporations in the United States have their own captive operations in other parts of the world. As a result, profits generated by the offshore company also directly benefit the U.S. corporations.

Transfer of jobs in one area creates jobs in other areas. Labor released from one function requires retraining to reskill them in another function. Hence, demand for trainers and those who can teach the new skill increases. Typically, employees have moved from more routine and boring jobs to higher skilled and more interesting jobs. Also, industries that benefit from the export demand may require additional talent.

Transfer of work to overseas locations also releases capital tied up in real estate and other assets in the U.S. This capital can be redeployed to provide greater returns to shareholders through other investments.

As noted earlier, the Bureau of Labor Statistics is forecasting a significant growth in the job market throughout this decade—and the growth continues to be in the service industries.

Just as today no one questions China as the place to manufacture most products, similarly, tomorrow no one will question services provided from other parts of the world.

So what kind of services are being outsourced? Texas Instruments has had an offshore operation developing software in Bangalore, India, for over 30 years. It created a process that addressed the limitations of the communication infrastructure in India at that time by creating a direct satellite link and yet found it more economical to develop the software in India. Similarly, Burroughs created a joint venture many years ago with Tatas (a major diversified business house of India), and the joint venture not only developed internal software for Burroughs' locations worldwide, but also was responsible for its implementation and support. In the case of the early adopters of this model, most of the outsourcing (or offshoring) has been in the area of software development and technical support.

Since that time, local Indian business houses have developed significant expertise in the software development arena, with many of them having Capability Maturity Model (CMM) level 5 (highest level) quality certification. Many of their employees have trained in the United States and developed an understanding of the U.S. client needs and expectations. The Indian external service providers (ESPs), like TCS, Infosys, Satyam, and Wipro have been leading the charge and have been competing with the best in the world. Their commitment to provide the service at a lower cost and at least of the same quality is further supported by formal service level agreements (SLAs). The actual performance on the SLAs typically exceeds their commitment to their clients. This has provided comfort to the U.S. corporations to consider moving more of such activity offshore to become more competitive in the marketplace.

Now the larger U.S. companies are opening offices in India and providing the services to their U.S. clients from India. Today, many large traditional global service providers like Accenture, BearingPoint (formerly KPMG Consulting), Ernst & Young, Deloitte, Cap Gemini, Computer Sciences Corp., Electronic Data Systems, and Keane are aggressively redefining their own global delivery strategies, which include offshoring option for their clients. They are formidable contenders in this space and serving the U.S. clients from India and other nations around the world.

This is a win for U.S. customers. Today they have an option to have work performed in any of the four global delivery models:

· Onsite (work is performed at customer's location)
· Onshore (work is performed in the United States but not at customer's location)
· Nearshore (work is performed outside the United States at a remote location but in a country that is not too distant, e.g., Canada or Mexico)
· Offshore (work is performed outside the United States at a more distant location, e.g., Ireland, Philippines, etc.)

Undoubtedly, each has a different price tag attached to it, and the selection will based on the particular need. It is not uncommon for many projects to have a hybrid approach, which includes two or more of these options.

The low cost countries (LCCs) that had expertise in software development are now gradually moving into other related areas. The North American market is increasingly willing to transition the management of networks, systems, and applications to offshore centers in LCCs. These vendors monitor networks and systems remotely and provide solutions for any problems they encounter. Their value proposition is that they can manage more IT elements—more efficiently and effectively, and at a significantly reduced cost.

Kevin Francis, President and Chief Executive Officer of CenterBeam, recently shared his views on the impact on globalization and said, "High technology has wrought many benefits. Perhaps the most important benefit has been to make the world a neighborhood of people we know. Technology removes barriers and extends our reach. Millions of people are communicating with each other—anytime, anywhere—as quickly as they can dial their telephone.... The Internet, and its corollary technologies, erase the lines of latitude and longitude that separated us one from another and make obsolete governments that hold back technology's irresistible progress."

As the companies around the world have provided comfort to the U.S. client their capability to deliver quality services and significantly lower costs, the span of services has continued to expand. From technology, it is now extending itself to other services and is being termed as business process outsourcing (BPO). Gartner Inc., a technology research and consulting firm, predicts that the global BPO market will be worth more than $240 billion by 2005.

## BUSINESS PROCESS OUTSOURCING

What makes BPO an attractive option? In services, the compensation (or labor) component is a very high proportion (50% to 75%) of the total cost of service. As the compensation costs have continued to rise, these costs are much lower in many other parts of the world. The U.S. firms that have global operations recognize this differential and use it as an opportunity to reduce their total cost and be competitive in the market.

Let us consider labor costs in India. Chartered Accountants—India's version of certified public accountants (CPAs)—make about one-seventh of the annual salary of their counterparts in the United States. A similar compensation ratio holds true in the call center industry. When you compare compensation with some of the other LCCs this differential increases even more—labor costs in the United States can be ten to twelve times higher than those in countries like Philippines and China.

Recognizing the labor arbitrage with foreign resources that can provide labor at one-sixth to one-tenth the cost, firms are seeking other opportunities to take advantage of this cost differential. In every organization, many processes are labor intensive. Human intervention, along with decision-making skills, are required to perform tasks such as interpreting, validating and editing, translating, transliterating, analyzing, or transforming information. These back-office processes are prime candidates for being migrated from the United States and other Western countries to other parts of the world. However, the labor cost should not be the only criterion to evaluate when considering your offshoring options—you need to evaluate the labor market holistically, as well as the political, economic, and regulatory environment.

The advent of technology is making firms more virtual and increasing their dependence on their partners in the supply chain. Carter, who received the 2000 CFO Excellence Award for Implementing Best Practices in Finance, a 19-year veteran of Motorola Inc., joined Cisco in 1995. He wanted to reduce Cisco's month end financial closing cycle from 14 days to one day—something Motorola had achieved—while cutting costs in half. Today, Cisco boasts that its back office is even more automated than its front office. More than 85% of customer orders arrive via the Internet, and the majority of those are automatically farmed out to subcontractors, who ship finished products directly to customers—never touched by a Cisco employee. Could such a supply chain be located beyond our boundaries? Absolutely. When the technology is providing us the capability to transcend beyond borders, why are we limiting our solutions and opportunities to maximize the potential to increase shareholder value to artificial borders that do not exist?

The LCCs are becoming prime hubs for BPO. Among financial institutions—ABN Amro, Citibank, GE Capital, HSBC, JP Morgan Chase, Standard Chartered—each has its own back office processing centers in India. Other global companies such as Accenture, Cap Gemini, EDS, Ernst & Young, PeopleSoft, and Hewitt Associates have either set up or expanded operations in India in the last year. This direction is allowing these businesses to expand their business and grow—an opportunity that would not come their way otherwise. It would be wrong to conclude that this expansion is at the cost of American jobs.

**Goldman Sachs' perspective.**    Goldman Sachs met with 19 participants in the offshore industry during a field trip to India in September 2003. These leading offshore companies were both pure-play firms and in-house captives of multinational corporations (MNCs).

Their meetings spanned Indian pure-play IT services firms (Birlasoft, Cognizant, HCL Perot Systems [HPS], Infosys, iFlex

Solutions, Satyam, and Wipro), Indian pure-play BPO firms (Daksh, MphasiS BFL Group, and Spectramind), offshore operations of U.S. IT services firms (Accenture, ACS, CSC, the offshore BPO operations of GE Capital, Keane, and Sapient), and customers (CSCO and GE and VCs). Their main conclusions were as follows:

1. The offshore has fully moved into the mainstream from both a demand and a delivery capability perspective, and demand will likely remain very robust.

2. Pricing had stabilized and there were pockets of improvement. For perspective, prevailing average wage for mid-level managers was $9,000–$12,000 annually versus $8,000–10,000 at the entry-level. Although there is potential for wage inflation in this segment, it probably will not have a big impact on the financials of most service firms.

3. The labor supply was more than keeping pace with employment trends, limiting wage inflation at the entry level. Inflation at the mid-level may pick up as U.S. companies embark aggressively on recruitment plans. The National Association for Software and Services Companies (NASSCOM) indicated that, annually, 250,000+ students graduate with engineering degrees (both three-year and four-year degrees). Another 500,000 or so graduate with nonengineering degrees, suggesting a total pool of tech-ready or "tech trainable" graduate students of over 750,000.

4. The offshore model poses a potential risk to the data center industry. While the hardware isn't going overseas, the offshore firms are beginning to win portions of the monitoring, network management, applications management, and operating system maintenance segments.

5. In BPO, offshoring is a necessity, not a luxury. To date, ACS and Accenture have made the most progress in this area, but Convergys and Exult are moving in the same direction.

A report published by Forrester Research in December 2002 predicts that over the next 15 years, 3.3 million U.S. service industry jobs and $136 billion in wages will move offshore to countries like India, Russia, China, and the Philippines.

There are many different business models for BPO, including offshoring. There are three primary candidates for your business:

- A U.S. BPO vendor (e.g., EDS, IBM, and Accenture)
- A foreign BPO vendor (e.g., Infosys, Satyam, and Wipro)
- Own captive office in a foreign country (e.g., GE, American Express, HSBC, and Hewitt Associates)

While these three options exist, you must reach a critical mass before it makes economic sense to create your own captive BPO center located offshore. (In discussion with Mr. Anupam Prakash of Hewitt Associates, who is a leading consultant providing consulting services to foreign corporations desiring to locate their BPO operations in India, he said that you probably need to be able to keep a group of 100 people busy regularly before you consider setting up your own captive operation in India.) For a size of operation lower than that, you are probably better off partnering with one of the ESP firms based in the offshore market.

**American Express.**    American Express (Amex) has an interesting model—a variation on the basic theme. Strictly speaking, Amex isn't engaged in business process outsourcing. Rather, its divisions around the world send accounting tasks to an American Express-owned and operated transaction processing center in India. American Express is among the pioneers when it comes to establishing a back-office operations facility overseas. The company opened its Financial Center East (FCE) in New Delhi in 1994. FCE is one of three major transaction processing centers for Amex worldwide, with the others located in Phoenix, Arizona and the United Kingdom. FCE's workforce has reached about 900 workers in three shifts, and the facility handles functions that include accounts payables, incoming payments, general accounting, and financial forecasting.

Approximately 38% of work done at FCE is for American Express' Asia-Pacific operations, 20% is for European operations, 24% is for U.S. and Canadian operations, and 8% is for Latin-American operations. The reason to push financial functions to India include not only lower cost, but higher quality as well. Processing transactions out of India costs between 25% and 40% of what it does in the United States or the United Kingdom. That's true partly because of the way technology and telecommunications have improved in recent years. The cost of high-bandwidth telecom links to India have fallen, while imaging software has improved, making it easy for anyone to scan invoices and zap them electronically to Delhi.

**World Bank.**    World Bank (WBO) transferred back-office operations from Washington, D.C. to India over a year ago. The World Bank achieved significant gains through this action. As per Chuck McDonough, director of the World Bank's accounting department:

· First, the WBO slashed its costs by 15% in the lower-wage country.
· Second, the organization significantly reduced its processing backlog of accounts receivable and expense forms from hundreds of items to just a handful.

The World Bank's shift is part of a growing practice of sending back-office work from developed nations to developing countries. Thanks to a growing willingness to outsource operations, lower labor costs, and technology improvements, firms are deciding to ship tasks such as customer contact, bill processing, medical transcription, and even animation to countries such as India, the Philippines, and Jamaica. Major corporations such as General Electric and British Airways pioneered this tactic with offshore operations.

The National Association of Software & Services Companies (Nasscom) in India says that the country's call center and BPO industry—what it calls IT-enabled services—grew by 70% during the 2001–2002 period to a total of $1.46 billion in revenues. Nasscom predicts Indian revenues in IT-enabled services should jump to $16.94 billion by 2008, capturing more than 10% of the global market.

**What tasks move to offshore locations?**    Looking at examples of certain companies that have been active in the BPO arena for some period provides an indication of the types of jobs that move offshore. The first temptation is that it allows companies to have a 24/7 operation that has great value for U.S. corporations in terms of enhancing customer service and gaining competitive edge.

Some examples of service activities that are being offshored include:

- Information Technology
  - Software development
  - Software implementation
- Contact Centers
  - Inbound call centers (customer support)
  - IT help desks (technical support)
  - Outbound call centers (calling customers and prospects)
  - Email responses
- Document processing
  - Data entry into electronic forms
  - Conversion of raw data into electronic format (digitizing)
  - Document storage
  - Claims processing
- Accounting
  - Accounts payable
  - Expense processing

- Intercompany reconciliation
- Tax preparation and filing
· Human resources management
- Payroll
· Marketing and sales
- Telesales

As companies become more comfortable with offshoring, they are more willing to move up the value chain in identifying opportunities that can be performed remotely. At the lower end of the scale are functions such as data entry and processing of forms where there is no client contact. Then there are contact centers, which can include inbound or outbound calls from customers and employees. Finally, there are functions that require the application of certain decision criteria such as processing insurance policy applications and claims processing.

**Where are the tasks moving among offshore locations?** General Electric (GE) is one of the pioneers in its commitment to transitioning the back-office operations on a global basis. GE Capital has operations based in India, China, and Mexico. Their facilities, also known as "Global Processing Centers," provide around-the-clock services. GE's financial services unit has grown over the last few years, and its operations in India have now reached 15,000 employees. The Mexican facility alone processes more than 3.5 million documents a day, with turnaround times of as little as eight minutes.

Some of the other large corporations that have back-office operations in India include:

· British Airways, which created links with World Network Services, a data management unit.
· Conseco, the Indianapolis-based financial services company, which has operations in Jamaica and India, where it purchased business process outsourcing firm Exlservice.com in April 2002.

- Oracle Corporation, the leading database software firm, which also plans to set up a back-office unit in India.
- Affiliated Computer Services, a Dallas-based IT and business process outsourcing firm, has operations in India, Jamaica, the Dominican Republic, Barbados, Mexico, and Ghana.

According to Gartner, other countries with business-process outsourcing facilities include the Philippines, Czechoslovakia, and Hungary.

**Why are tasks moving to offshore locations?**    In an interview suggests Wharton professor Ravi Aron: "Companies start out for cost, stay on for quality, and then realize that they get a lot of managerial initiative." This is what lets corporations move up the value chain.

Cost: Even though the wages in many locations are a much lower proportion, the total cost reduction is more around 50% depending on the operation being offshored and its complexity. Call centers in India can result in a 30–40% cost savings.

Productivity: Turnover in the U.S.-based call centers can run as high as 100%, while 40% is quite common. Absenteeism can reach 14%, causing performance and customer satisfaction issues. These problems are significantly lower in the lower cost countries as these jobs represent well paid and prestigious opportunities in those countries. Some BPO operations claim to have achieved productivity levels in excess of 150% of the levels in the United States. Another factor that could contribute to higher levels of productivity is the education level of the workforce, which typically consists of at least an undergraduate college degree.

Quality of work: The high level of employee education also results in a higher level of work quality. Conseco has found that its operations in India meet or exceed service level standards 95% of the time. The more seasoned operations are willing to offer 99% quality compliance standards. Another ancillary benefit of this process of moving work overseas is that it gives you an opportunity to redefine the process and improve it—because you are forced to find solutions to many workaround processes that have developed over a period of time and increased your processing costs. This is where applying a Six Sigma standard to the process can be very beneficial in ensuring that the correct quality standards are applied. Few corporations in the United States have moved toward adopting Six Sigma in this process. Those like GE that have adopted it, have reaped significant benefits.

## New Technologies

New technologies bring enormous benefits with them. Probably the most important one has been to make the world a smaller place and a network of people we know (even though we may have not met them). Technology brings down distance barriers and extends our reach to a wider audience. Due to the relatively low cost of the Internet, millions of people are communicating with each other—anyone can communicate anytime and from anywhere.

The advent of the Internet and a proliferation of new advanced technologies has introduced new ways of doing business. New business models are evolving, and it is also helping companies move up the value chain.

In the electronics industry, there is a lot of talk about e-diagnostics, the basic foundation of which is information. By connecting directly to clients' hardware and software, one can receive real-time data that can be analyzed and provided to customers to help them to make more informed decisions and increase their productivity and throughput. The same information can also be used to provide them enhanced service. IBM has been performing remote preventive maintenance of its customers' various elements of computing equipment for over 20 years. From a centralized location, IBM service representatives can diagnose the performance of your computer system and anticipate upcoming problems before they are encountered and impact your uptime or performance of your system. Corporations like BASF monitor the level of their customers' liquid chemical inventory in storage tanks and ensure it is replenished in time before they fall below safety stock levels. This kind of use of information allows vendor corporations to provide value-added services to their customers and also create a competitive advantage.

The more creative companies have used the new technologies to redefine the business model of the entire industry itself. Jeff Bezos, CEO of Amazon, used Web-based technology to redefine the book retailing industry. Creating a virtual model that initially eliminated the need to carry inventory, he offered customers millions of books to select, and the convenience to shop from anywhere, and at anytime. This has forever changed the way books are retailed. Through the use of intelligent software, Amazon also has been able to provide personalized suggestions for books based on an individual's interests and purchasing habits. Similarly, Dell has helped redefine the personal computer retail industry, offering each customer an opportunity to have a custom-designed computer to fit his or her needs without having to pay an exorbitantly higher price. Dell provides a selection of standard configuration based on the needs of a typical home user, student, or business user, which can be shipped within 24 hours. For users with special needs, however, it just takes a little longer to ship the product. All these changes are driven by ease with which information can now be shared. As the trend is toward greater sharing of information, the need to have accurate, secure, and timely information will continue to increase. Cisco is acclaimed as a leader in hands-free information processing, whereby a large majority of its customer orders get processed without anyone in the firm having to handle a piece of paper.

These new technologies help to reduce costs, as well as allow information to be distributed quickly, easily, and cheaply. They remove the limitations of distance and offer innovative and cost-effective solutions to increase efficiency and customer satisfaction.

## GLOBALIZATION

The demand for services is only going to grow. The need for services is also a function of disposable incomes of individuals. With the deflationary trends that have been experienced in the

United States over the last few years, products are costing less, and hence, the consumer has more disposable income to spend on services. We can thank the global movement, which transferred manufacturing activity to lower-cost locations around the world, for this trend. The following are examples of on-demand patterns and price trends for products in the United States between 1998 and 2001:

- U.S. imports of household cooking appliances more than doubled to $640 million; however, prices of these items have declined by almost 50% during the same period.
- Demand for televisions and audio equipment in the United States has grown at 13% annually to $6.1 billion; however, retail prices of these items have declined by almost 9% annually during the same period.
- Imports of tools and metal implements grew 23% annually to reach $1.5 billion; however, their retail prices declined by almost 1% annually in that time.

Let us also consider pricing trends in certain broad product lines typically used by consumers (see Table 1.5).

TABLE 1.5. Decling Price Trend in Consumer Goods

| SECTOR | PRICE CHANGE |
| --- | --- |
| New autos | −3.8% |
| Personal computers | −20.9% |
| Appliances | −2.7% |
| Apparel | −2.2% |

The transition of manufacturing operations to China is a major factor in this stability or decline in product prices. Assuming this trend will continue, the demand for services will continue to grow.

**The China factor.**    China poses an interesting challenge for all countries around the world. China became the leading country for foreign direct investment (FDI) in 2002 when it attracted $50 billion in foreign investments. In the United States, which until then had held the leadership position, FDI dropped from $124 billion in 2001 to $44 billion in 2002.

In the '90s, we saw manufacturing operations move from the developing countries in Europe and North America to China. However, now it is starting to affect developing countries as well. Since December of 2002, it is estimated that over 500 factories employing over 223,000 people have moved from Mexico to China. Many Asian countries, previously considered hubs of future manufacturing prowess, are also facing a similar fate.

## GREATER ACCOUNTABILITY

The unprecedented fall of a number of giant U.S. corporations like Enron, Global Crossing, Lucent, MCI, and Tyco—resulting from inappropriate accounting and management practices—prompted the U.S. Congress to personally hold CEOs and CFOs responsible for the accuracy of the information they share with the investing public.

The Sarbanes-Oxley Act (SOX) is becoming a major driver of growth for service firms. Corporations listed on the U.S. stock exchanges must comply with their requirements starting October 2004. Many foreign nations are considering similar requirements to protect their stakeholders. This legislation will drive demand for services in the coming years. Management Consultant International estimates that U.S. corporations may be spending over $75 million annually in SOX compliance with focus on the following:

· Enterprise resource planning (ERP)/supply chain systems—To further automate acquisition, monitoring, and reporting financial information with the aim of reporting any material changes to external guidance as early as possible

· Data warehousing—To ensure all relevant information is collected and to enhance the quality of information captured

- Business intelligence—To gain better understanding of the global business environment
- Change management—To increase awareness of SOX requirements and to foster cultural change to ensure compliance going forward
- BPO and process consulting—To ensure processes are in compliance with the new requirements

As we are largely dealing with financial data, this initiative will be highly focused on "transaction" activities. This will trigger a significant opportunity for providing additional services to organizations in the United States and around the world. The CEOs and CFOs can potentially face criminal penalties in case of noncompliance. As a result, compliance is not a choice. When the need for compliance is so critical, the use of tools offered by Six Sigma methodology, which focuses on quality and accuracy, is an ideal application for this situation.

As mentioned earlier, the average U.S. company operates at three Sigma (i.e., approximately 66,800 defects per million). Given the penalties and potential business impact of incorrect financial information, a significantly higher standard is needed for these data.

**Transaction versus nontransaction-based services.**   Given the focus of this book, it is worthwhile distinguishing between transaction-based services and nontransaction-based services. Almost every industry has both transaction-based services and nontransaction-based services, so a particular industry cannot be categorized as one or the other. Table 1.6 identifies typical characteristics of both types of services.

Information processing is an example of a transaction-based service, while a customer service call is an example of a nontransaction-based service. Given the characteristics in Table 1.6, the following attributes are important in a transaction-based service:

- Accuracy
- Authenticity

· Security

· Speed

Typically, transaction-based services are *low value, high volume*, while the nontransaction-based services are *high-value, low volume*.

TABLE 1.6    Characterists Distinguishing Transaction-Based
            Services

| Characteristics | Transaction-Based Services | Nontransaction-Based Services |
|---|---|---|
| Volume | High | Low |
| Frequency | High | Low |
| Cost | Low | High |
| Process | Automated—low touch | Manual—high touch |

## SIX SIGMA IN THE SERVICE ECONOMY

Let us review the magnitude of savings that has been experienced by some of the pioneers as a result of their efforts in this field. Motorola, GE, and AlliedSignal have claimed astronomical savings as a result of implementing Six Sigma.

Motorola claims more than $15 billion in savings in manufacturing alone since its development of the Six Sigma program in the 1980s. In one example, the company has reduced the manufacturing time for pagers from 40 days to less than one hour. At Motorola Lighting, there has been an estimated 70% improvement in quality and defects over the past eight years as a result of the program.

GE is frequently recognized for its Six Sigma implementation excellence. Introduction of Six Sigma usually requires a significant up front investment. GE started the program in 1996. Even in the first year of the program, it achieved break even in most of GE's 12 business units. In the second year (1997), the savings exceeded GE's $300 million investment, and by a whopping $100 million. GE Aircraft Engines division had projected annual savings of approximately $70 million, for a total of $500 million by 2000. Similarly, the plastics business saved $400 million in funding requirements for investments in the first year and another $400 million by the year 2000. In addition, it added 300 million pounds of new production capacity in the first year.

AlliedSignal has claimed that it saved over $2 billion since 1991, through Six Sigma activity. In addition, it experienced a boom in productivity and an increase in operating margins.

In the preceding examples, the huge numbers result largely from material and labor cost saved as a result of the Six Sigma program. However, in addition to the material cost savings, discussions with these companies have indicated that they are also enjoying other benefits. These include:

· Increase in capacity
· Larger output
· Faster throughput
· Higher customer satisfaction
· Streamlined process flow
· Enhanced productivity
· Shorter cycle times
· Reduction in inventory and work-in-progress

In a service economy, the advantage stemming from savings in material costs are not there, but all the other elements are still very relevant, and they are even more important when one is faced with ever increasing labor costs.

While Six Sigma was originally developed for the manufacturing sector, it has significant relevance to service functions as well. In today's information economy, service functions within almost every sector are also using Six Sigma methods to boost performance.

Six Sigma was originally developed to address processes with high volume and with a high degree of standardization. To that end, it does not matter whether it is manufactured product processing or information processing. Its goal is to eliminate waste by achieving near-perfect results (Six Sigma-level quality means no more than 3.4 defects per million).

After having exhausted the opportunities in the manufacturing processes, companies are using Six Sigma to shape up such nonmanufacturing processes as accounts payable, accounts receivable, sales orders, and forms processing. Dow Chemical has applied the Six Sigma methodology to environmental, health, and safety services and claims to have saved the company $130 million in two years. It has launched other initiatives in corporate R&D, finance, information systems, legal, marketing, public affairs, and human resources processes. Considering the success of this methodology in service functions, financial institutions, consumer product companies, and health care firms are all jumping on the Six Sigma craze.

Six Sigma came to even greater prominence when former GE executives James McNerney and Bob Nardelli took charge of 3M and Home Depot, respectively, and introduced Six Sigma to those organizations. Service institutions like Bank of America and Merrill Lynch have also adopted Six Sigma in the recent times.

**Six Sigma is critical to transaction-based services.**    As found in earlier examples of GE, American Express, and other companies, the opportunity for realizing hard savings is significant. When considering transaction-based processing, any defect will create diversion in the process and typically require manual intervention. The statistics in Figure 1.16, from the banking industry, will help one realize the importance of error-free transactions. The potential of the Internet to reduce back-office and transaction costs is significant, as demonstrated by the banking industry experience. Figure 1.16 shows that the cost per transaction is over 100 times greater when a customer walks in the door inside a bank branch because of the face-to-face contact with the client.

When you multiply the cost per transaction by the number of transactions, the resulting impact on financial results is enormous. The Federal Reserve System recently sponsored

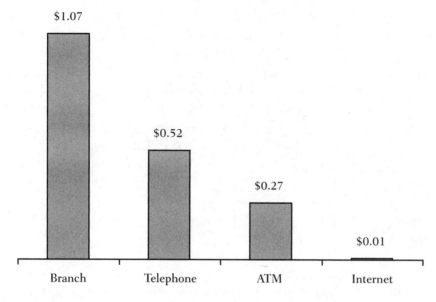

Source: Gartner Inc.

**FIGURE 1.16.** Unit Cost per Transaction by Channel in Banking

research on retail payments. Highlights of this research are provided below. This is of particular importance considering the high volume of transactions processed.

# EXAMPLES OF HIGH TRANSACTION-BASED INDUSTRIES

## DEPOSITORY FINANCIAL INSTITUTION CHECK STUDY ON PROCESSING OF CHECK PAYMENTS

Let us consider an area with very high volume of transactions—check payments. The recently published Depository Financial Institution check study on processing of check payments, sponsored by the Federal Reserve System, provides some interesting insights (see Figure 1.17). It is estimated that people write 42.5 billion checks each year in the United States, representing $39.3 trillion in payments. An interesting fact is that despite the advances in technology, the study estimates

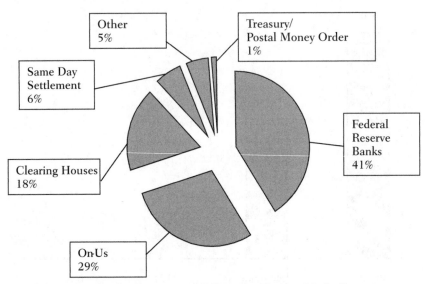

**FIGURE 1.17.** Check Volumes: 42.5 Billion (by Clearing Method)

that the check volume is still increasing at the rate of 1.2% annually.

The study also identified that 251 million checks are returned, representing 0.6% of the total volume. About half the checks are written by consumers, while businesses receive half of all checks.

**Categorization of check payments by purpose.**    Check payments are categorized into six purpose categories:

· Casual
· Income
· Remittance
· Point of sale (POS)
· Remit/POS
· Unknown

In Figure 1.18, commercial payment (remittance and POS check payments combined) represents over half of all check payment volume (57%), while income and casual check pay-

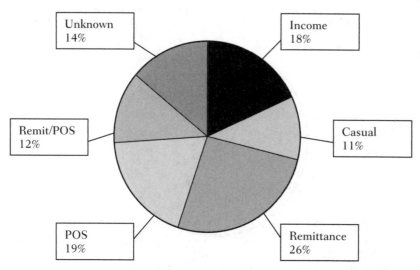

**FIGURE 1.18.** Check Payment Volumes: 42.5 Billion (by Purpose)

ments (those received by consumers mainly) represent about 30% of all check payment volume.

Looking at check payments by dollar value, POS checks (largely written by consumers) make up only 9% in terms of the total value of checks, while they represented 19% of total volume. On the other hand, the Remittance/POS category checks (high-value checks largely issued between business or government payers and payees) make up 25% of the total value of checks, while representing only 12% of the total volume (see Figure 1.19).

Further analyzing these transactions allows us to see where the opportunities may exist to further streamline these processes.

As shown in Table 1.7, the largest segments of check payments are business and/or government income payments to consumers (17.8%) and consumer remittance payments to business and/or governments (17.7%). It is surprising that 18.3% of the check transactions are BG2BG payments (payments from business or government payers to business or government payees). These payments are generally larger and represent 49% of the total value transacted through checks.

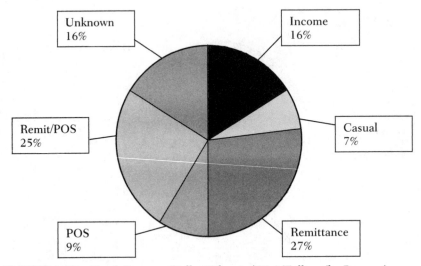

**FIGURE 1.19.** Check Payment Dollar Volume: $39.3 Trillion (by Purpose)

**TABLE 1.7**    Distribution of Check Volumes—Between Consumers "C" and Business and/or Government "BG"

| PURPOSE | C2C | C2BG | BG2C | BG2BG | UNKNOWN |
|---|---|---|---|---|---|
| Income | | | 17.8% | | |
| Casual | 11.2% | | | | |
| Remittance | | 17.7% | | 7.9% | 0.1% |
| POS | | 14.1% | | 4.9% | 0.1% |
| Remit/POS | | 6.4% | | 5.5% | |
| Unknown | | | | | 14.5% |
| Total—# of checks | 11.2% | 38.1% | 17.8% | 18.3% | 14.7% |
| Total—$ value | 6.7% | 11.5% | 16.5% | 49.0% | 16.2% |

## SIGNIFICANT FINDINGS: ELECTRONIC PAYMENT INSTRUMENTS STUDY

Despite the surprisingly large volume of check payments, a substantial volume of transactions are conducted electronically. The Electronic Payment Instruments Study (EPIS) estimated the volume and dollar value of electronic payments. During calendar year 2000, 29.5 billion electronic payments were originated in the United States, with a value of $7.3 trillion.

As shown in Table 1.8, the majority (51%) of electronic payment transactions were made using credit cards, but 78% of payment dollars were handled through the automated clearing house.

TABLE 1.8     Total Estimated Volume and Dollar Value of Electronic Payments

| ELECTRONIC PAYMENT INSTRUMENT | TRANSACTION VOLUME (MILLIONS) | DOLLAR VOLUME ($MILLIONS) | AVERAGE PAYMENT VALUE |
|---|---|---|---|
| General purpose credit cards[1] | 12,300.2 | $1,072,555 | $87.20 |
| Private label credit cards[2] | 2,748.6 | $162,819 | $59.24 |
| Offline debit (signature-based)[3] | 5,268.6 | $209,980 | $39.85 |
| Online debit (PIN-based)[4] | 3,010.4 | $138,151 | $45.89 |
| Automated clearing house (ACH)[5] | 5,622.0 | $5,674,851 | $1,009.40 |
| Electronic benefits transfer (EBT)[6] | 537.7 | $13,744 | $25.56 |
| **Total** | **29,487.5** | **$7,272,100** | **$246.62** |

*Source:* Retail Payments Research Project—Sponsored by the Federal Reserve System
*Notes:*
1. General-purpose credit cards include cobranded credit cards, charge cards, cobranded charge cards, secured credit cards, T&E cards, commercial cards (including business, corporate, and purchasing), and new payment technologies that route transactions through the card associations' networks.
2. Private-label credit card programs include those run by individual retailers or gas companies, third-party fleet-card issuers, and third-party receivable owners.
3. Offline debit refers to debit card transactions that require the customer's signature (rather than PIN) as a means of authentication. Visa and MasterCard have the only two networks for offline debit transactions. Visa's offline debit statistics also include the new hybrid online/offline debit card offered by Visa, the Visa Check Card II.
4. Online debit card transactions require the customer to enter a four-digit personal identification number (PIN) as a means of authentication (as opposed to signing a sales receipt). Online debit card transactions are originated, cleared, and settled over EFT networks (the same networks that process ATM transactions) as opposed to Visa or MasterCard's networks.
5. The ACH is an electronic payments network that allows credits and debits to be processed between financial institutions. ACH transactions between financial institutions are processed by one of four ACH operators who were surveyed. ACH transactions that are On-Us (a DFI is both the originator and receiver of a transaction) have been estimated using data from National Automated Clearing House Association (NACHA).
6. EBT is an electronic system that allows a recipient to authorize transfer of his/her government benefits from a federal account to a retailer account to pay for products received. EBT is currently being used in many states to issue food stamps and other benefits.

## Key observations on each type of transaction

*General purpose and private label credit card.* General purpose and private label credit cards were the most common electronic payment instrument used in the United States during the year 2000: 15.0 billion transactions were originated, with a value of $1,235 billion. The average transaction size for general purpose credit cards was much larger than that of private label cards: $87.20 vs. $59.24. Credit cards accounted for 51% of all electronic payment transactions and 17% of the dollar value. Eighty-two% of credit card transactions and 87% of transaction value came from general purpose credit cards.

*Online and offline debit cards.* Debit cards represented the second most common form of electronic payment, accounting for 8.3 billion transactions and a dollar value of $348 billion in 2000. On average, each debit transaction was $42, compared with $87 for the average general purpose credit card transaction. In 2000, 64% of transactions and 60% of the value was contributed by offline debit (i.e., signature-based); 36% of transactions and 40% of value were from online debit (i.e., PIN-based).

*ACH.* Although ACH was the third most commonly used electronic payment instrument, with 5.6 billion transactions, it dominates on a dollar value basis, accounting for 78% of the monetary value. The average transaction volume was more than 11 times larger than that of general purpose credit card transactions ($1,009 vs. $87).

*EBT.* EBT volume has increased dramatically due to initiatives at the federal level and significant efforts by state governments to electronify both food stamps and cash assistance payments during the 1990s. Nevertheless, EBT was the smallest volume payment instrument, with 500 million transactions and $13.7 billion in value.

*Emerging payments.* The survey of emerging payments involved companies that provide services in such markets as electronic bill payment, person-to-person payments, stored value, Internet currencies, and other emerging technologies. In general, emerging payment volumes for the payment instruments studied were quite small in 2000. Organizations participating in the survey reported 76.2 million transactions involving $12.7 billion. However, these numbers represent only a small portion of the total emerging payments.

## IMPLICATIONS AND OVERALL OBSERVATIONS

As the Federal Reserve, like many other payment processors, looks for ways to make the payments system more efficient, it is vital to understand where opportunities exist for migrating check payments to electronics.

The data from this research show that remittance and point-of-sale payments written by consumers offer the most significant opportunities for substitution, as these are the largest categories of checks written today. This implies that the ACH and credit and debit cards are poised for meaningful growth in the near term.

Given the large number of checks still being written in the United States, and the increased usage of electronic forms of payment, businesses and financial institutions are going to have to maintain multiple channels for the foreseeable future. While checks, we believe, will account for a decreasing portion of total payments, they will continue to be around for some time to come.

Despite an annual volume of 42.5 billion checks, it appears that Americans are changing their historical, conservative use of payments. The fact that 30 billion electronic payments were initiated in 2000 indicates clear acceptance by consumers and businesses. And given that debit and credit card payments and ACH transactions collectively have grown exponentially in the last 20 years—some 500% since 1979—these electronic forms of payment will become more prevalent and increasingly a requirement of doing business for U.S. companies and financial institutions.

The data from the study are important because they offer factual evidence to financial institutions, the financial services industry, and the Federal Reserve System on the volume and value of payments. From this, industry stakeholders may make inferences about the migration of the payments system and where prudent opportunities for investments in payments system technology exist. Going forward, the Federal Reserve plans to repeat this research and establish a trend line that will enable both industry stakeholders and the Fed to measure the progression of the payments system and the migration of paper payments to electronics. This study is just a picture—a snapshot—of the continuing evolution of the payments system.

## KEY TAKEAWAYS

- Service sector represents 80% of the U.S. GDP today and is continuing to increase in importance.
- Growth in the service sector is three times the rate of the nonservice sector.
- Approximately 93% of the employment is being created today is in the service sector.
- Service is the key to our competitive advantage moving forward.
- Trends such as new technologies, globalization, and outsourcing are significantly impacting current service models
- Six Sigma can help us make service more customer focused and efficient so that we can increase customer satisfaction and deliver services more cost effectively.

# QUALITY IN SERVICES AND TRANSACTIONS

## EVERY ORGANIZATION IS A SERVICE ORGANIZATION

E very organization is a service organization—some deliver products as well!

In any corporation, whether it is part of a service industry or a nonservice industry, components of service are present. In fact, this desire to categorize corporations as to whether they are service organizations, has created many misconceptions among management and leaders of these corporations.

What benefit is this distinction providing? How do you decide whether you are a service organization? Is it some criteria determined by what component of the total cost is represented by labor cost? Or, is it that product cost should be below a certain percentage of the total operating cost? As we ponder over this question, we start to realize that it is very subjective and academic. No matter what your organization does, you provide a service to your customer. Every organization is a service organization—some deliver products as well!

## SERVICE COMPONENTS IN A CORPORATION'S VALUE CHAIN

The value chain in a corporation can vary from industry to industry. Even within the service industry, some service industries deal with physical products (e.g., retail and wholesale), while others do not deal with a product, but simply information (e.g., banking or insurance). Depending on your particular value chain, the proportion of the service component may vary based on your business model. Let us consider a typical value chain in the retail industry (see Figure 2.1).

A service component is in each link of the value chain. Performance within each link of the value chain will impact the overall service and price to the customer. Let us consider each link in the value chain and how it may impact the customer.

**FIGURE 2.1.** Value Chain in the Retail Industry

**Demand forecasting.**    Demand forecasting relates to projecting and anticipating customer demand for your products and services. Poor demand forecasting will result in high inventories or shortages—in one situation, the customer ends up paying more for the product; in the other, the customer is not able to get the product.

**Supply contracting.**    Once an organization has a handle on what their customers are going to need, it is important to translate that demand into the resource components required to service that demand. The next step would require the organization to reach out to its suppliers and secure required products and services. Contracting with the vendors is a critical component of the price, product, and quality provided to the customer.

**Procurement.**    Having contracted with the suppliers, procurement deals with timely ordering and receipt of goods and services based on the contracted terms. If the goods are not received in time, the customer is dissatisfied. If the wrong goods (in terms of specification or quality) are received, the customer is again dissatisfied.

**Inventory management.** Inventory ties up capital—you do not want to tie up more capital than required. How you manage the inventory and its distribution again impacts your customer—empty shelves hurt you in sales and overstocking increases costs.

**Marketing.** Marketing effectiveness is key to positioning the product correctly, reaching the right customers, and setting the right customer expectations. It also requires a good understanding of competitors and their strategy. Effective marketing ensures that sales at least achieve the forecasted demand level. Poor marketing will result in lower sales and increase the overall cost of operation. Once an organization is under cost pressures, service suffers.

**Sales management.** Sales management is important to ensure that the channels of distribution are functioning effectively. It helps determine which channel to use for which customer segment and provide the right level of resources to support them. Today, we have a more diverse selection of service channels. Historically, the emphasis was on direct sales. Today, we have agents and distributors, third-party alliances, value-added resellers, and the Internet, to name a few. From the customers' perspective, the service should be of the same quality no matter which channel they use—unless their expectations are managed appropriately and different channels are being used to position the product differently in the market.

**After sales support.** Once product or service is delivered, customers continue to require support—to ensure they are satisfied with the quality and performance. In the past, this has been considered a pure service function. This is where many times the pulse of the customer's satisfaction with the organization resides—it is typically the first point of contact when the customer has a problem. While on one hand, this link does not add to the initial sales, on the other hand, a poor after-sales service can result in a permanent loss of customer and loss of repeat business.

As you can realize, every link in the value chain has risks associated with it that could dissatisfy the customer. It becomes extremely important to optimize the entire value chain to provide the best service to customers and gain their loyalty.

An interesting example is cited in *IndustryWeek's* "European Best Plant: Siemens AG, Medical Solutions Computed Tomography." Siemens tracks first pass yield, cycle time, and on-time delivery for each of the following processes:

· Order management
· Scheduling
· Materials logistics
· Assembly
· Testing
· Shipping
· Installation

Editorial Research Director David Drickhamer reports: "Knowing the costs of each of these subprocesses... Siemens CT managers can optimize the total process to best serve the customer... [while] delivering higher economic value added" for the company. He cites the company's decision to deliver by airfreight all the X-ray systems shipped to customers outside Europe. It's more costly, but "it accelerates the total cycle time, decreases the amount of capital tied up in inventory, and speeds cash flow." Such a decision could not be intelligently made without detailed knowledge of how each process affects cost, cycle time, and quality of the finished, delivered product.

Many companies do not track their performance on various elements of the value chain. Those that do track performance, track it on selected elements of the value chain. The better the knowledge across the entire value chain, the more powerful is the information to enable you to make customer-centric profit optimizing decisions. Six Sigma tools can be of great assistance in analyzing and optimizing the overall value chain.

Most U.S. companies operate at three Sigma (i.e., approximately 66,800 defects per million). Six Sigma quality measures bring this down to 3.4 defects per million. Is that the right target? It will depend on the criticality of the information in terms of its use and business impact—but three Sigma standard also may not be acceptable.

**What is service?**    Having established the importance of the service sector in our economy, let us define what we mean by service. Does service mean direct contact with the customer? Does it require being physically interactive with the customer? Does it mean providing a consulting report or a product? What is service?

Looking at the industries that are classified as service industries, it is clear that service industry could include delivery of a product—as in retail or wholesale industry. However, the service in itself is not a product—it is an experience. A consultant delivers a report, but the bound paper and ink are not the service. The service component is in the skills and knowledge that is provided by the consultants and how well they are able to address the problem they were tasked to solve.

We go to a restaurant to have a meal. The meal is the product we consume; however, the service is in how we are treated on arrival, how long we have to wait to be seated, how appealing is the ambiance, how comfortable are the seats, what menu selection is provided, how courteously and quickly the meal is served, how appetizing does the meal look and smell, and the like. The food we consumed was the product, but there is no such equivalent in service. Service is not something physical, and as a result, it cannot be possessed by the receiver—it can only be experienced. We would claim that every organization is a service organization—some deliver a product as well (see Figure 2.2)!

Similarly, when you go on a vacation, the travel agent, airline, hotels, and car rental agency are all providing services that contribute finally to your vacation experience. The airline carrier contributes to your experience through services such as:

· Effortlessness of making reservations
· Courtesy of their ground staff

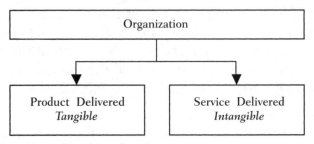

**FIGURE 2.2.** Key Deliverables of an Organization

· Ease and time it takes you to check in at the airport
· Punctuality of their flight arrivals and departures
· Décor and cleanliness of their aircraft
· Appearance and behavior of the flight attendants
· Comfort during the flight
· Speedy and careful handling of your baggage

All these elements are part of the service to the customer.

**Service does not require physical presence.**   A common misconception is that service must be delivered by an individual and through physical presence. Certainly that is one way to deliver service. In the earlier examples, the consultant is delivering service to the client management, the waiter is attending to the guests in the restaurant, and the airline staff is assisting the passengers. However, many times, various components of service do not require the physical presence of an individual (e.g., a call center operator assists you from a remote location). More important, many components of the service do not require any human interaction by the server at all! The ambiance of the restaurant, the décor of the aircraft, the comfort of the seats—none require personal service.

The trend today—largely driven by advances in technology and the need to lower costs to be competitive—is to provide many services that previously required personal human interaction through machines or electronic devices. For example, your airline ticket is provided to you at the airport by a machine that recognizes your identity, in the form of an e-ticket, or you are able to withdraw funds from an ATM machine rather than having to go inside a bank and ask a teller. Your calls are directed to the right individual or department through an automatic voice response (AVR) system rather than a telephone operator.

**How to determine what services to provide.** Corporations that are known for their good service are frequently willing to provide services above and beyond. In today's competitive environment, keeping customers satisfied is critical. However, providing services above and beyond usually means that it costs the firm in terms of resources to provide the additional service.

Was the additional investment desirable? Many times this investment is warranted for strengthening the relationship—it keeps the customer happy and does not cost the firm much in resources. At other times, the service provided does not resonate with the customer, and while you incurred the cost, it does not increase the satisfaction of the customer.

How do we ensure that the investment we make in satisfying the customer is valuable to the customer? To start, we need to better understand the customer—what is valuable to them. If we are providing services that they do not value, it does not matter how well we deliver them; they will not increase customers' satisfaction. On the other hand, we need to be honest about the services we provide and determine what we are good and not good at providing.

If we try to plot all the services we offer in the following four quadrants, it will help us to determine the right strategy for maximizing customer satisfaction and creating a competitive advantage (see Figure 2.3).

· *Quadrant 1*—The customer places high value and it is your strength as a Service Provider.

You have hit a bull's eye! Here are a set of services that are your sweet spot. The customer is excited about these services because they are important to him or her and add value to the organization. You have the domain knowledge, and you know you are good at delivering these services. This is what gives you the edge over your competition.

|  | Customer Values | Customer Does Not Value |
|---|---|---|
| Strength | Emphasize | Discontinue (or find a customer that values these services) |
| Not a Strength | Consider alternative delivery strategy, e.g. partners | Discontinue |

**FIGURE 2.3.** Service Evaluation Matrix

· *Quadrant 2*—Customer places high value, but it is *not* your strength as a Service Provider.

If it is important to the customer, it has to be important to you. However, if you are not good at delivering these services—do not mess it up! The customer will not forgive you easily for not delivering well on something that was important to him or her. No organization can be good at delivering every service—so why not find an organization that is good at it and work with it in partnership. The whole concept of outsourcing is based on the premise that each organization is good at something and hence needs to focus on its core competency.

According to The Outsourcing Institute, the top two reasons for outsourcing are:

1. Improve the focus of the company
2. Gain access to best-in-class capabilities

Outsourcing allows you to focus on your sweet spot, and your partner can enhance your image with your customer by providing the other services important to your customer in an efficient and effective way.

Alternatively, if, due to high value placed by the customer, you wish to make this a core competency, then you may wish to follow an acquisition strategy.

· *Quadrant 3*—The customer places *low* value, but it is your strength as a Service Provider.

The question is very simple. If you are not doing the right things, what is the value of doing things right? If customers do not want a particular service, then you are only increasing your cost by maintaining that capability. So, sell that capability to someone who needs that service. With the advent of computers, mechanical typewriters have become obsolete. Most of the typewriter operations have closed down or have gone into Chapter 11. For instance, Smith Corona, a leader in its heyda, is currently in Chapter 11 and now trying to turn around to leverage its capability in the small office equipment arena.

· *Quadrant 4*—Customer places low value, and it is also not your strength as a Service Provider

The answer here is pretty obvious—stop providing these services! These services will only take away precious management time from things that are important to your customer. You probably have blocked investment in these services—training, resource material, certifications, people, space, that could be better utilized focusing on your strengths.

## CONCEPT OF QUALITY IN SERVICE

Defining quality has never been easy in the service sector. Quality has its roots in the manufacturing environment. However, the concept of quality is associated less with the product and more with the customer—it is not what the engineer would define as acceptable, but what the customer expects for the product to perform reliably. This desire to reduce variability and improve the reliability resulted in corporations adopting quality control techniques such as Six Sigma, total quality management (TQM), and statistical process control (SPC).

If we adopt the broader, customer-centric view of quality, we will see that there are minimal differences between a service environment and a manufacturing environment. In their article in *Harvard Business Review*, C.K. Prahlad and M.S. Krishnan talk about how Disney defines its quality of service. "Disney's complex amusement rides have to work flawlessly all the time. But the Disney magic isn't just about defect-free rides. It is about managing emotions, expectations, and experiences." The situation becomes even more complex when you have a variety of customers, there are wide variations in their expectation, and each consumer may have different tolerance levels. Disney sends out live Disney characters for crowds to see and play with when there are long lines for favorite rides. It creates a diversion and entertains the crowd.

A few years ago, while traveling in India as a tourist, we came across a quaint restaurant called Tomatoes in Ahemdabad, India. Well known for its pizza, it faced a very similar problem to Disney—long waiting queues of customers during peak hours. With creative minds at work, the management hired students from the local university to entertain the waiting customers with magic tricks. It truly was a novel experience—totally unexpected—and fun for all the waiting guests.

Charles Schwab customers while on hold on the telephone line can select to listen to financial news, stock quotes, and other news, to make the wait more interesting and less onerous. This is much more client-friendly than the majority of companies that insist you listen to a description of their services and use it for a sales pitch, even if you have no interest in it. Some companies will provide you with an estimate of the time you may have to wait—part of managing your expectations.

One firm that has successfully embedded the service culture into the organization is Hewitt Associates. Through the vision of "one-firm firm" under the leadership of its CEO, Dale Gifford and his leadership team, the firm operates as a single service organization, globally. If you are a client, it does not matter where in the world you are located, and what portfolio of services of the firm you utilize, you are able to tap into the resources of other consultants and quickly leverage their expertise to obtain world-class solutions. Given that Hewitt offers a broad spectrum of services from HR consulting to technology-driven outsourcing, this is a critical differentiator for Hewitt and it provides them a strong competitive advantage.

When focusing on quality, many product-based operations focus on the specifications of the product, engineering, function and features, operational performance (breakdown, defects), and the like. The service aspect is considered less important, and the service element comes typically in the after-sales services' part of the value chain. It is considered an unnecessary cost and primarily a feeder to the product design team to develop a better product. On the other hand, for service firms, the mantra appears to be to please the customers at all costs with little focus on economic considerations. Clearly, we cannot ignore service quality in either situation. Figure 2.4 describes the drivers of the service component and their importance in any organization.

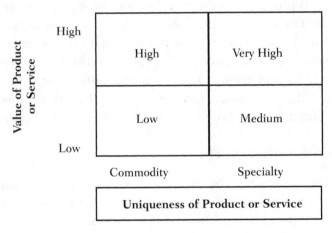

**FIGURE 2.4.** Service Expectation Matrix

In the figure above, the y-axis is defined as the value rather than the price. The reason is that the value of a product varies by customer while the price is fixed (same) for all customers. For example, an organization that does not use consultants may place high value (and expectations) on a $100,000 assignment, while another organization that spends millions of dollars on consultants may consider it a low value assignment and have much lower service expectation.

**Commodity products.**    A customer's expectation of service is "low" for purchase of a low-value "commodity" product or service (e.g., you do not expect much service when you purchase a school notebook at a departmental store). However, if you were going to purchase a low-end automobile, even though it is a commodity, you expect someone to attend to you, explain all the functions and features, take you for a test drive, and help you make the right selection (if your mind is not already made up). These service elements will increase if you are purchasing a luxury vehicle—you may expect free servicing along with a courtesy vehicle, free roadside assistance, and free car washes.

**Specialty products.**   On the other hand, if you want a custom-made bouquet of flowers, depending on the value you are paying, you may be willing to pick them on your own or have someone help you make the selection. You would still need someone to put it all together in a nice arrangement for you to gift it. On the other hand, if you wish to have an outside contractor build your basement or a home, which is a much larger investment, your expectation of service would be much greater. You would expect them to prepare multiple designs and help you select the right one, purchase all required materials, manage the construction process, and obtain all required permits.

## SERVICE COMPONENTS

As we start to analyze the delivery of service, we can categorize it into three major components—process component, environmental component, and personal component (see Figure 2.5 and Table 2.1).

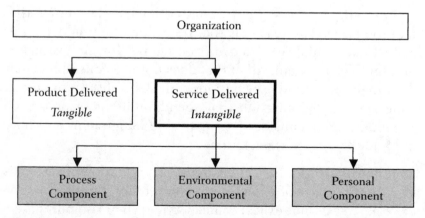

**FIGURE 2.5.** Components of Service Delivery System

**TABLE 2.1**    Examples of Process, Environmental, and Personal
Component in an Industry

|  | PROCESS COMPONENT | ENVIRONMENTAL COMPONENT | PERSONAL COMPONENT |
|---|---|---|---|
| Retail Store | • Selection/ variety | • Display and layout | • Welcome on entry |
|  | • Ease of purchasing | • Feeling while shopping | • Responsiveness of sales associates |
|  | • Ease of delivery | • Smartness of associates | • Mannerisms of associates during interaction |
| Restaurant | • Waiting and seating process | • Ambiance | • Attention on entry |
|  | • Ordering | • Comfort of seats | • Service from the waiter |
|  | • Serving the meal | • How the food smells and looks | • Dealing with special requests |
| Airline | • Reservation | • Décor and cleanliness of aircraft | • Courtesy of check in and flight staff |
|  | • Check in | • Appearance of flight attendants | • Service during flight |
|  | • Departure and arrival at destination | • Comfort in flight | • Handling |

**Interaction and transaction.**    Another dimension for categorizing service delivery is where there is direct "interaction" with the customer, and where there is not, "transaction." So, any given service delivery element can have an interaction component and a transaction component. For example, in the retail store situation, the services element "ease of purchasing" is interactive with the customer with respect to the customer being able to make the selection, place the order, and pay for it. The "transaction" component to that process includes:

· Updating the inventory records for sale of the item
· Preparing an invoice ensuring the correct price

- Checking the credit worthiness of the customer
- Processing the payment

In today's digital world, the customer does not see this aspect of the service; however, the speed and accuracy with which it is performed is critical to the satisfaction of the customer.

A customer becomes upset if:

- The checkout process takes long.
- He or she is charged the wrong price.
- Credit card takes forever to get approved.
- The computers are not working.
- Tape runs out in the register.

Eventhough, these are invisible components of the process, they are very important.

These transactions can be visible to the customer too. In fact, before the proliferation of the computer, they were performed in front of the customer—and in many parts of the world, they are performed that way even today. The important aspect is not whether they are "visible" or "invisible," but that they are performed quickly, efficiently, and accurately while minimizing the waiting time for customers, as these are all nonvalue-adding activities for them.

Thus, the process component clearly has "interaction" and "transaction" elements, but what about the environmental and personal components? Even environmental and personal components can have a "transaction" element to them—but they can vary by element.

In the preceding example of the retail store, a good display is important to the environmental component. You use logistics (that are transaction-based) to constantly replenish the goods that are sold and have them neatly displayed throughout the day.

Let us take another example of environmental component, this time in the airline industry, regarding the décor and cleanliness of the aircraft. The décor may be static for a given flight in a given section, but certain airlines change the décor to be more pleasing to their customers as they fly across different international sectors. As for cleanliness, that requires significant planning ahead to ensure the aircraft is cleaned by a crew everywhere it makes a landing.

Finally, it would seem that the "personal" component would have little in terms of "transaction" or noncustomer facing elements. A closer look would suggest many "transactional" elements exist here too. The last time I took my car for an oil change, the new service manager addressed me by my name as I walked up to him. I was impressed. It took me a moment to figure out how he knew my name when I had never met him before. Here is what he did: As I walked from my car to the main entrance, he got a glimpse of my car's number plate, keyed it in, and instantly knew who I was, when I had last been there, and for what service.

Similarly, due to the importance of the customer interaction, a high level of training and feedback loop is required that is very transaction based.

If you accidentally lose your baggage during a flight, you can register your complaint with an airline official. However, numerous transactions follow to record the information in the database, send out baggage tracer messages, and record the status of events until resolution of the problem. If a payment has to be made, it will also trigger such requests to the appropriate department and maintain the entire history related to the claim. Hence, each of the three components—process, environmental, and personal—can have both "interaction" elements, which deal directly with the customer, and a "transactional" element that focuses on preparation, sharing of information, and tracking progress.

Let's use the retail store situation again and consider the interaction versus transaction components (see Table 2.2).

As shown in the preceding example, while the "interaction" with the customer is a critical component of satisfying the customer, the support by the "transactional" elements is equally critical. Particularly in today's environment, our dependence on "transactional" elements has increased tremendously. A sales associate at the checkout counter can be extremely polite and courteous; however, if the computer system was not functioning, in many stores, they would not be able to process the sale, as it would inhibit:

· Verification of credit
· Processing of credit
· Recording of the sales transaction
· Updating of the inventory record

This would potentially result in a dissatisfied customer.

TABLE 2.2    Retail Store: Interaction versus Transaction

|  | PROCESS COMPONENT | ENVIRONMENTAL COMPONENT | PERSONAL COMPONENT |
|---|---|---|---|
| Activity | · Ease of Purchasing | · Display of Merchandise | · Price Check for an Item |
| Interaction | · Product selection | · Ability to look and feel merchandise | · Courteousness and responsiveness of the sales associate |
|  | · Checking out the selection and making the payment | · Create an attractive display of merchandise |  |
| Transaction | · Accurate scanning | · Tracking of inventory on display | · Ability to quickly and accurately check the price of an item |
|  | · Verification of credit<br>· Processing of payment | · Communication to warehouse to replenish display merchandise |  |

**Relationship with service component.** In the chart describing the service expectation matrix (Figure 2.4), the service expectation varied from "low" to "very high." As the service expectation increases, the "interaction" element of the service component also increases. In the example of purchasing a luxury automobile, you will get much greater attention and time from the salesperson. You will also be dealing with more experienced and trained professionals. You will get additional services like courtesy car during service, free roadside assistance, free car wash, and the like.

The "transactional" element in the service component does not increase proportionally. A minimum level of "transactional" support is required. Whether you purchased an economy car or a luxury car, the "transactional" elements are largely the same.

## STAGES OF EXCELLENCE

Most organizations that have a product offering struggle to define themselves as service organizations as well. Organizations that have realized that the service is the "differentiator" have moved to a different level—they do not segregate product from service—they are part of a single offer to the customer.

In the world today, products are becoming a commodity; hence, service allows organizations to differentiate themselves.

Dell is the leader in the PC business. Its product is not unique, however. You could purchase a PC from many vendors, and most of the components will be the same or similar. No other vendor has been able to replicate Dell's business model successfully, and Dell continues to distinguish itself in its service.

Southwest Airlines is another leader in its industry. After the September 11 terrorist attack on the World Trade Center, all U.S. airlines reported significant reduction in their passenger volume and an increase in operating losses. Southwest, well known for its friendly services, continued to operate profitably even in those difficult times and to provide its customers the usual high-quality service.

Another point to note is that higher quality service does not mean lower profits. Both Dell and Southwest are highly profitable in industries that are considered highly commoditized, and with low margins.

**Table 2.3    Stages of Excellence in Service**

|  | Stage 1 Traditional | Stage 2 Emerging | Stage 3 Leading | Stage 4 World Class |
|---|---|---|---|---|
| Strategy | None | Local strategy development | Central strategy development | Differentiated based on portfolio—firm wide |
| Leadership Practices | Reactive | Recognition of service as a differentiator | Proactive | Forward-looking |
| Service Organization | Isolated function Low ranking Limited resources | Mid-ranking focus across the organizations Some impact on decisions | Centralized | Leadership focus |
| People | No training and development | Limited training and development | Well educated staff Formal training | Highly educated staff program development |
| Process | No customer feedback process | Annual customer satisfaction survey | Active and periodic solicitation of customer feedback | Customers' input embedded in the design of service |
| Technology | Independent systems—not integrated | Basic and independent systems in place, some integration | Well integrated systems | Standardized, integrated, and modular systems |
| Measurement | No measures, no incentive | Few measures, little incentive | Quantitative and qualitative evaluation, performance-based incentive | … same and highly linked to compensation |

Table 2.3 categorizes the growth of services within an organization into four stages. It does not really matter whether you are a product-based service organization or a pure-play (nonproduct based).

**Strategy.** This refers to the understanding of the market and the competition to develop a differentiation for long-term competitive advantage. The Stage #1 organizations really do not have a defined strategy, while the leading firms have a differentiated strategy to serve each market segment.

**Leadership practices.** Many companies struggle to even recognize that service can be an excellent differentiating factor; the leaders leverage it extensively to create more value. The Stage 1 organizations consider service an overhead cost and minimize resources devoted to it—they constantly react to customers' service needs. The leading firms have a planned and proactive approach that addresses any potential issues even before the customer faces them.

**Service organization.** Importance of the service organization correlates to the service maturity of an organization. The Stage #1 organizations struggle to invest in the service organization, considering it a less important function. The more mature organizations look at the service organization as their customers' pulse and the voice that helps shape future products and services to be delivered.

**People.** Quality of people—their qualifications, training, and attitude—all are reflected in the service. The Stage #1 organizations that give service a low priority, also do not invest in quality of the resources supporting that organization. The Stage #4 firms invest heavily in the education and training of its human resources supporting the customer.

**Process.** This reflects the importance of obtaining feedback from customers to determine how well their needs are being met. Early-stage organizations do not see the need for customer feedback, while the advanced stage organizations embed that as a routine process that encourages customers and provides them with the customers' views of the organization.

**Technology.** This refers to how much technology to use and how efficient and user-friendly it is for the customer. While the cost of technology has significantly come down, the rate of innovation has resulted in this remaining a high cost of operation. Also, advancements in technology have the potential to increase the service level at a lower cost. The mature service organizations have invested in the technology that adds value to the customer and is integrated with the organization's own processes.

**Measurement.** This is a reflection on accountability and how the service provider measures its performance. The non-progressive firms do not see the need for customer feedback and do not recognize as necessary their customer service personnel for a high level of customer satisfaction. The progressive world-class organizations not only create qualitative and quantitative scorecards, but also link performance to pay.

## TRANSACTIONAL SERVICES

Considering that the performance of each link in the service value chain (Figure 2.1) is critical to the overall service and satisfaction of the customer, Table 2.4 summarizes some of the transactional elements in each link of the value chain.

It is apparent that there are a significant number of transactional elements in any service value chain. For the customer, who typically is not faced with many of these transactional elements, the assumption is that these transactional operations are working perfectly and efficiently. Whenever these transactional elements fail to function and the customer becomes aware of the flaw (because it affects him or her), it gives rise to irritation and dissatisfaction with the service.

These transactional elements are the easiest to fix and would significantly improve the quality of service delivered. Given that most organizations in the United States are operating at the three Sigma level of processing efficiency, there is significant room for improvement, and Six Sigma analytical tools and techniques could assist in that process.

Let us take an example of a supporting process like accounts receivable collections. If the receivables are not col-

TABLE 2.4    Transaction Elements in the Value Chain

### TRANSACTIONAL ELEMENTS

| | |
|---|---|
| Demand Forecasting | • Product enhancements and new product pipeline |
| | • Customer demand forecast by product and by location |
| | • Conversion of customer demand into product specifications and components to be ordered |
| | • Product scheduling |
| Supply Contracting | • Supplier database by product |
| | • Supplier capability, capacity, and pricing negotiation |
| Procurement | • Purchase orders |
| | • Delivery scheduling |
| | • Terms and conditions |
| | • Supplier and product performance |
| Inventory Management | • Goods received versus goods approved and available for sale to customer |
| | • Warehousing and inventory recording |
| | • Defective/damaged goods |
| | • Inventory tracking by product and by location |
| | • Accounts payable |
| | • Payment processing to suppliers |
| Marketing | • Customer profiling |
| | • Advertising effectiveness tracking |
| Sales Management | • Sales and backlog tracking by product and by location |
| | • Customer data management |
| | • Customer credit approval |
| | • Recording of sales transactions and credit sale (accounts receivable) |
| | • Rebate management |
| | • Inventory recording |
| | • Inventory replenishment request to warehouse |
| | • Delivery management |
| | • Cash transaction processing |
| After-sales Support | • Goods returned—customer cause tracking |
| | • Product performance tracking |
| | • Inventory tracking by product and by location |
| | • Warranty management |

lected on time and efficiently, they could significantly increase the investment in the business. Higher investment translates into lower return and decreasing shareholder value for the corporation. A service provider had committed to collect receivables and manage the overdue bills within certain parameters. However, the situation was getting out of hand. High transaction processes relating to billing and collections were involved.

The problem was given to a team of Six Sigma specialists to analyze and find a solution. On an overdue accounts receivable base of over $6.1 million, the objective was to reduce it by approximately 5% ($0.3 million). Through a detailed analysis, the team was able to strike at the heart of the problem and deliver results far exceeding the expectation. Against a target of 5%, they were able to obtain a 23% improvement that was almost five times the target! It resulted in reduced investment of $1.5 million in the business in addition to other benefits.

The following results were achieved:

- Collections increased (investment reduced) by 23%.
- Accounts going to an outside collection agency were reduced by 35%.
- Collection effort and related expenses (e.g., number of calls) were reduced by 16%.
- Follow-up turnaround within 24 hours reached 100% achievement.
- Accuracy in updating information improved from 98% to over 99%.

In addition, there was improvement in first-time resolution of queries, and the workforce productivity and morale also improved.

As you can see, the benefits are far-reaching and not focused entirely on the task at hand. More details are provided in the summary of the case study (see Table 2.5 and Figure 2.5).

## A    C A S E    S T U D Y

*Six Sigma Quality Improvement Project for Accounts Receivable—*
*Collections Process*

**TABLE 2.5**    Use of Six Sigma in the Collections Process

| | DESCRIPTION |
|---|---|
| Problem Statement | Delinquent receivables (past due in excess of 60 days) as on March 31 for the business unit were $6.1 million. If not addressed, it could potentially result in missing the commitment under the SLA by the service provider. |
| Goal Statement | Reduce delinquent receivables as at March 31 by 5% within four months. |
| Start Point | Customer accounts being sent to the external collection agency by the business unit. |
| End Point | Monthly analysis of delinquent receivables for the business unit. |

FIGURE 2.5. Defining the Process

The Six Sigma team defined various other components—input, process, and output in the following measures:

Defining input measures:

- Timely receipt of Customer Report
- Completeness of Customer Report

Defining process measures:

- Number of calls per full-time equivalent (FTE) per day
- Accuracy of information to be updated
- Number of customer queries resolved
- Number of no-response calls
- Follow-up action not taken within 24 hours
- 100% coverage of "follow-up list"

Defining output measures:

- Receivables amount
- Amount forwarded to collection agency
- Number of invoices over 30 days per FTE

Improvement achieved in process:

- All calls are made during the morning
- Staffing is maximized in the morning
- High-dollar calls are made by experienced callers
- Research and follow-ups are done in the afternoon
- High focus on accuracy in updating information

Results achieved:

- Collections increased (investment in accounts receivable reduced) by 23%
- Accounts going to an outside collection agency were reduced by 35%
- Collection effort and related expenses (e.g., number of calls) were reduced by 16%
- Follow-up turnaround within 24 hours resulted in 100% achievement
- Accuracy in updating information improved from 98% to over 99%
- Improved query resolution
- Improved productivity and morale of workforce

Considering the improvements in the process that were achieved, will it result in higher customer satisfaction? Was it worth the investment in the Six Sigma effort? There is no doubt that it will increase customer satisfaction. The investment in overdue accounts receivable adds no value to the customer. Once this investment is released, it is available for initiatives that will be attractive to the customer—that add value to them. Improved efficiency would potentially lead to lower costs for them too. Improving the quality of the transaction processes is an area where Six Sigma methodology and tools could be very effective and provide a significant return on investment.

# KEY TAKEAWAYS

· Every organization is a service organization—some deliver products as well.

· Drive your service model based on what is valued by the customer.

· A customer's expectation of service varies depending on the value they place on a particular service. Different customers can place a different value on the same service.

· A service has three components—process, environmental and personal—each component can potentially have an interaction and a transaction dimension—each component plays a critical role in determining the overall satisfaction of the customer.

· Organizations go through different stages of excellence based on the characteristics of their service philosophy and can mature to World Class.

· Use of Six Sigma methodology to solve business issues can provide breakthrough results and improvement in service quality that are above and beyond expectation.

# SERVICE PERFORMANCE INDICATORS

## IF IT IS IMPORTANT, IT MUST BE MEASURED

## WHY MEASURE?

Many management gurus state that if you cannot measure it, you cannot manage it. Others will tell you, "What gets measured, gets done." Measurement is nothing new. From the early days of management by objectives (MBO), which required a goal to be specified, we focused on measurement. You had to define a goal in some quantitative terms so that one could determine if it was actually achieved or not.

Today, the literature talks about SMART goals. The acronym SMART represents the characteristics of the goal:

- Specific
- Measurable
- Achievable
- Relevant
- Time-based

The "M" stands for the importance of being measurable. Measurability is also a prerequisite to accountability. If you cannot measure performance, how can you hold anyone accountable for it?

## MEASURING PERFORMANCE OF SERVICES INDUSTRY

To enable us to measure the performance of any corporation, one of the key sources of information is their financial statements. If one compares the financial performance ratios over a period of time, one can determine if the performance has improved or deteriorated. Figure 3.1 shows the performance of Fifth Third Bank over the last five years. If we look at measures such as "revenue growth" and "operating margin," we can comment on the historical performance. One of the best forward looking measures is the share price, which is based on the future economic potential of the firm.

The Fifth Third Bank is part of the banking industry. If we look at its performance in isolation to the industry's performance, it provides only a partial picture.

If the banking industry was growing much faster than the growth experienced by Fifth Third Bank, then its growth was insufficient and did not keep pace with the industry. As a result, it would lose market share. To enable such comparison, certain organizations measure and track the performance of the entire industry. This helps give a benchmark against which to compare. The Dow, Nasdaq, and S&P 500 are all indices that track performance of all their constituents in a consolidated manner, as well as on a more selective basis.

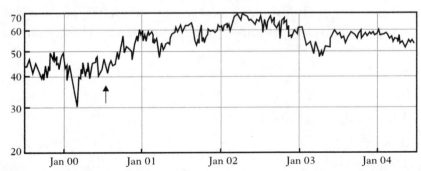

**FIGURE 3.1.** Performance (Share Price in Dollars) of Fifth Third Bank Shares over Last Five Years

Let us review the growth of Fifth Third Bank compared to the Nasdaq Bank index over the last five years.

As shown in Figure 3.2, the Fifth Third Bank's performance was better than the Nasdaq Bank index for almost three years. Since January 2003 (the last year and a half), however, it has slowed down tremendously. Hence, even if the Fifth Third Bank's performance may have shown improvement over the last year and a half, it has still not performed well compared to its peers in the industry.

Figure 3.3 is the comparison to a broader selection of companies, the Nasdaq composite index.

Hence, the industry indices, though based on publicly traded firms, still provide a useful benchmark to compare performance for any corporation.

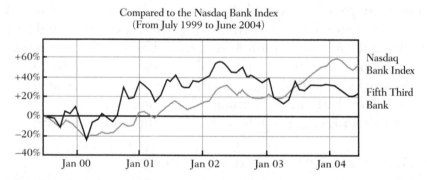

FIGURE 3.2. Relative Performance of Fifth Third Bank over Last Five Years

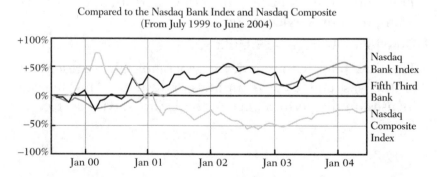

FIGURE 3.3. Relative Performance of Fifth Third Bank over Last Five Years

Measuring service performance is not easy. For publicly traded companies, we at least have share price as a comprehensive measure of overall performance of the corporation. For private companies even that measure is not available. Let us look at various organizations and institutions that help us measure services in more detail.

# DOW, NASDAQ, S&P 500, JD POWERS

## PERFORMANCE OF SOME SERVICE INDUSTRIES LISTED IN DOW INDUSTRIALS

Three major Dow Jones Averages are quoted—the most widely followed is the first one, "Dow Jones Average—30 Industrial." It is prepared and published by Dow Jones & Co. and is one of the oldest and most widely quoted of all the market indicators. The Dow Jones Industrial Average is comprised of 30 stocks that are major factors in their industries and widely held by individuals and institutional investors. These 30 stocks represent about one-fifth of the $8 trillion-plus market value of all U.S. stocks and about one-fourth of the value of stocks listed on the New York Stock Exchange (see Table 3.1).

Two averages are totally focused on the service sector:

· Dow Jones Average—20 Transportation

The Dow Jones Transportation Average represents 20 stocks in the airline, trucking, railroad, and shipping businesses. This measure focuses on the performance of the Transportation service sector.

· Dow Jones Average—15 Utilities

The Dow Jones Utility index is geographically representative of the gas and electric utilities industries. This measure focuses on the performance of the Utility service sector.

TABLE 3.1    Components of the Dow Jones 30 Industrial Average

| SYMBOL | NAME | SECTOR | INDUSTRY |
|--------|------|--------|----------|
| **Primarily Service Industry** | | | |
| AIG | AMER INTL GROUP | Financial | Insurance (Prop. & Casualty) |
| AXP | AMER EXPRESS CO | Financial | Consumer Financial Services |
| C | CITIGROUP | Financial | Money Center Banks |
| DIS | WALT DISNEY CO | Services | Broadcasting and Cable TV |
| HD | HOME DEPOT INC | Services | Retail (Home Improvement) |
| JPM | JP MORGAN CHASE | Financial | Money Center Banks |
| MCD | MCDONALDS CORP | Services | Restaurants |
| SBC | SBC COMMS | Services | Communications Services |
| VZ | VERIZON COMMS | Services | Communications Services |
| WMT | WAL-MART STORES | Services | Retail (Department and Discount) |
| **Other Industries** | | | |
| AA | ALCOA INC | Basic Materials | Metal Mining |
| BA | BOEING CO | Capital Goods | Aerospace and Defense |
| CAT | CATERPILLAR INC | Capital Goods | Constr. and Agric. Machinery |
| DD | DU PONT CO | Basic Materials | Chemicals – Plastics & Rubber |
| GE | GENERAL ELEC CO | Conglomerates | Conglomerates |
| GM | GENERAL MOTORS | Consumer Cyclical | Auto and Truck Manufacturers |
| HON | HONEYWELL INTL | Capital Goods | Aerospace and Defense |
| HPQ | HEWLETT-PACKARD | Technology | Computer Peripherals |
| IBM | INTL BUS MACHINE | Technology | Computer Hardware |
| INTC | INTEL CORP | Technology | Semiconductors |
| JNJ | JOHNSON&JOHNSON | Healthcare | Major Drugs |
| KO | COCA COLA CO | Consumer Noncyclical | Beverages (Nonalcoholic) |
| MMM | 3M COMPANY | Conglomerates | Conglomerates |
| MO | ALTRIA GROUP | Consumer Noncyclical | Tobacco |

**TABLE 3.1**    Components of the Dow Jones 30 Industrial Average
    (*Continued*)

| SYMBOL | NAME | SECTOR | INDUSTRY |
|--------|------|--------|----------|
| MRK | MERCK & CO | Healthcare | Major Drugs |
| MSFT | MICROSOFT CP | Technology | Software and Programming |
| PFE | PFIZER INC | Healthcare | Major Drugs |
| PG | PROCTER & GAMBLE | Consumer Noncyclical | Personal and Household Products |
| UTX | UNITED TECH CP | Conglomerates | Conglomerates |
| XOM | EXXON MOBIL | Energy | Oil & Gas—Integrated |

Many of the corporations that would not consider themselves in the service industry have a significant component of their operation related to services.

## PERFORMANCE OF SOME SERVICE INDUSTRIES LISTED IN NASDAQ

Developing prominence through the high-tech companies, the Nasdaq became very popular. It has many indices, including a number of them focused on the service industry.

**NASDAQ Composite Index.**    The NASDAQ Composite Index measures all NASDAQ domestic and international-based common type stocks listed on The NASDAQ Stock Market. Today the NASDAQ Composite includes over 4,000 companies, more than most other stock market indices. Because it is so broad-based, the Composite is one of the most widely followed and quoted major market indices.

On February 5, 1971, the NASDAQ Composite Index began with a base of 100.00.

In addition to the composite index is the Nasdaq 100 Index and a portfolio of sector-specific indices (see Table 3.2).

**NASDAQ-100 Index.**    The NASDAQ-100 Index includes 100 of the largest domestic and international nonfinancial companies listed on The Nasdaq Stock Market, based on

TABLE 3.2    Sector-Specific NASDAQ Indices

| SYMBOL | NAME | INDEX VALUE (AS OF MAY 20, 2004) |
|--------|------|----------------------------------|
| IXFIN | NASDAQ Financial-100 | 2,472.80 |
| IXID | NASDAQ Industrial | 1,588.37 |
| IXTR | NASDAQ Transportation | 1,684.48 |
| IXBK | NASDAQ Bank | 2,784.42 |
| IXTC | NASDAQ Telecommunications | 172.17 |
| IXIS | NASDAQ Insurance | 2,930.35 |
| IXCO | NASDAQ Computer | 862.54 |
| IXFN | NASDAQ Other Finance | 3,124.88 |
| NBI | NASDAQ Biotechnology | 740.47 |
| NIN | NASDAQ NM Industrial | 649.12 |

market capitalization. The Index reflects companies across major industry groups including computer hardware and software, telecommunications, retail/wholesale trade, and biotechnology. It does not contain financial companies, including investment companies. The NASDAQ-100 Index is calculated under a modified capitalization-weighted methodology.

On January 31, 1985, the NASDAQ-100 Index began with a base of 250.00.

On January 1, 1994, the NASDAQ-100 base was reset by division by a factor of 2.00 to 125.00.

**NASDAQ Financial-100 Index.** The NASDAQ Financial-100 Index includes 100 of the largest domestic and international financial organizations listed on The Nasdaq Stock Market, based on market capitalization. The Index contains bank and savings institutions and related holding companies, insurance companies, broker dealers, investment companies, and financial services.

On January 31, 1985, the NASDAQ Financial-100 Index began with a base of 250.00.

**NASDAQ Bank Index.** The NASDAQ Bank Index contains all types of banks and savings institutions and related holding companies, establishments performing functions closely related to banking, such as check cashing agencies, currency exchanges, safe deposit companies, and corporations for banking abroad.

On February 5, 1971, the NASDAQ Bank Index began with a base of 100.00.

**NASDAQ Biotechnology Index.** The NASDAQ Biotechnology Index contains companies that are classified according to the FTSET™ Global Classification System as either biotechnology or pharmaceutical, which also meet other eligibility criteria. The NASDAQ Biotechnology Index is calculated under a modified capitalization-weighted methodology.

On November 1, 1993, the NASDAQ Biotechnology Index began with a base of 200.00.

**NASDAQ Computer Index.** The NASDAQ Computer Index contains computer hardware and software companies that furnish computer programming and data processing services, and firms that produce computers, office equipment, and electronic components/accessories.

On November 1, 1993, the NASDAQ Computer Index began with a base of 200.00.

**NASDAQ Industrial Index.** The NASDAQ Industrial Index contains companies not classified in one of the other subindices, including agricultural, mining, construction, manufacturing, retail/wholesale trade, services, public administration enterprises, health maintenance organizations, and companies that do not meet the NASDAQ Biotechnology Index criteria.

On February 5, 1971, the NASDAQ Industrial Index began with a base of 100.00.

**NASDAQ National Market Industrial Index.** The NASDAQ National Market Industrial Index is a subset of the NASDAQ Industrial Index and consists of all companies included in the NASDAQ Industrial Index, which are listed on the NASDAQ National Market tier of The Nasdaq Stock Market.

On July 10, 1984, the NASDAQ National Market Industrial Index began with a base of 100.00.

**NASDAQ Insurance Index.** The NASDAQ Insurance Index contains all types of insurance companies including life, health, property, casualty, and brokers, agents, and related services.

On February 5, 1971, the NASDAQ Insurance Index began with a base of 100.00.

**NASDAQ Other Finance Index.** The NASDAQ Other Finance Index includes credit agencies (except banks and savings institutions and related holding companies), security and commodity brokers, exchanges and dealers, real estate, and holding investments companies.

On February 5, 1971, the NASDAQ Other Finance Index began with a base of 100.00.

**NASDAQ Telecommunications Index.** The NASDAQ Telecommunications Index contains all types of telecommunications companies, including point-to-point communication services and radio and television broadcast, and companies that manufacture communication equipment and accessories.

On November 1, 1993, the NASDAQ Utility Index was renamed the NASDAQ Telecommunications Index. The former NASDAQ Utility Index was reset to a base of 200.00, using a factor of 5.74805.

**NASDAQ Transportation Index.** The NASDAQ Transportation Index contains all types of transportation companies, including railroads, trucking companies, airlines, pipelines, (except natural gas), and services incidental to transportation, such as warehousing, travel arrangements, and packing.

On February 5, 1971, the NASDAQ Transportation Index began with a base of 100.00.

Given the importance of service sector, many of the NASDAQ indices track performance of specific service industries. This not only allows these service firms to have a benchmark to compare against, but also provides them with a feedback on how the market perceives them, and who they consider to be their competitors. This information, along with customer feedback can be helpful in improving the service quality to the customer and to gain competitive advantage.

## PERFORMANCE OF SOME SERVICE INDUSTRIES LISTED IN S&P 500

Widely regarded as the standard for measuring large-cap U.S. stock market performance, this popular index includes a representative sample of leading companies in leading industries. The S&P 500 is used by 97% of U.S. money managers and pension plan sponsors. Some $626 billion is indexed to the S&P 500. The performance of many mutual funds is compared to S&P 500 index as a benchmark.

## J.D. POWER'S VIEW OF SERVICE INDUSTRIES

J.D. Power and Associates has established itself as an organization that has monitored the quality of products and services for some time. Through its research and surveys it provides credible, meaningful, and easily accessible customer-based information to enable better decisions and results for business and consumers.

When Chairman J.D. Power III founded the firm in 1968, his goal was to provide an unbiased source of marketing information based on the opinions of consumers. Each year, they survey millions of consumers and business customers to understand their opinions and expectations regarding the products and services they purchase. Today, its clientele include numerous Fortune 500 firms, although it is best known for its coverage of the automotive industry. However, they cover a number of other industries, including:

· Commercial vehicles
· Finance
· Healthcare
· Marine
· Office products
· Professional services
· Real estate

- Retail
- Sports and entertainment
- Telecommunications
- Travel
- Utilities

The firm offers a variety of services to help businesses, as follows:

- Understand B2B and B2C customers
- Improve product and service quality
- Increase customer satisfaction levels
- Enhance the customer experience, including retail and dealer sales processes
- Increase operational efficiency

J.D. Power and Associates also conducts syndicated research based on the assessment of an industry via a survey of the customers in that sector. The purpose of syndicated customer satisfaction research is to:

- Establish competitive benchmarks for quality and customer satisfaction in an industry.
- Identify the strengths and weaknesses of individual companies in an industry with regard to product quality and customer satisfaction.
- Provide specific recommendations on how individual companies can improve their quality and customer satisfaction levels.

J.D. Power and Associates also conducts custom research. In addition, it measures and tracks customer satisfaction for individual companies on a proprietary basis. Proprietary services include:

- Customer satisfaction tracking research (to measure improvement over time).

- Quality tracking studies.
- Retail sales tracking.
- Dealer service tracking.
- Retailer quality audits.
- Certification programs to ensure compliance with predetermined industry benchmark standards.

It then provides clients access to weekly, monthly, or quarterly data on demand, via a customized extranet, to identify issues and enable quality and customer satisfaction process improvements.

J.D. Power and Associates also awards rankings for quality and service based on responses from consumers and business-to-business customers who have used the products and services being rated.

The firm's awards program provides consumers with product quality and customer satisfaction data that enable them to make more informed purchasing decisions. Companies that rank highest in the firm's studies may license the use of the J.D. Power and Associates name for advertising claims.

The following is a list of studies and surveys that are conducted by them in the automotive industry alone:

- Initial quality
- Dealer maintenance and repair services
- Collision repair and satisfaction
- Vehicle performance and features satisfaction
- Sales satisfaction
- Brand and owner loyalty
- Dealer attitudes
- Long-term dependability
- Supplier component quality reports
- Interior quality
- Audio quality

- Acoustic quality
- Seat quality
- Brake and handling quality
- Transmission quality
- HVAC quality
- Component branding
- Satisfaction with vehicle features
- Navigation features
- OEM website evaluation
- Online buying services
- Online shopping satisfaction
- Financing satisfaction—consumer
- Financing satisfaction—dealer
- Insurance satisfaction
- OE tire satisfaction
- Replacement tire satisfaction

As you can see, the review is very extensive and focuses on the "service" aspect of the industry. It not only assesses customer satisfaction with many specific components of the product that the customer can assess, but also with the service provided at various points of contact with the customer—the dealer attitude, the sales process, financing, insurance, and maintenance and repair services.

**Impact of measuring quality.** As an example, the *Initial Quality Study* measures a broad range of quality problems, heavily weighted toward defects and malfunctions, quality of workmanship, drivability, human factors in engineering (i.e., ease of use or convenience), and safety-related problems. Among these categories, the area that accounts for the greatest product improvement since 1998 are defects and malfunctions, down from 61 parts per 100 to 40 parts per 100. Further, those that are related to safety show a 44% improvement—from 25 parts per 100 down to only 14.

This industry benchmark study for new-vehicle initial quality has been conducted for 18 years, and their assessment has led to significant improvements by the product manufacturers and an increase in the satisfaction of the consumers.

# MEASUREMENT OF SERVICE QUALITY BY ORGANIZATIONS OTHER THAN REPRESENTATIVES OF THE STOCK MARKET

## 100 BEST PLACES TO WORK IN IT

Measurement is not limited to these few organizations. Organizations have multiple ways of measuring quality. Recently, Computerworld surveyed thousands of employees from various companies to determine the 100 Best Places to Work in IT in the United States. Its survey-collected data measured:

· Satisfaction with training and development programs
· Base salary
· Bonuses
· Health/life balance
· Morale
· Career growth
· Management's fair and equal treatment of employees

Based on the preceding criteria, it was able to analyze the information to establish rankings for the top 100 organizations in the United States.

## BEST COLLEGES IN THE UNITED STATES

U.S. News publishes a ranking of universities and colleges around the United States. It has developed a three-stage process that starts with categorization of the colleges, then collects information from a large sample that includes a college's student body and its faculty. In addition, they evaluate a college's financial resources and resulting performance—how well the institution does its job of educating students. Information from each university is collected through a survey and then analyzed to determine the overall ranking.

**Categorization.**    The categorization of colleges is based on their mission. In 2004, the selection included:

- 248 national universities that offer a full range of undergraduate majors, plus master's and doctoral degrees, and emphasize faculty research.
- 217 liberal arts colleges that focus almost exclusively on undergraduate education. They award at least 50% of their degrees in the liberal arts.
- 573 universities that offer a full range of undergraduate degrees and some master's degree programs.
- 324 comprehensive colleges that focus on undergraduate education, but grant fewer than 50% of their degrees in liberal arts disciplines.

**Data collection.**    Next, they gather data from each college on up to 15 indicators of academic excellence. These are based on input from education experts as reliable indicators of academic quality and the *U.S. News'* view of what is important in education.

These indicators of academic quality fall into seven categories (see Table 3.3).

**Analysis.**    Each measure is rated on a scale of 1 to 5, where 1 is marginal and 5 is distinguished. Each measure is weighted (see Table 3.3) based on judgment about the relative importance of the measure. This facilitates computation of a composite weighted score for each college.

To arrive at a school's rank, they calculate the composite weighted sum of its scores. The colleges in each category are then ranked against their peers, based on this composite weighted score. The final scores are rescaled, with the top

**Table 3.3** Measuring Academic Quality

| | Category | Additional Components/Description | Weight |
|---|---|---|---|
| 1 | Assessment by administrators at peer institutions | Assessment by top academicians—presidents, provosts, and deans of admission at peer institutions | 25% |
| 2 | Retention of students | Six-year graduation rate and freshman retention rate | 20–25 percent |
| 3 | Faculty resources | Average class size, faculty salary, proportion of professors with the highest degree in their fields, the student–faculty ratio, and the proportion of the faculty who are full-time | 20% |
| 4 | Student selectivity | SAT or ACT test scores, the proportion of freshmen who graduated in the top 10% of their high school classes, and the acceptance rate | 15% |
| 5 | Financial resources | Average spending per student | 10% |
| 6 | Alumni giving | Percentage of alumni who made contributions to their school | 5% |
| 7 | Graduation rate | Difference between a college's real six-year graduation rate for the class that entered six years ago and the predicted rate for that class | 0–5% |

school being assigned a value of 100 and the other schools' weighted scores recalculated as a proportion of that top score.

The preceding examples illustrate how even complex services like the service an employer provides its employees and the service of a university to its customer (the students) can be measured using various measures that represent what is important to an employee or the quality of the education received by a student.

## Best Business Schools

The *U.S. News* also publishes a ranking of all (377) master's programs in business accredited by AACSB International in the United States. Using a survey, it collects the data needed to calculate rankings based on a weighted average of the eight quality indicators and then determines the overall ranking in a manner similar to that for colleges.

Table 3.4 provides the ranking criteria that the U.S. News uses to evaluate the master's programs in business schools around the country. This is to contrast how similar processes can have very different measurement needs.

TABLE 3.4    U.S. News Ranking Criteria for Business Schools

|   | CATEGORY | ADDITIONAL COMPONENTS/DESCRIPTION | WEIGHT |
|---|----------|-----------------------------------|--------|
| 1 | Quality Assessment | By business school deans and directors of accredited programs (60%) and by corporate recruiters and company contacts (40%) | 40% |
| 2 | Graduate employment | Mean starting salary and bonus (40%) and employment rates for full-time M.B.A. program graduates, computed at graduation (20%) and three months later (40%) | 35% |
| 3 | Student selectivity | Mean GMAT (65%), mean under-graduate GPA (30%), and the proportion of applicants accepted by the school (5%) | 25% |

As you can note, some differences include:

· Input from corporate recruiters and company contacts, and
· The entire category of graduate employment.

On the other hand, there are some similarities too (e.g., in peer assessment and in student selectivity). Therefore, based on the desired outcomes, the quality criteria can be different.

## HOW EFFECTIVE ARE THE PERFORMANCE MEASURES IN A SERVICE ENVIRONMENT?

If it is important, it must be measured. Undoubtedly, performance measures are required to:

- Create accountability
- Create focus
- Provide direction
- Monitor progress

We would be lost without them—they help create a business map with a visible destination. If there were no measures, you would not be able distinguish between progress and deterioration. You would not be able to drive efforts for improvement to satisfy a target, as there would be no specific goal to achieve. Without performance measures, it would be very difficult to gain agreement between parties, to determine whether each fulfilled its part of the bargain.

Service Quality Institute based in Minneapolis suggests measuring and tracking certain measures monthly, quarterly, and annually. These include:

- Financial—Sales, market share and profits
- Customer related—Defection rate, customer count, customer satisfaction and repeat business
- Employee focused—Turnover, tardiness, absenteeism, grievances
- Others, such as, shrinkage, theft, claims, etc.

It is important to measure, but even more important is what we do with the information. Do we use it positively or negatively? Do we use it steer in the right direction or is it used to place blame? More positively we use the measures, greater the acceptance of the measures in direction setting.

Yet today, there is an increasing level of discomfort with performance measures. Given the importance of the stock market and the importance of the share price to the shareholders, there is disproportionate weight given to the financial measures. A statement by Donald Lehmann, a marketing professor at the Columbia Business School, captures this growing sentiment: "Customer awareness, customer satisfaction, and market share are metrics, and they are nice to know about. But the CEO [is more concerned with] shareholder value, market capitalization, return on assets, and return on investment."

What drives this kind of behavior? We can find volumes of text written on the importance of the customer, and rarely will you find a company that does not state that people are its most important asset—yet it seems that financial measures dominate the evaluation of a corporation's performance—to an extent driven by the stock market. The focus on quarterly earnings and on meeting stock analysts' expectations seems to be driving the culture in many organizations. When our CEOs start getting rewarded based on the shareholder value they create, it serves as the "icing on the cake." Even if they cared about other factors, the link between a measure and compensation will drive the behavior.

The corporate compensation committees have a major role to play. If a CEO's compensation was based on multiple measures, as promoted by the "Balanced Scorecard" concept of Kaplan, you would have a more balanced approach. If the CEO's compensation was determined by a combination of factors that, in addition to financial measures, included "customer satisfaction," "employee engagement," and "process efficiency," we would see a different result, a different behavior. The truth is that people like measures, but do not wish their compensation to be adversely affected if they are not achieved. A few examples illustrate this point.

In a recent article in the *Wall Street Journal*, it was stated that the CEO of a major wireless service provider had "subscriber growth" as one of the key elements in its performance target for 2002. In mid year, however, it became apparent that the wireless industry was growing at a slower pace than expected. As a result, the target was changed to maintaining market share instead of increasing it—so that the CEO could still be rewarded. In 2003, the market share goal was eliminated as a bonus criterion. And, even though the company lost market share in 2003, the senior executives received a significant increase in bonuses.

Similarly, there are additional examples which point toward a prevalence of executive compensation practices that suggest that if performance falls short of the target, adjust the target!

When performance measures get linked to compensation, they drive behavior. And, as we have seen in the recent past, if the measures are not properly balanced, they can also drive the wrong behavior. In the recent past, there are many examples of corporation leaders (such as Ken Lay of Enron), who, because of the compensation incentives they received, were motivated to make undesirable decisions.

Does that mean we should not have measures? Not at all; we need to identify the right measures and set the right targets. Otherwise, we are embarking on a journey without the right compass and without knowing our destination. We need to understand why we are measuring what we are measuring. And, when we plan to link measures to compensation, we need to realize that they will impact behavior. Hence, do we have the right set of measures (including compensating measures) so that they do not result in undesirable behavior?

## How Does Six Sigma Help Identify the Right Measures and Enhance Performance?

Six Sigma can play a significant role in helping on both accounts—identifying the right set of measures and enhancing performance. When we look at the analytical tool set offered by Six Sigma, the analysis provides much needed insight into the processes. While we could determine the high-level business objective without the help of Six Sigma, the analysis identifies additional supporting measures required to achieve the ultimate objective. Let us review the case study example at the end of Chapter 2. The overall objective or the measure was to maintain delinquent receivables (defined as overdue accounts over 60 days overdue) below a certain dollar limit. As the Six Sigma team analyzed the process because of the need to improve the performance, they identified numerous supporting measures required to achieve the primary measure, such as:

**Input measures:**

· Timely receipt of Customer Report
· Completeness of Customer Report

**Process measures:**

· Number of calls per full-time equivalent (FTE) per day
· Accuracy of information to be updated
· Number of customer queries resolved
· Number of no-response calls
· Number of follow-up actions not taken within 24 hours
· 100% coverage of follow-up list

**Output measures:**

· Receivables amount
· Amount forwarded to collection agency
· Number of invoices over 30 days per FTE

It is important to realize that just setting a target for delinquent receivables to be below a certain dollar limit is a result of a number of other activities being measured and meeting their target performance. By tracking the performance of these submeasures, management is better able to understand where the problem lies so that it can take appropriate action. Without a deeper understanding of these additional supporting measures, management may make undesirable decisions to achieve the target.

For example, to be extreme, the management decides to send highly threatening letters to the customers that are categorized as "delinquent." Receiving such threatening letters, some customers may pay up, but you may lose them as customers for the future. Some of them may have called to resolve disputes and it may be the corporation was slack in responding or resolving the queries—do you know that you are not behind in dispute resolution? Similarly, some of them may have paid their dues, but the corporation's cash receipts processing may be slacking behind—do you know that you have the most current stats at hand? Tracking these additional measures helps make the right decisions when exceptions occur.

Six Sigma is known for delivering breakthrough performance. Due to the structured and detailed analysis performed, it is able to deliver significant improvements in performance, as it is able to pinpoint most potential sources of error. Once the sources of error are monitored for satisfactory performance, the end result will always be more than satisfactory.

Let us continue to examine the preceding example of delinquent receivables in greater depth. The Six Sigma team identified two input measures and six process measures. Let us assume that each one of these activities was performing at 90% efficiency. In most business environments, where we apply the 80/20 rule of Pareto's law, this would be more than satisfactory performance. Would this result in achieving 90% of your target for the overall performance? Would you be shocked if it resulted in only 43% of your targeted performance? Simple math will tell us that is exactly what we should expect (Table 3.5).

TABLE 3.5    Cumulative Impact of 90% Efficient Activities in a Process

| PROCESS | ACTIVITY EFFICIENCY | CUMULATIVE EFFICIENCY OF PROCESS |
|---------|---------------------|----------------------------------|
| Activity 1 | 90% | 90.0% |
| Activity 2 | 90% | 81.0% |
| Activity 3 | 90% | 72.9% |
| Activity 4 | 90% | 65.6% |
| Activity 5 | 90% | 59.0% |
| Activity 6 | 90% | 53.1% |
| Activity 7 | 90% | 47.8% |
| Activity 8 | 90% | 43.0% |

If we only receive 90% of the reports on time and only 90% of them are complete, we are already down to 81% efficiency (90% x 90%). And by the time we multiply that six more times by 90% for efficiency of the supporting activities identified in process measures (i.e., $81\% \times 90\% \times 90\% \times 90\% \times 90\% \times 90\% \times 90\%$), the process will be performing at 43% efficiency. To convert this into a formula—if $E$ was the efficiency of each activity, and we had $n$ activities supporting a process, the overall efficiency of the process can be represented by $E^n$. The Six Sigma methodology helps identify the supporting activities that are relevant and that will impact the overall result of the process.

## SMALL SERVICE INDUSTRIES

**Importance of small service industries.**    If you recollect from Chapter 1, out of 6.7 million businesses in the U.S. economy, over 95% of the businesses are small firms that employ less than 50 workers (see Figure 3.4).

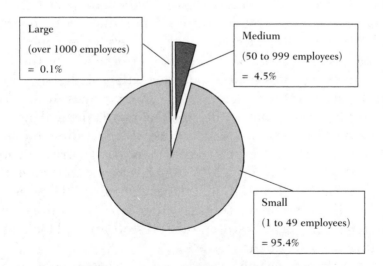

Large
(over 1000 employees)
= 0.1%

Medium
(50 to 999 employees)
= 4.5%

Small
(1 to 49 employees)
= 95.4%

Total number of businesses: 6.7 million

**FIGURE 3.4.**  Distribution of Business in the United States Based on Number of Employees

Further, 90% of these firms are in the services sector and account for 90% of jobs at small firms. When it comes to job creation, over 90% of the jobs (about 350,000) are created in the services sector through these small firms. So, the small service firms are a significant component of the service economy.

**Service sectors for small business.**    It is difficult to segregate industry sectors that are uniquely suited for services. Even the strong manufacturing sectors need to be supported by service. The question is not which sectors are more appropriate for service, but how can service add more value in any value chain.

## HOW PERFORMANCE MEASURES OF SERVICE FACTORS VARY BETWEEN SMALL AND LARGE FIRMS

Research has shown that the performance measures do not significantly change between small and large firms. However, the emphasis, or the importance placed, can vary between the two types of organizations.

For example, public corporations measure their success in terms of shareholder value. Shareholder value is at least one of *the* most important measures if not the most important measure. We have different stock markets where the public companies are listed. Based on the performance of the company (past and projected), and expectation of the shareholders, the share value is determined. And, changes in the value are visible every moment the stock market is open. This is an appropriate measure of success, as the investing public is looking for a return on their investment. Large private corporations use a similar measure for themselves even though they may not be listed on a public stock exchange. They usually fund their requirements through other financial sources (e.g., private placements); however, the investors are still looking for a return on investment.

What determines shareholder value? The three major factors that have the most significant impact on shareholder value are:

· Revenue growth—Greater the growth, greater the value it can generate

- Profit margins—Higher the margins, better the returns
- Investment required to generate profits—Lower the investment required, higher the returns to the investors

Would this situation be any different for small business owners? Would they not want to grow the business? Would they not want it to be profitable with the least amount of investment? They too would like to achieve these results. However, their emphasis and the time frames in which they may like to achieve them could be significantly different. As one entrepreneur said, "I don't think it's all about money. If it were all about money, I wouldn't be doing what I'm doing."

Typically, small businesses offer a unique selling proposition in a niche market. Many small business owners want to make a difference. A lot of them want to do well by doing good in their particular area of expertise, so they're driven by passion. So, do they measure their success in financial terms? They generally like to have financial security and stability—but they are not driven to maximize financial returns.

Most entrepreneurs would like to determine their own measures of success. When you work for a business you own, it is purely a matter of what success looks like for you. No one else even matters. As an entrepreneur, you want to have an advisory board (consisting of people whom you can trust), but you are not answerable to them or the shareholders. As a result, you have a certain amount of freedom to do what is right (in your view) at any given moment. Hence, the measures of success can vary significantly from one individual to another. The following examples provide how some entrepreneurs measured their success:

- New product development—Measured success by the number of patents they filed.
- Research in specific area—Measured success by the number of research projects and research grants they received.
- Passion for certain process—Measured success by throughput speed (automating the process and elimination of manual intervention).

- Innovation—Measured success in continuously driving to miniaturize certain medical devices.
- Desire to help the community—Measured success by number of community boards he was on.
- Support a particular cause—Measured success by the number of operations he could sponsor for the blind.
- Provide opportunity to members of the family—Measured success by the number of his kids who joined the business— gave each one their own small business to operate.

As you can see, the measures of success can vary significantly—but to survive, all businesses must achieve a certain level of financial success as well.

## KEY TAKEAWAYS

- If it is important, it must be measured.
- Goals must be SMART.
- Service quality is difficult to measure, but the stock market and other organizations have found creative ways to measure it.
- A process with eight activities each being performed at 90% efficiency will result in 43% efficiency for the entire process—90% quality rating of an activity is unacceptable.
- Six Sigma can provide a robust set of tools to measure quality.

# THE SERVICE CRISIS

## DEFINING SERVICE IS A CONTINUOUS PROCESS

L ittle has been written about the service industry (compared to the nonservice industry). This is not because there is lack of interest, but more because the customers' needs keep changing and vendors are continuously trying to add more value for their customers through more innovation. As a result, customers' expectations are continuously changing, and the service offers are redefining themselves (see Figure 4.1).

This service cycle consists of:

1. Defining or redefining customer requirements and understanding them.
2. Innovating a solution by leveraging knowledge and talent.
3. Delivering service.
4. Satisfying customer expectations.
5. Driving further improvement for additional value and adapting to the changing business environment.

Most service providers stop at the fourth step—once they have satisfied the customer. However, the last step is also a critical step in this cycle, as it requires service providers to con-

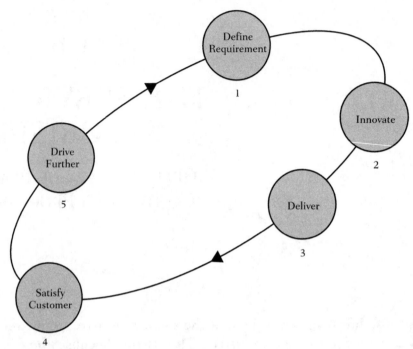

**FIGURE 4.1.** Customer Service Cycle

tinuously find new ways to add value while providing the service more efficiently and cost-effectively. Companies that sit on their laurels after satisfying a customer will eventually be left behind.

Today's business environment provides numerous challenges that a business must address. The only constant in business today is change. A service provider needs to constantly be aware of these changes and respond to them. Some of the major challenges include:

· Alignment with customer expectations
· Changing threshold
· Globalization
· Outsourcing
· Measuring in a state of constant change
· Cost and productivity

These challenges are described in more detail below, along with opportunities to address them.

## KEY CHALLENGES

**Alignment with customer expectations.**   It is rare to find service organizations that fully meet all the expectations of their customers. Why? It is very common even in today's customer-oriented environment for service firms to describe their value proposition in terms of their capabilities, their expertise, their capacity, and their experience. These are very important factors in providing good service; however, they do not present themselves in terms of the needs of the customer.

As an example, a customer has little to gain by knowing whether the vendor has its customer support center staffed by 100 or 200 persons (except for providing some assurance that it is above the critical mass required to impact service if a customer support representative [CSR] takes a day off). What they are interested in is, when they call the support center, how quickly will they get through to a CSR, how frequently will that person be able to solve their problems rather than having to refer them to a more qualified person, and how long will it take them to provide a solution?

The origins of service firms like FedEx are found in strong alignment with customer needs. In the early days, when the use of large expensive mainframe computers was growing, the need to deliver spare parts quickly in case of a breakdown was a critical customer need. FedEx guaranteed a 24-hour delivery and created a leading courier service.

**Changing threshold.**   Developments in technology continuously make business faster and less expensive. The pace of innovation has consistently increased, and we see new and technologically more advanced versions being introduced in the marketplace at ever increasing speed. The form and function that would "Wow!" the customer yesterday is something the customer starts to expect as a *given* today. This continuously changing threshold and raising of the bar requires service providers to constantly upgrade their offer and continuously find innovative ways to service the customer.

Let us look at Amazon as an example. The concept of shopping online for books started as:

- Providing the convenience of ordering books without having to go to the store
- Having a large selection of millions of books available to customers—not initially provided by a typical bookshop
- Offering a discounted price for books—sharing a large component of the margin provided by the publisher to the retailer with the customer

The preceding value proposition, when coupled with free shipping, attracted the attention of the consumer.

As the concept of online ordering became more popular, however, the model was copied by many of the other retailers, like Borders, Barnes and Noble, and WalMart. For them, online shopping was an additional distribution channel to their regular operation. They had to provide it because some of their customers were now looking for the convenience of online shopping—changing customer expectations.

As these retailers started to close the gap for Amazon, they had to raise the bar. Their web site evolved over time in form and functionality to include:

- Confirmation of order placed
- Tracking of the order
- Keeping history of customers' purchases
- Providing a selection of books to the customer based on their previous selections of interest
- Providing Table of Contents of books to let purchasers have some feel for the contents of the book before purchasing it
- Including the Introduction or a chapter of the book to let purchasers have a chance to browse the book before purchasing it
- Sharing comments from other readers about the book

- Increasing purchasers' involvement by allowing them to comment on the book
- Providing an opportunity to sell the book once customers have finished reading it, etc.

Every time Amazon would add functionality to its website to make the customer experience more enjoyable, it was only a matter of months in which the competition would also add a similar functionality. As Web sites can be accessed by anyone (except for any restricted areas), any competitor can view the content and functionality. As a result, one has to continuously innovate to stay ahead of the competition and keep raising the bar. It is important in the service industry to continuously find more and better ways to serve the customer—that is what provides the competitive advantage.

Products typically have protection available to them through patent and copyright laws. Such protection does not normally apply to service processes. In particular, web-based processes are relatively easy to copy and reverse engineer. There is little protection available to prevent competition from substantially copying form and functionality of the process. As a result, as discussed in the above example, there is a constant pressure to innovate to maintain the competitive advantage.

**Globalization.**    Globalization has meant different things to different people. The most common understanding is the concept of operating in multiple markets worldwide. Even in this concept there are different ways to operate. For example:

- Do you provide one standard product or service to any customer worldwide?
- Do you have a standard offer from a central location, but customized for the local needs of each country (e.g., language, culture, tastes, etc.)?
- Do you set up an operation in each country?

Depending on the business model, the challenges will vary in complexity.

While the above largely focused on a customer's perspective, there is another concept from a delivery perspective—setting up core operations in different parts of the world. This is more commonly known as "global sourcing" today.

Traditionally, corporations like IBM had their worldwide operations compete for manufacturing products at a given location for distribution around the world. Today, services are also being set up at central locations to serve customers around the world. Services like call centers, technical support, software development, and research centers are being set up to serve the needs of customers worldwide. While such global offers promise efficiency and lower cost, they create new challenges. Some of the traditional challenges include language, regulatory compliance, and culture.

- *Language*—While English is spoken in many parts of the world, not everyone speaks it. So how do we communicate with customers who would like to be served and prefer communication in their own language?
- *Compliance with local laws*—Knowledge of the local laws and regulatory requirements is critical to doing business in any country.
- *Culture*—Again, knowledge of culture is important as certain works, phrases, or mannerisms may be quite acceptable in one country, but highly offensive in another. For example, pork is a very common food in the United States but not in any Islamic country. McDonald's burgers in India are not made from beef as they are in the rest of the world because cows are considered sacred in India.

Today, new challenges in the form of terrorism, business continuity, and ethics are emerging.

- *Terrorism*—U.S. businesses and personnel are being targeted by terrorists. Safety of our people has become a very important issue when operating globally.
- *Business continuity*—Two business trends have raised the concern for business continuity. First, the increase in outsourcing

and off-shoring results in part of the operations being managed by someone else. In most situations, these are noncore processes, yet they are critical to the functioning of the corporation. Second, with increasing global unrest and terrorist attacks, many organizations are more concerned about business continuity plans as a backup for their operations.

- *Ethics*—The recent wave of corporate scandals and overnight collapses of seemingly healthy businesses as a result of unethical practices has raised the concern with ethics to new heights. Introduction of Sarbanes-Oxley reforms and legislation has attempted to increase disclosure and accountability with corporate executives. Considering that many of the unethical practices have been discovered in a nation ranked very high in ethical standards, it is worrisome to think what practices may exist in countries that rank much lower on the scale of ethical standards.

**Outsourcing.**   We discussed previously the trend toward business process outsourcing. This was a means of focusing on core competencies, taking advantage of specialization by others and a strategy to reduce costs. While these advantages are clear, it is difficult to ensure compliance with:

- Privacy laws relating to an individual's personal information
- Security of information that may be confidential or proprietary
- Protection of intellectual capital through copyright and patent laws

These issues continue to plague businesses with their outsourcing and globalization decisions.

**Measuring in a state of constant change.**   The importance of performance measurement is undisputed. There is a popular belief that "what gets measured, gets done." But a critical component of measurement is being able to compare against a target. For example, if a corporation achieves a profitability of 10% (as a percentage of investment), is it good? Well, the answer is that *it depends*. If the industry is averaging 8% profitability, then 10% profitability may be good. But, if the industry is averaging 12% profitability, then the 10% does not seem so good any more.

Another point of reference may be last year's performance, or a target set for the current year. In the preceding example, if we achieved profitability of 10% last year as well, we are maintaining status quo. If cost reduction efforts should have increased profitability to 11% this year, however, then we are not tracking to target.

In today's business environment, which is constantly changing, measurement has become a challenge. Regulations are changing, as are the customer's expectations and behavior. Challenges of globalization and economic uncertainty have made it difficult to determine what is a realistic target. One can make assumptions about economic factors, but reality will be discovered only after the time has passed. Given today's volatility, chances are that reality differs from your assumptions, and performance measures are not helping in managing the business proactively.

**Cost and productivity.**   If you have tried to call customer service or the help desk of one of the services you use, the chances are that you went through significant aggravation in punching numbers for various options, but could not speak to a live representative. Finally, when you were able to find a way to reach a representative, you were asked to call back, as the waiting times were too long. Corporations are trying to reduce their cost of service by providing self-help capability and supporting them with automated technologies like automated voice response systems. Customers are getting more and more frustrated, however, as many of the efforts are too time consuming for the customers and not leading to a speedy resolution of their problems. Typically, as productivity increases (through a greater use of technology), the cost of service should go down (while maintaining quality) (see Figure 4.2).

It appears that the quality of service has also decreased in the process, and now service providers are starting to substitute technology with service from lower cost locations. Finding the right balance between cost and quality and technology remains a challenge.

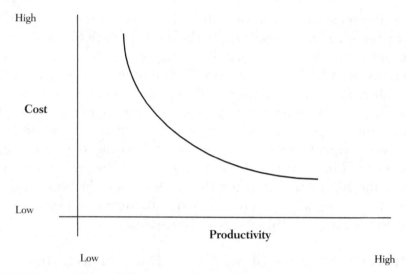

**FIGURE 4.2.** Effects of Productivity Improvement on Cost of Service

## OPPORTUNITY FOR IMPROVEMENT

It is believed that most American companies operate at a processing efficiency of three Sigma. Hence, it provides a significant opportunity to improve efficiency—from three Sigma to Six Sigma. Undoubtedly, improving the quality will result in some increase in processing cost as new controls, checks, and balances are put in place. The benefit of moving to Six Sigma will depend on the criticality of the process (see Figure 4.3).

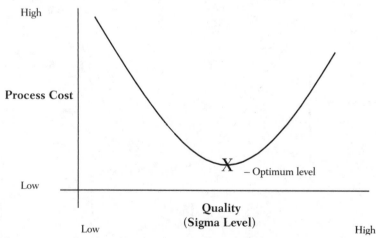

**FIGURE 4.3.** Optimizing Process Cost and Service Quality

An organization that operates all its processes at Six Sigma may find itself uncompetitive in the market, particularly, if consumer needs do not require Six Sigma quality. Service organizations need to clearly understand the cost of improving quality and benefit to the customer and then determine the optimum quality level. As shown in Figure 4.4, the optimum quality level for nonstrategic services is likely to be at point $X_1$, while for mission critical services it would be $X_2$ (as the higher defect rate would be unacceptable). As per Rajeev Grover, who manages the BPO operations for Hewitt Associates in India, striving for a target between 4.5 and 5.0 Sigma is probably a desirable range for many service processes.

## PLAN OF ATTACK—SOME POTENTIAL APPROACHES

Given the significant opportunity for improvement and various challenges involved, a concerted and well-planned effort is required to improve service. Six Sigma, through its highly structured analytical approach, helps to initially define the problem. Unless the problem is correctly defined, the solution may not eradicate the problem. For example, a vice president of a leading outsourcing firm, while reviewing the Six Sigma analysis with respect to employee turnover, discovered that many of the employees were leaving not to join a competitor for higher salary, but to undertake higher studies. Retention solu-

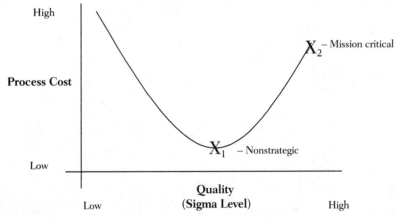

**FIGURE 4.4.** Process Criticality Drives Quality Level

tions as a result were focused on providing assistance with respect to further education rather than increasing salaries.

Firms in various countries around the world are achieving significant productivity gains. TBM Consulting Group, in a recent survey of firms across five countries—United States, United Kingdom, Germany, Mexico, and Brazil—found that the top two initiatives of the executives in these countries were:

· Continuous process improvement
· Work flow/procedures process improvement

In today's competitive environment, where it is difficult to differentiate between competitors, quality is a key factor. As a result, every company is starting to differentiate itself based on quality. But how do you communicate your quality to differentiate yourself? Six Sigma again provides a universally accepted methodology and measure. Companies that are serious have made a commitment of resources to the extent that they have even appointed a very senior level head of Six Sigma. It is recognized that Six Sigma is a notional concept and a target that you may rarely wish to achieve. But its rich tool set and rigorous methodology lends itself to a highly structured analysis. In addition, any communication that defines your quality in terms of Sigma provides a very concise and clear message—not subject to misinterpretation.

Companies more known for their manufacturing prowess than service focus have Six Sigma leaders, such as:

· Textron Inc.
· Raytheon Company
· 3M Company
· Johnson Controls, Inc.
· Dupont

Service organizations like American Residential Services (subsidiary of ServiceMaster) and ITT Industries have also appointed a leader of Six Sigma.

The challenges in providing excellent service to a customer continue to increase at an ever-faster pace. This is driven not only by our changing environment, but the customers' demands are also changing more rapidly. Being close to the customer and understanding their needs more thoroughly could help us serve them better and maintain a competitive advantage.

## KEY TAKEAWAYS

· Defining and improving service are both distinct continuous processes.

· When the only constant is change, it creates additional challenges in providing quality service.

· Unlike products, services do not have the protection of patents. That makes it easy for competition to replicate the service, and harder for you to sustain the competitive advantage.

· Six Sigma tools and framework can be very helpful in clearly defining the quality standard required by the customer and in communicating it to others.

· Every process needs to be evaluated to determine the desirable quality level based on the importance of the process. Unless a process is mission critical, Six Sigma quality level is unlikely to be the optimal level.

# TRANSACTIONAL SIX SIGMA

# INTRODUCTION

Six Sigma for transaction service takes a fresh look at Six Sigma applications in the service sector. Transaction services as defined in Part One of this book are high-volume services. Evidently, economy of scale plays a greater role in high-volume services. The following chapters discuss the material, keeping in view the total system perspective of a service. The analysis is not limited to transaction services alone, however.

Part Two of this book describes Six Sigma in a broader service economy context. Chapters are organized as follows:

- Chapter 5 introduces the concept of Six Sigma and its application to the transaction processes. The chapter also introduces the Six Sigma methodology to be used. The methodology is described as Define, Measure, Analyze, Innovate, and Embed (DMAIE).
- Chapter 6 describes the define phase of Six Sigma methodology. Define phase includes strategic alignment, process analysis, blueprinting analysis, and overall performance management.
- Chapter 7 describes the measure phase of Six Sigma methodology. This phase includes performance indicators and baseline and goal setting.
- Chapter 8 combines analyze and innovate phases into one chapter. Techniques like TRIZ (Russian acronym for the Theory of Solving Inventive Problems) and mind mapping are described in this chapter.

· Finally, Chapter 9 of Part Two describes the embed phase. This phase describes training, projects, and continuous improvement steps. A template is provided, which can be used readily for most service quality situations.

The discussion keeps in mind the following observations:

· The service sector is a dominant sector in the majority of industrialized nations and is increasingly becoming a dominant sector in developing economies.

· Six Sigma is often considered a reactive approach. We wish to provide a proactive approach to continuous improvement.

· Six Sigma methodology is thought to consider only "operational" aspects at the process level. We wish to apply the methodology at the corporate level—the methodology takes a system approach, considers feedback and delays, and is aligned with corporate goals.

· Service quality discussion is often dominated by "marketing" oriented concepts and frameworks. We wish to provide a balanced framework, as marketing and operations are intimately connected and overlap greatly in service delivery design.

There are excellent textbooks on service operations management. Some of these textbooks provide an excellent overview of service quality dimensions. However, no text has provided a comprehensive discussion on Six Sigma application in services. A textbook by Michael L. George (2003) titled *Lean Six Sigma for Service, How to Use Lean Speed and Six Sigma Quality to Improve Services and Transactions* provides some examples of Six Sigma applications in the service sector.

## OVERVIEW OF SIX SIGMA

Most organizations simply understand Six Sigma to be a measure of quality that strives for near perfection. More specifically, Six Sigma is a disciplined, data-driven approach and methodology for eliminating defects by driving toward six standard deviations between the mean and the nearest specification limit. It is applicable to any process from manufacturing to transactional, and from product-oriented to service-oriented businesses. Six Sigma methods integrate the principles of business, statistics, and engineering to achieve tangible results.

The fundamental objective of the Six Sigma methodology is the implementation of a measurement-based strategy that focuses on process improvement and variation reduction through the application of Six Sigma improvement projects, accomplished through the use of two Six Sigma submethodologies. Some common methodologies are Define, Measure, Analyze, Improve, and Control (DMAIC) and define, measure, analyze, design, verify (DMADV). The DMAIC process is an improvement system for existing processes falling below specification and looking for incremental improvement. The DMADV process is another system used to develop new processes or products at Six Sigma quality levels or improve a current process that requires more than just incremental improvement.

The majority of submethodologies for Six Sigma were developed for manufacturing-based tangible products. We need to further improve the submethodology to make the most impact in the service setting. We will describe the DMAIE process in greater detail later in this chapter.

Six Sigma in general refers to a quality improvement method to increase efficiency and reduce errors to 3.4 defects per million operations. The program is similar to other quality programs in that it is customer-focused and uses analytical methods. In fact, when a company implements Six Sigma, it learns the same lessons learned in the 1980s with total quality management (TQM). Lessons revisited were working closely with suppliers on design and manufacturing and avoiding variations in a design or production process. Six Sigma is more structured and profit-oriented than TQM. The principle of TQM was to fix the problem without worrying about the cost.

When Six Sigma is skillfully applied by an organization, the company is generally able to reduce costs by 50% or more through a self-funded approach to improvement, reduction in the waste chain, understanding the customer's requirements, improving delivery and quality performance, providing critical process inputs needed to respond to changing customer requirements, developing robust products and processes, and driving improvements rapidly with internal resources; in short, the possibilities are endless.

## APPLICATION TO THE TRANSACTION PROCESSES

The service sector needs Six Sigma and quality management tools as much as any other sectors do. Various success stories have proven the value of quality initiative in a service setting. As far back as 1992, Edwards Deming (Deming, 1992) proposed the idea of using his "methods for improvement" in the service sector. One chapter is devoted to this topic in his 1992 edition of the text. Actual applications differ by the business context. The service sector has a whole spectrum of business context that could be sliced and diced along various dimensions. Fitzsimmons and Fitzsimmons (2004) have provided various ways to look at the business contexts. Some of these dimensions are based on:

- Nature of the service act
- Relationship with customers
- Customization and judgment
- Nature of supply and demand
- Method of service delivery
- Extent of customer participation

A service process matrix based on the degree of labor intensity and degree of customization is presented in Figure 5.1. Four types of service businesses can be identified here: service factory, service shop, mass service, and professional services. Service factory is analogous to transaction services providing standardized services. Professional services provide individual attention and the highest degree of customization, and employees are often highly trained business specialists.

Degree of Interaction and Customization

| | Low | High |
|---|---|---|
| Low | Service factory:<br>• Airlines<br>• Trucking<br>• Hotels<br>• Resorts and recreation | Service shop:<br>• Hospitals<br>• Auto repair<br>• Other repair services |
| High | Mass service:<br>• Retailing<br>• Wholesaling<br>• Schools<br>• Retail aspects of commercial bank | Professional service:<br>• Doctors<br>• Lawyers<br>• Accountants<br>• Architects |

Degree of Labor Intensity (left axis label)

Reprinted with permission from Fitzsimmons & Fitzsimmons, 2004

**FIGURE 5.1.** The Service Process Matrix

While describing Six Sigma, we need to briefly discuss ISO application in the service sector. The literature review offers varying opinions on ISO 9000 for services, and there is lack of research on ISO application for professional services. A recent survey of professional quantity surveyors (McAdam and Canning, 2001) showed that companies have registered for marketing reasons. However, these same companies have experienced internal improvements.

To use Six Sigma submethodologies, we should be able to identify and quantify relevant measurements. Some measurements in the service setting are readily quantifiable. However, other performance indicators (for example, courtesy and professionalism) are harder or at times impossible to quantify. A few myths regarding Six Sigma in services are presented here.

**Myths about use of Six Sigma in services.**   Myth 1: Six Sigma is a toolkit containing various tools for problem solving. The approach includes a philosophy about continuous business improvement. The philosophy includes customer perception and includes measurement aspects.

Myth 2: Six Sigma is only good for fixing the problem and is useless for preventing the defects/problems: Design for Six Sigma tools and methodology precisely includes tools like quality function deployment (QFD), failure mode and effects analysis (FMEA), and poka yoke. These tools enable defect-free design from the customer' perspective and prevent problems from occurring.

Myth 3: Six Sigma is about having 3.4 defects per million opportunities: We need not take Six Sigma too literally. For some business processes, the goal might be higher (for example, landing/take off of aircrafts), and for some business processes, the goal might be lower (for example, baggage claims).

Myth 4: Statistical process control is not applicable to service businesses. As evident from literature on focused factory and examples like Shouldice Hospital and E-bay, etc., we have use for techniques like statistical process control. Various measures are quantifiable in the service business and these measures also have natural variances. Other measures will have no specification limits and lack a notion of process capability. We could call this a human side of statistical process control. Situations may occur where companies have tried to put some loose specification limits around these "human" parameters.

Myth 5: Six Sigma cannot be applied at the corporate level. As evident from the recent publication of the book titled *Six Sigma Business Scorecard*, the approach deserved merit even in tracking the corporate business performance.

Myth 6: Six Sigma is the universal answer to all business problems. We need to have an appropriate perspective for Six Sigma. The technique and philosophy is not an answer to all business problems, and a similar statement is valid for similar approaches like reengineering and business improvement.

## MANUFACTURING, SERVICES, EXPERIENCES, AND TRANSACTIONAL PROCESSES

Customer perception of the service quality is usually immediate, unlike the perception of the quality of a product. Customers will usually form an opinion about the quality of their car over a few months to a few years. However, quality at a restaurant or at a bank is different. Often, customers immediately form an opinion about the quality of services.

## How are services different?

*Services are produced and consumed at the same time.*   The customer is often present, and the server and customer together, with enabling assets, produce the offering. The service or offering is consumed at the same time. Why is this relevant? Quality is much more important in this scenario.

Therefore, in services, we encounter the sensitive issues of rework and employee training while customers are present. Customers will observe if a new employee fails to fill the form out correctly at a mortgage institution. The customer might lose confidence and will have lower trust in the services of this particular company. Often, irreversibility is attached to the service (for example, hair cuts—once hair is cut, there is no way to reverse it). The job has to be done right the very first time.

*Perishable.*   Services cannot be stored for later use. Consider an empty room at a hotel or an empty seat in an airplane or an empty ride in a park. Products can be stores for future use, but services cannot be stored. These systems are called fixed capacity systems, and firms use the following strategies to manage the profit:

- Price discrimination—Charging different pricing depending upon the time of purchase
- Reservations or appointments
- Price incentives—Price discounts for nonpeak use

At times, service capacity could be increased by hiring part-time employees—for example, in a retail store during Christmas season. However, quality implications are associated with these strategies. For example, companies need to have an excellent training program to train temporary employees. Temporary employees might not be very happy working at these firms.

*Intangible nature of service.*   Customers are able to touch and feel a physical product. However, customers can only feel the service ideas and concepts. Service acts are intangible in nature. Quality perceptions will also vary with the customer's state of mind and the employee's performance at the time of delivery (often called "moment-of-truth").

These characteristics provide another dilemma for the service provider—since service acts are ideas and concepts, these acts do not qualify for a patent by the patent office. Other companies will be able to copy these concepts readily.

*Heterogeneous nature.*    Each customer is different and, therefore, quality perception is different. This characteristic affects the extent of diversity of design for a service. For a physiotherapist, no two patients will ever be exactly alike. Even two consecutive sessions with the same patient will be different. The reasons could be a lack of homogeneity in customer segment, customer participation, and customer involvement in the process. Customers also enter the system with preconceived notions and perceptions about what they will or might expect at this service encounter.

With these characteristics, we have a difficult time standardizing the process used in service design. Lack of process is often debated. Even finding customer standards and requirements of what customers really want is a challenge.

**Production line and empowerment approach.**    The service delivery system and hence, quality management systems, heavily depend on the type of service act and the way to deliver this service act to customers. Bowen and Lawler (1992 and 1995) have used degree of empowerment of employees and presented various types of the service delivery systems. Employees have the least empowerment in a production line system. Managers design and control the process—employees merely follow a predetermined process with minimal or zero input into the system. The advantage of a production line approach is that the organization controls the system and leaves nothing to the discretion of the employee. The system could be the easiest one to control from a quality control perspective. We could also call these systems a transaction service. The other end of the empowerment spectrum is a total empowerment approach that allows employees to participate in the process, as managers and owners to decide the strategic direction of the firm. Employees provide input into the process, design the process and use discretion to make decisions, in turn satisfying immediate customers' needs (Bowen and Lawler, 1995).

Employees' participation and involvement in an empowerment system contributes to the commitment and cooperation of staff in the evolving work environment and thus reduces staff turnover levels. In this context, the continual development of all aspects of service delivery should become the firm's top priority for all levels of staff.

Being in the back room makes it much easier to manage. Quality control in services in the front room is mostly limited to process control. However, we want to illustrate in this section the usefulness of techniques for back room and front room operations.

## RETURN ON QUALITY

Studies show that customer satisfaction and service quality directly impact customer retention, market share, and profitability. The challenge is to provide operational methods for measuring and optimizing this link.

Authors (Roland et al., 1995) present a managerial framework that can be used to guide quality improvement efforts. The framework is based on the following four assumptions:

1. Quality is an investment: Treat quality investment like any other investment.
2. Quality efforts must be financially accountable.
3. It is possible to spend too much on quality.
4. Not all quality expenditures are equally valid.

The authors have presented various examples of companies where rising costs associated with quality initiatives caused a backlash from stakeholders, resulting in financial difficulties or in their quality program being dismantled.

The manufacturing sector has an economy of scale and often the process is a standard process. Lessons from such a sector can be applied to standard services and transaction services. However, customized services will not have an economy of scale, thus resulting in higher cost of quality. In these situations, quality programs will improve revenue stream or top line, but will not reduce cost due to the lack of economy of scale available. Therefore, services need to justify all quality programs and need to analyze the return on each quality program. In an era of cost cutting, quality initiatives need to be made financially accountable. The framework is represented as a chain of events starting with an improvement effort. Improvement efforts lead to higher perceived quality and customer satisfaction (and perhaps reduced cost). Increased customer satisfaction leads to a higher level of customer retention, leading to increased revenues and market share. New customers are also attracted by this positive cycle.

Roland et al. (1995) present quality improvement as a five-stage process:

- Stage 1: Preliminary information gathering: Customer surveys, business intelligence, internal capability analysis, and management intuition.
- Stage 2: Opportunities identification: Identify the best opportunities based on various business indicators.
- Stage 3: Limited testing of improvements to determine effectiveness.
- Stage 4: Financial projections based on hard data.
- Stage 5: Full rollout of quality improvement efforts.

With a brief understanding of Six Sigma in services, understanding the special nature of service acts and understanding return on quality, Six Sigma methodology is presented in the following section.

# TRANSACTIONAL SIX SIGMA METHODOLOGY (DMAIE)

The improved Six Sigma problem solving approach (DMAIE) employs the following five areas:

## 1. Define (D)

· Identify their customers and culture of the firm or an institution.

· Determine a Six Sigma approach for customer needs and feedback as well as meeting business objectives and goals.

· Identify the qualities that customers consider to be most important.

## 2. Measure (M)

· Determine how to measure service processes and performance.

· Identify the key internal processes that influence service quality and identify the defects in these processes.

## 3. Analyze (A)

· Use tools to determine the source of likely defects. Six Sigma methodology is rich in tools. We present additional tools to analyze the source of defects and identify improvement methods.

· Understand why the defects occur and the key variables that directly contribute to these processes.

· Confirm that key variables and effects on quality in services that customers feel are most important.

· Identify acceptable ranges of the defects in these specific quality service areas.

· Develop a system for measuring and monitoring the deviations in these variables.

## 4. Innovate (I)

· Make changes to the processes in order to stay within these acceptable ranges at all times.

- Think about ways to further improve the process.
- Apply innovative techniques to improve the service delivery system.
- Identify ways to remove defects from these processes.

## 5. Embed (E)

- Apply concepts and manage the project.
- Train employees and gain employee loyalty first.
- Gain customer loyalty.
- Continuously review the process.
- Fine tune the process.
- Manage projects.
- Continuous improvement: Determine how to continuously improve these processes.

## EXAMPLES OF MEASUREMENT IN A SERVICE SECTOR

Three examples in this section illustrate the application of various Six Sigma methodology. We will only introduce these examples; full examples using methodology will be presented in Chapter 9. Three examples presented here are a restaurant, a government library, and a call center.

**Restaurant.**  A restaurant strives to provide excellent food with appropriate ambiance while customers are dining in-house. The whole process can be split into four parts—making a reservation, arrival process, dining process, and billing and leaving the restaurant. We could follow a measurement variable to track the performance of a restaurant (Kimes et al., 1988):

- Measure related to the arrival process:
  - Internal measures: Forecasting and overbooking
  - External measures: Guaranteed reservations, reconfirm reservations, and service guarantee
- Duration measurement:

- Internal measures: Menu design, process analysis, labor scheduling, communication systems
- External measures: Pre-bussing, check delivery, coffee-dessert bar, visual signals, reduce time between customers, process analysis, communication system

A restaurant operator has to estimate the length of time a party stays once it is seated. Predicting meal duration affect reservation decisions and estimates of wait times for walk-in guest. Some authors have suggested controlling the length of time customers are occupying their seats. The approach is very attractive to a particular segment of customers-time-pressed professionals who have other meetings lined up at the meal. Appropriate operations levers are reducing the uncertainty of arrival, reducing the uncertainty of duration, or reducing the length of time between customers meals.

In this case, restaurants are not selling meals but selling time in the form of meals of predictable length—for example seating parties every two hours, with a reminder to leave when the time is up.

We might need to redesign the service delivery system to improve the operational efficiency while maintaining or improving service quality. An example is having a dessert and coffee bar where guests can move to chat after a meal. Appropriate operations levers are reducing the uncertainty of arrival, reducing the uncertainty of duration, or reducing the length of time between customers meals.

**Government services: Library services.** A public library is a not-for-profit entity and strives to provide the best service to its stakeholders. Authors Hernon and Altman (1998) suggest quality managers examine the following element of the library services:

- Resources: Investments available, staff time, materials, services, and supplies
- Physical environment: Lighting, temperature, humidity, and other factors

- Functions and processes: Typical functions of a library are identification, selection, acquisition, organization, preparation, storage, interpretation, utilization, dissemination, and management
- Team or various units: These teams could be cross-functional in nature

The next logical question is about the measurement. What should be measured and how should we measure these at a typical library? At a library, performance measures could be divided into categories based on ownership or who decides these measures:

- Library control: How much to invest, how many volumes and allocation of resources, how economical, and how promptly to acquire these volumes.
- Library plus customer decide together: How valuable are the resources (volumes), how reliable is the service, and how accurate is the service?
- Customers decide: How courteous, how responsive, and how satisfied?

Examples of some typical measures are:

- Number of satisfied customers, and number of dissatisfied customers
- Number of graduating students who are borrowers in relation to the total number enrolled
- Donations per customers and total donations received

The above examples provide a glimpse into the usefulness of Six Sigma methodology in different business contexts. The following chapter will provide an introduction to define and develop phase of Six Sigma methodology.

# KEY TAKEAWAYS

- Six Sigma methodology is considered both proactive and reactive approach. The approach should be aligned with corporate strategy and goals.

- There is a lack of comprehensive literature on Six Sigma applications in transactions and services.

- Application of Six Sigma methodology in transactions and services should take into account the inherent differences between services and manufacturing operations.

- DMAIE (Define-Measure-Analyze-Innovate-Embed) approach for Six Sigma implementation is introduced.

# DEFINE AND DEVELOP

This chapter describes the define and develop step of Six Sigma methodology. The step is key to setting up the entire methodology. This chapter describes the strategic alignment, definition, and development phases.

## CHOOSING A METHODOLOGY

A first step for a firm is to select a methodology or a set of methodologies. Service firms will often choose more than one methodology to systematically improve their performance. Broadly speaking, one can choose from the following methodologies:

- Total quality management (TQM): Includes basic quality tools such as flowcharting, fishbone diagram, and Pareto diagram.
- ISO 9000: Process-oriented and next level of sophistication. The method could be used to bring standardization into the picture.
- The Malcolm-Baldrige award model: The model is used to improve the overall competitive business performance of a firm.
- Six Sigma model: The topic of this section of the book concerns overall performance of the business, including innovation.

No matter which combination of methodologies a firm chooses to adopt, an alignment with the corporate vision and strategy is necessary and a required condition for the initiative to succeed. Service firms' operations strategies should be in alignment with their corporate strategies. Operations strategy will depend on business context, life cycle of the offering, and many other contextual variables. The operations strategy of a service firm will be based on what the firm decides to offer its customers in terms of price, quality, response time, and level of customization. Adoption of an operations strategy has internal implications for a firm. A firm needs to make capacity, channel, degree of customization, service recovery, and information decisions.

Next, we will compare existing Six Sigma methodology, DMAIC, with the Six Sigma methodology presented in this section, called DMAIE. Table 6.1 shows a side-by-side comparison between the two. Earlier steps are similar and the difference only occursat the later stages.

## Strategic Alignment

The first step in the *define and develop* phase is to revisit the strategic alignment of a service firm. The strategic direction of a service firm has a direct bearing on the operations strategy of a service firm (Soteriou and Chase, 2000). Competition among service companies often boils down to service quality. Rust et al. (1994) noted that 93% of all large U.S. corporations have initiated some form of service quality initiative. These initiatives include ISO standards and the Baldrige award criteria. Quality investment decisions should be looked at like any other investment decision. According to Rust et al. (1994), "Spending on quality is like any other resources allocation decision; it is expected to produce returns that are greater than the costs." An overview of service quality is presented by Harvey (1998).

The concept of "focused factory" was discussed in a 1974 article by Skinner (Skinner, 1974). Focused operations: make a focused factory, and hence, we can apply many tools and techniques. The idea is to concentrate on a narrower range of services for a particular market sector. Van Dierdonck and Brandt (1988) have also described the application of this concept.

TABLE 6.1    Six Sigma Methodologies Comparison

| METHODOLOGY PHASE | TOOLS USED | METHODOLOGY PHASE | TOOLS USED |
|---|---|---|---|
| **Define:** Identify the problem, define requirements, set goals | Project selection, definition, stakeholders' analysis, and others | Define | Project selection, definition, service operations audit, stakeholders analysis, and others |
| **Measure:** Refine goals, measure key steps/input/outputs, and validate | Data collection, Pareto chart, control charts, process cycle efficiency | Measure and baseline | Data collection, benchmarking, service recovery |
| **Analyze:** Develop hypothesis, identify key root causes, and validate hypothesis | Fishbone, regression analysis, hypothesis testing, FMEA | Analyze | Fishbone, FMEA, regression, and hypothesis testing |
| **Improve:** Develop ideas to improve, test ideas, and make solutions standard, and measure the outcome | Poke-yoke, FMEA, benchmarking | Innovate | TRIZ, brainstorming, and mind mapping |
| **Control:** Establish standard measures to maintain performance and correct problems as needed | Control charts, training, project control plan | Embed | Project implementation, training, and monitoring and feedback |

The focused factory concept in a service environment means that a firm has to make a decision about its offerings. The firm has to limit its offering and strive to make standard processes. Reducing offerings reduces variation in terms of the demand of the product. Reducing demand helps us match supply with the demand. Hence, reducing variability has a net positive effect in terms of allocating resources and training employees.

A significant difference between a manufacturing firm and a service firm is employee participation. Employee satisfaction and participation is an intricate part of any service business. Service businesses often struggle with the question of employee empowerment. Bowen and Lawler (1992) presented a method to identify the degree of empowerment based on the business context.

The term "empowerment" means that employees have power to make decisions. Lower level empowerment or a production line approach system provides almost no empowerment to employees. Empowerment comes with a cost to the company, but could provide additional customization to customers and employees. The contingencies of empowerment presented by Bowen and Lawler (1992) are presented in Table 6.2.

As per an article (Wysocki, 2002), Dr. Berwick, a member of the Harvard Medical School faculty, considers the healthcare system rife with errors, waste, and delays. Dr. Berwick studied the theories of quality-control gurus such as W. Edwards Deming and Joseph Juran. His group also studied the operations of several quality–conscious companies, such as Motorola Inc. and Toyota Motor Corp. We could consider these wastes according to the lean manufacturing concept of Toyota Production System. One example presented is to abandon one-on-one office visits to the doctor's office. He argues that 50% to 80% of such visits are "neither wanted by the patient nor deeply believed in" by the doctor. The system should be designed around the need of patients as the end consumer ("patient-centered") and not around the need of doctors. Insurers have legitimate concerns, and doctors often have a conservative attitude toward any change process. His contention is that systemic problems, rather than the failing of individual institutions or individuals, are at the core of many hospital problems. Some of these problems include poor internal communications and lack of coordination. Dr. Berwick argues for a two-pronged approach:

1. Patients should get immediate one-on-one access to doctors and nurses when they really need it. In one example,

**Table 6.2**  Contingency for Employee Empowerment

| Contingency | Production Line Approach | Circle the Number | Empowerment |
|---|---|---|---|
| Basic Business Strategy | Low cost, high volume | 1 2 3 4 5 | Differentiation, customized, personalized |
| Time to Customer | Transaction, short time period | 1 2 3 4 5 | Relationship, long time period |
| Technology | Routine, simple | 1 2 3 4 5 | Nonroutine, complex |
| Business Environment | Predictable, few surprises | 1 2 3 4 5 | Unpredictable, many surprises |
| Types of People | Theory X managers, employees with low growth needs, low social needs, and weak interpersonal skills | 1 2 3 4 5 | Theory Y managers, employees with high growth needs, high social needs, and strong interpersonal skills |

STEPS:

Step 1: Score your organization based on the above five contingencies. Low score is appropriate if contingency sounds similar to production line approach, but a high score is appropriate if empowerment type approach is more valid.

Step 2: Find the total score.

Step 3: Decide based on the total score where the business context stands in terms of overall empowerment.

Reprinted from Bowen, D. E., & Lawler, E. E., III, (1992). The empowerment of service workers. Sloan Management Review, 3, 37, by permission of publisher. Copyright 1992 by Sloan Management Review Association. All rights reserved.

40% or more of each doctor's daily schedule was left unbooked to accommodate patients with immediate needs. Changing the system will have operations implications, as discussed later in the book. The service delivery system will look more like an emergency room, rather than a typical physician's office.

2. When patients don't need to see a physician face to face, doctors could team them in groups and could use other channels like email or telephone conferences (and chat forums) to answer questions. These are low-cost channels and will likely be more effective. One pilot project in the mid-1990s resulted in a cost reduction of as much as 30%. Shorter patient stays were also reported.

At a clinic, after implementing this open-access system, patient satisfaction was scored at 8.1 on a nine-point scale, up a bit from the previous year. To make this system work, health care providers will need to have a comprehensive electronic medical record of each patient. A patient should also have access to his/her comprehensive medical record.

Other pilot program's, according to the article, include reducing medical errors to zero and improving the quality of the health care. One hospital automated its ordering, dispensing, and monitoring medication system to reduce errors.

## REVIEW OF EXISTING SERVICE QUALITY FRAMEWORKS

After providing a context for services, next we discuss the available service quality framework presented in literature. The list is not exhaustive by any measure. The following frameworks will be presented:

- A service quality model, presented by Parasuraman and Zeithml
- Haywood-Farmer cube for service classification
- Service profit chain model
- Service chain design

**Service quality model: SERVQUAL.**   The Parasuraman et al. (1991) version of the model is presented here. The authors have contributed significantly to the service quality literature. They have tested the model in the following business contexts: banking, credit card, repair and maintenance, and long-distance telephone companies. The model uses a questionnaire to identify these gaps. A set of about 22 questions representing different service quality dimensions is administered. The survey asks questions about the firm and about its competition. Analysis of the survey provides gaps in a service delivery system, according to the model. The questionnaire has two parts: One part asks about the expectation of the service and a second part receives response about the actual service received. The final section has questions on the relative importance of quality dimensions. Based on the model, the following service quality dimensions are also presented:

· Tangibles: Physical facilities, equipment, and appearance of personnel
· Reliability: Ability to perform the promised service dependably and accurately
· Responsiveness: Willingness to help customers and provide prompt service
· Assurance: Knowledge and courtesy of employees and their ability to inspire trust and confidence
· Empathy: Caring, individualized attention the firm provides its customers

**Use of SERVQUAL questionnaire.**   Authors of SERVQUAL intended the unchanged version of the questionnaire to be used for all service settings, and this survey was supposed to cover all five "generic service quality dimensions." Studies have shown (Hope and Muhlemann, 1997, Chapter 13) that the unchanged version is much more relevant and successful in service systems, with the following features:

· Low customer contact

- Low customization or having a standard service offering
- Little need of professional judgment

The definition of each gap is presented as follows:

- Gap 1: Difference between consumer expectations and management perceptions of consumer expectations.
- Gap 2: Difference between management perceptions of consumer expectations and service quality specifications.
- Gap 3: Difference between service quality specifications and the service quality actually delivered.
- Gap 4: Difference between service delivery and what is communicated about the service to consumers.
- Gap 5: Difference between consumer expectations and perceptions.

After identifying gap(s) in the system, the model provides a direction for further improvement for the system.

Many researchers have adapted the questionnaire to fit the specific business context.

Generally, a step-by-step process is followed:

1. Hold a brainstorming session or a focus group consisting of users of the service to identify any missing dimension of service quality.
2. Change the questionnaire if conclusions from the brainstorming session indicate such direction. As a rule of thumb, avoid including negatively worded questions. Negatively worded questions have been known to distort the response.
3. Test the pilot-adapted questionnaire and make any modifications necessary.
4. Conduct the survey.
5. Analyze the response and conduct something analogous to a factor analysis to test the validity of service quality dimensions.

6. Identify gaps by finding the gap score. Gap score is the difference between customer expectation and customer perception of the actual service.

7. Prioritize managerial actions based on the gap score.

**Haywood-Farmer Cube classification.** Haywood-Farmer (1988) provided a classification scheme for service businesses. Rosen and Karwan (1994) suggested that the Haywood-Farmer (1988) service classification along the three dimensions of contact, customization, and labor intensity could be helpful when identifying the order of priority for service quality dimensions. Haywood-Farmer uses three dimensions to put every service into one of the cubes. The following dimensions are represented by the cube:

1. Labor intensity: High or low
2. Customization: High or low
3. Customer contact/interaction: High/low

Together, we have eight quadrants. The authors suggest that the original dimensions of SERVQUAL have a much better fit with service companies that fall in quadrant one and have the following characteristics: relatively low contact/interaction with customers, low level of customization, and low labor intensity. Table 6.3 shows the classification scheme using this model.

TABLE 6.3    Haywood-Farmer Classification of Services

| BUSINESS CONTEXT | POSITION ON CUBE |
| --- | --- |
| Medical services, accountant & lawyers, solicitors, travel agencies, pharmacies, hospitals, management information system | High labor intensity<br>High customization<br>High contact/interaction |
| Tire store | Medium labor intensity<br>Low customization<br>Low contact/interaction |

**TABLE 6.3**    Haywood-Farmer Classification of Services *(Continued)*

| BUSINESS CONTEXT | POSITION ON CUBE |
|---|---|
| Placement center | High labor intensity |
| | Low customization |
| | Low contact/interaction |
| Retail banking | Low labor intensity |
| | Low customization |
| | Medium contact/interaction |
| College/universities | Medium labor intensity |
| | Low customization |
| | Low contact/interaction |
| Medium price restaurant | High labor intensity |
| Hotels | Low customization |
| | High contact/interaction |
| Credit card company, product repair and maintenance, long-distance phone company | Low labor intensity |
| | Low customization |
| | Low contact/interaction |

**Service profit chain.**    Heskett et al. (1997) present a service profit chain model. A modified version of the model is presented in Figure 6.1. The model has three levels. Level one describes the process wherein a company chooses a certain strategic market position, and a service concept is the end result. The service delivery system needs to be consistent with the operations strategy of the firm, which in turn should be aligned with the corporate strategy. Level two refers to employees and customer satisfaction. Employees play a critical role in the process. Employee satisfaction and loyalty leads to customer loyalty and satisfaction. A firm needs to have a very good idea about the nature of work that it offers to its employees. Level three describes appropriate financial ratios or a profit model.

An audit developed by authors includes 44 dimensions, as presented in Table 6.4. The audit should be performed in five steps:

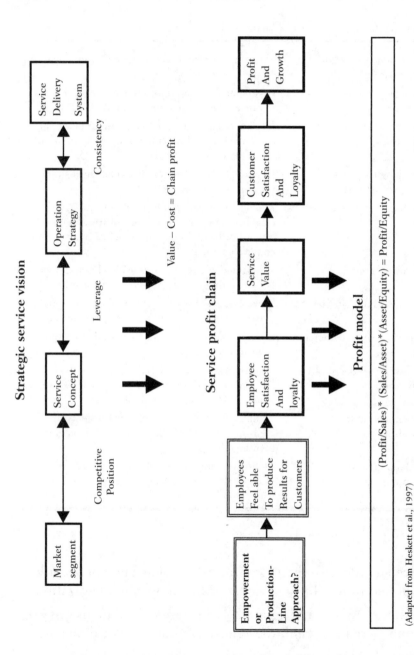

**FIGURE 6.1.** Service Profit Chain

**TABLE 6.4**    Developing a Strategic Service Vision

| MANAGEMENT PRACTICES | DIMENSIONS |
| --- | --- |
| Developing a strategic service vision | Targeting customers, business definition, operating strategy, focus of operations |
| Managing by the value equation | Customer segmentation, service design, service improvement, value enhancement |
| Practicing potential-based marketing | Lifetime customer value, effort to attract/retain customers, use of listening posts, proportion of customer potential realized |
| Managing the cycle of customer satisfaction | Primary goals, definition and goals, measurement, rewards and recognition |
| Building frontline capability | Management communication and action, establishing appropriate levels of frontline capability, frontline employee selection, frontline employee support |
| Improving processes | Extent and scope of effort, leadership and sustained effort, getting buy-in, objectives of the effort |
| Managing multisite networks | Service delivery system, standardization in multisite networks, international networks, franchising |
| Achieving total customer satisfaction | Eliciting customer feedback, latitude to respond, service recovery, guarantee |
| Measuring service profit chain progress | Measurement of profit chain components, determination of component relationships, use of balanced scorecard, use of measures |
| Reengineering the organization | Establishing the need for change, leadership for change, selection of the vehicle for change, implementing change |
| Leadership service profit chain management | Identifying and communicating core values, nature of core (shared) values, use of core values, and leadership behavior |

Adapted from Heskett et al., 1997, 260–269.

- Step 1: Identify the organizational unit—The audit could be applied to one business unit or to the whole organization.
- Step 2: Assess the relative importance of dimensions—Let managers force rank the relative importance of dimensions.
- Step 3: Assess current practice in the marketplace.
- Step 4: Identify and measure the gaps.

· Step 5: Establish priorities and take actions.

## NEW FRAMEWORK FOR SERVICE CHAIN DESIGN

The next section provides a theoretical framework to design a service operation. A manager has the following internal decisions:

· How much total service capacity should we keep?
· Where should this capacity reside?
· How much flexibility should be provided in terms of capacity?
· Should we outsource this service capacity?
· What channels should a firm use for delivery of the service?
· Why will customers go a particular channel, and how much does it cost us per customer?
· What is the degree of customization?

**Service operations chain design.**    The following section is taken from a white paper published by the author at Northwestern University (Chawan et al., 2003). Services have traditionally been viewed as being fundamentally different from products. The framework is still in its early stages. The goal of this section is to build a framework for service chain design that parallels supply chain frameworks. The concepts are then used to draw inferences for service chain design.

To be successful, product or service delivery chains must be designed to support the strategy of a firm and fill the customer needs targeted by the firm. In a service context, the strategic position of a firm may be defined in terms of the following customer needs:

· Speed of response
· Variety/customization
· Level of interaction
· Willingness to pay
· Quality

In a service context, we make the same point. For any service, both demand and service times are uncertain. Variability is influenced both by the strategic position of the firm and the nature of the service itself. A firm trying to be very responsive in providing service must be prepared to face a much higher degree of variability than one providing a lower level of responsiveness. An emergency room trying to provide service within half an hour of patient arrival has to be prepared to deal with much higher variability in the demand for service than an emergency room that will let patients wait for longer. The ability to let patients wait longer allows the emergency room to delay the patients who arrive during a period of heavy demand and thus level the demand for service on the medical facilities. Guaranteeing a half hour response, in contrast, does not allow this opportunity for leveling, leading to a demand for service that is almost as variable as patient arrival.

Similarly, as customization in the service offered increases, the service chain has to deal with greater demand as well as service time variability. The variability is also affected by the type of service offered. For example, a dentist's office faces less variability because most visits are scheduled and service time is predictable. In contrast, an emergency room faces much higher variability of both demand and service time. As with product supply chains, a service chain facing low variability must exploit the low variability to lower costs and be efficient, whereas a service chain facing high variability must be responsive to changing customer demand.

Service chain capability is defined in terms of its *responsiveness* to customer needs and changing demand, and the *cost incurred* to provide service. A manager's goal is to design a service chain that provides the appropriate level of performance in both dimensions. To identify the drivers of service chain performance, consider the main decisions a service chain designer must make. She must decide the level, location, flexibility, and ownership of service capacity. She must identify the steps of service to be completed before a customer arrives and those that are finished afterwards. She must decide if the customer will get the service from a person or by direct access to a machine. She must decide on how information will be used to support the providing of service. She must also make similar decisions with regard to the service recovery chain. The four key drivers of service chain capability are:

1. Service capacity network
2. Service inventory
3. Service delivery channel
4. Information

A manager must clearly articulate the customer needs that the service chain aims to fill and design service capacity, service inventory, service channels, and information to effectively serve the customer's needs. The basic framework linking the drivers of service chain performance and customer needs is shown in Figure 6.2.

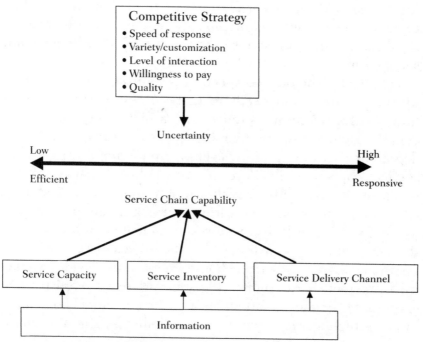

**FIGURE 6.2.** A Framework for Service Chain Design

## SERVICE DESIGN PRINCIPLES

**Service capacity network.**   The service network is the set of resources allocated to serve customers. The service capacity of the network is defined as the number of customers that can be served per unit time. A manager must design both the network and allocate service capacity at different locations. Key service capacity-related decisions are:

· Level: How much capacity to have?
· Degree of centralization: Should the capacity be centralized or decentralized?
· Flexibility: How flexible should the capacity be?
· Ownership: Should the capacity be internally owned or outsourced?

**Level of capacity.**   Level of capacity is the amount of capacity the service chain carries relative to the number of customers served. The level of capacity required is influenced by the desired speed of response and the variability of demand and service time.

The faster the desired speed of response, the more capacity the service chain needs to carry. A restaurant promising lunch within 15 minutes must carry a higher level of cooking capacity than a restaurant aiming for a slower response. As the degree of customization or level of interaction increases, the service capacity will have to be higher because of the increased service time variability. For example, a fast food restaurant offering limited variety needs a much lower capacity compared to a full-service restaurant, if both are to serve the same number of customers. As the price a customer is willing to pay increases, a firm should increase the level of capacity it provides. For example, it is appropriate for Nordstrom to carry more service personnel because its customers are willing to pay a higher price than a customer shopping at a discount store.

**Degree of centralization.** Centralization pools demand service time variability, thus decreasing the level of capacity required. Our goal should be to centralize capacity whenever feasible. Decentralized capacity close to the customer is appropriate when customers require a high speed of response or want a high level of interaction. For example, Charles Schwab has opened offices in many neighborhoods as it is trying to increase the financial advice part of its business, which requires a higher level of customer interaction than its previous focus on simply executing trades. Given the high variability and decentralization, the extra capacity the service chain has to carry will be high. As a result, the margin per customer must justify the increased speed of response or higher level of interaction demanded by the customer.

Decentralized capacity is also appropriate when demand for the service at each location is large and predictable. Conversely, when services have a lower volume of highly unpredictable demand, a centralized model is more appropriate, to be able to pool the variability. Thus, the greater the variety/customization offered, the more effective centralized capacity will be. For example, demand for dental services is more stable and predictable than demand for medical services. This justifies the greater consolidation in medical service providers compared with dental providers. Within medical services, primary care services with more predictable demand are more decentralized compared to specialist services where demand is lower and less predictable. Another good example is the change Delta is planning for airport check-in. Travelers with a simple check-in will be able to access local agents or self-service check-in machines. Customers with more complicated requests such as ticketing changes will have access to phone banks to speak to telephone reservation agents who are centralized. Travelers using the phone can obtain boarding passes from a nearby printer. Delta will thus decentralize capacity for routine requests while centralizing capacity for complicated, high-variability requests.

Another example that illustrates the appropriateness of both low and high degrees of centralization is the human resources (HR) function of a large firm. One of the tasks performed by HR is recruiting employees, from entry level to senior management positions. Demand for entry level candidates is large and predictable at each plant. It is thus appropriate to decentralize the hiring function for lower level positions to each location. The need to fill senior management positions, however, is less frequent and less predictable. Thus, it is better to centralize senior recruiting at the headquarters.

**Capacity flexibility.**    Managers could use the following steps to match supply with the demand:

· Use part-time or temporary employees: Most firms use temporary staff during the peak season.

· Outsource the capacity to a third party: A firm may decide to outsource the capacity at a short notice.

· Cross train employees within the company. Cross training employees creates more flexibility within the company.

· Schedule downtime or activities not touching the customers in a low demand period. A 7-Eleven store might want to get deliveries in early morning or late in the morning or in low demand periods.

Capacity flexibility defines how easily the available capacity resources can provide multiple services or be reallocated across time. Flexibility allows a firm to aggregate variability and reduce the service capacity it carries because flexible capacity can provide whatever service there is demand for. Capacity flexibility is very valuable to a firm aiming for a high speed of response or providing high variety/customization. In both cases, the required service capacity would be very high in the absence of flexibility. Call centers have successfully provided flexibility by distinguishing between services requiring quick response (answering a call), and those that can tolerate a slower response (answering e-mail). Cross training the staff allows them to switch from answering e-mails to answering inbound calls as the inbound call volume increases. This provides better service to customers calling in and improves the utilization of available capacity.

Capacity flexibility over time enables a firm to adjust the service capacity to meet current market requirements. Capacity flexibility over time is obtained by using a part-time workforce or by shifting the workforce from one task to another based on need. For example, a grocery store varies the number of lanes that are open at a given time based on customer traffic through the store. To achieve this flexibility, the grocery store uses some part-time workers and moves others from stocking shelves to the checkout counter as needed.

**Degree of ownership.** The degree of ownership refers to the level of control that a firm exercises over each business function. Economists have considered several factors, such as transaction cost and incomplete contracts, that influence the outsourcing decision. From a supply chain perspective, economies of scale and variability of demand/service time are two factors that play a key role in decision making for degree of ownership. We look for situations where a third party can provide the same level of responsiveness at a lower cost than the firm itself. When faced with low demand and high variability, a firm gains by outsourcing because the third party can pool demand/capacity across multiple firms, including competitors, and achieve lower variability and better economies of scale. Mid-size television stations having newscast service often do not provide in-house weather reporting. These stations outsource weather reporting to another company that specializes in this service and achieves economies of scale. Most firms outsource recruiting for very high level managers because demand at this level is low and highly unpredictable. This is appropriate because the third party recruits for several firms and is able to pool the variability and achieve economies of scale. Outsourcing high-level recruiting to a third party thus allows the HR service chain to operate with a lower level of capacity.

**Service inventory.** In a product supply chain, inventory is defined as the number of units of the product in the supply chain waiting to be sold. Inventory may be held in various forms, from raw materials to finished goods. In most service chains, for example a call center providing customer service, no product inventory is waiting to be sold. Thus, we have to broaden the definition of inventory to capture its role in services. We do so by considering the following key inventory-related decisions in a supply chain:

- In what form should the inventory be held (degree of completion)?
- Where should the inventory reside (distance from customer)?
- How much inventory should be carried?

It can be argued from a supply chain design perspective that the first two questions are very important, even though the definition of product inventory focuses on the third. We thus define service inventory as the degree of completion of a service offering at the point when a customer enters the service delivery chain. A high level of service inventory indicates that the service chain completes most of the steps in providing the service before a customer arrives. A low level of service inventory implies that the service chain performs most steps to complete service after a customer arrives. For example, a key service provided by an HR department is to fill open positions. The manager of an HR department can carry varying levels of service inventory as follows:

- Start advertising for the position only after the service request is received (no service inventory).
- Carry a database of potential hires and search the database after the service request is received (some service inventory).
- Carry a database of potential hires that can be sorted by skills and other preferences by the manager making the service request (more service inventory).
- Carry a group of people who are already hired by the firm and use one of them to fill the service request (very high service inventory).

The manager must decide the appropriate level of service inventory to carry.

Our definition of service inventory can be linked to the concept of the push-pull boundary in product supply chains. In product supply chains, push processes are performed in anticipation of a customer order, and pull processes are performed in response to a customer order. Like push processes, service inventory is the set of service processes performed in the push phase in anticipation of customer arrival.

**Degree of completion (service inventory form).**    Degree of completion is the portion of service finished before the customer enters the network. A firm can increase the degree of completion by explicitly completing some service steps or by providing a tool that speeds up service completion. For example, when a service chain allows an order taker to pull up all information on a customer based on the telephone number, it is increasing the level of service inventory. A section of frequently asked questions on the IRS Web site is a form of service inventory because it provides ready made answers to many questions people have.

The higher the service inventory a firm carries, the more capacity and expense it has already committed to completing part of the service ahead of demand. This will increase costs if demand for the service does not materialize, and the sunk cost is wasted. Thus, from Principle 3, when demand and service variability is high, it is better for a firm to limit the degree of service inventory and commit resources only after demand arises. This is equivalent to saying that it is better for a service chain to postpone service steps until after demand arises if variability is high. The less service inventory a chain builds, the more service capacity it requires to respond to customer requests. In contrast, when demand variability is low, a firm can increase the level of service inventory and carry less service capacity. Given the higher variability, it makes sense for a high-end restaurant to start cooking dishes from scratch, whereas a fast food restaurant facing lower variability can hold more service inventory by preparing some of the food in advance. The high-end restaurant will thus need a higher surplus capacity of cooks compared to the fast food restaurant. If the service being provided is standardized and has commonality across customers, a high degree of service completion is appropriate. If significant customization is required and commonality is low, then holding a lower service inventory is appropriate. For travelers with electronic tickets not requiring any change, Delta Airlines has provided service inventory in the form of self-service check-in machines where customers identify themselves using the credit card they used to purchase the ticket. This is appropriate because check-in requires the same procedure for all such travelers. For customers with a ticketing change, however, no service inventory is available, and they must contact a reservation agent.

Returning to the HR example, for low-level recruiting where variety/customization is low and demand is high, the firm may carry high service inventory as a pool of hired people. For midlevel managers, where demand is somewhat lower and less predictable, the HR department may carry lower service inventory by investing in a database of potential hires. For very high level managers, where the job description is very specific and demand is highly unpredictable, the HR department may carry no service inventory and start the search from scratch. In general, as the degree of customization required and the variability of demand and service time increases, it is preferable to carry less service inventory and increase service capacity to react to demand.

As the willingness of the customer to pay increases, it is appropriate for a service chain to increase the level of service inventory it carries. For example, the Ritz Carlton builds a database with the tastes and previous orders of its clients who are then offered services accordingly. It makes sense for Ritz to carry a higher level of service inventory, given the high willingness of its customers to pay. Investing in such a system would not make sense for a low-end hotel chain.

**Distance from customer (location of service inventory).** Distance from customer is the time and effort required by the customer to access service inventory. For example, customers can conduct trades themselves over the Charles Schwab web site. This service has a short distance for the customer having access to the Internet. When seeking advice from a financial planner, however, customers either visit or talk to the planner by phone. The distance of the service inventory from the customer is longer in this case.

If a customer desires a high level of responsiveness, then the distance between service inventory and customers must be minimized. Using the Internet and cell phones for stock trades are examples of bringing service inventory closer to the customer. Direct access is easier to provide if the service is standardized and a high degree of service inventory is maintained. However, if significant customization or level of interaction is required, direct customer access is difficult to provide. For example, financial service firms provide a variety of calculators that customers can use to analyze their portfolio. In all these instances, the underlying service inventory, in the form of an algorithm that does the analysis, is standardized. A customer who needs a more customized analysis of his or her portfolio cannot access the service directly and will need to go to a professional. Similarly, frequently asked questions are a form of service inventory that serves standardized requests. Customers are provided direct access to the service inventory in this case. For more complex requests, customers do not have direct access to the service inventory and need to go through intermediaries.

**Service delivery channel.**   The service delivery channel is the method(s) by which a firm provides services to its customers. Customers interact with a service chain when making a service request and when receiving the service. When designing its service chain, a firm can select from among the following service delivery channels:

· Person-to-person direct

· Person-to-person remote

· Person-to-machine

· Machine-to-machine

Cost implications of channel selection were discussed in Chapter 1 for a typical bank. Consider, for example, a financial services provider such as Charles Schwab. Schwab serves a variety of customers, from those investing a few thousand dollars to those investing much larger amounts. The firm needs to decide which channels it should use to serve its customers. Should all transactions be through face-to-face meetings with a financial advisor? Should a hybrid combination of channels be used? If so, which customers should have online access? Which customers should be able to access an agent for advice on the phone?

These channels vary in the cost of providing service as well as their ability to fulfill customer needs in terms of speed of response, customization, and level of interaction. The level of customer participation also varies with the channel selected. As a result, the choice of service delivery channel(s) has a significant impact on cost and customer service provided. We now discuss the strengths and weaknesses of each channel.

**Person-to-person direct channel.** The person-to-person direct channel involves the direct interface between the customer and a service provider. The speed of response of the person-to-person direct channel is slower because it does require the customer and service provider to come together. For example, transferring funds through a bank teller requires a lot more time than transferring funds online. This channel has the highest cost of capacity because of loss of aggregation, especially if variability is high. For example, Charles Schwab has opened many local offices where financial service providers can meet with clients. This has required investment in many local offices that are often unused. Thus, the person-to-person direct channel is only justified for customers who are willing to pay for it. This channel is very effective if customers desire a high interaction with the service provider. For financial service providers, the person-to-person direct channel is suited for high net worth customers who have a variety of financial needs and are willing and able to pay for the service. This channel may also be justified if customers require some interaction, but variability is low because the increase in capacity due to disaggregation will be small.

**Person-to-person remote channel.** This channel has a remote interface by phone or web between the customer and the service provider. For example, most web hosting companies provide live web chats or phone contact to answer customer queries. Relative to the face-to-face direct channel, the remote channel provides a faster response. Relative to the direct channel, the remote channel can be operated with a much lower level of service capacity because of aggregation. The remote channel also allows capacity to be located at a low-cost site. For example, many providers of phone or online web services are locating in low-cost countries like India and China. The remote channel is most appropriate when cost reduction is important, demand and service times are variable, and customers require a limited amount of person-to-person interaction. Delta's decision to provide a phone bank for travelers with ticketing changes at check-in is appropriate because variability for such transactions is high and Delta wants to lower costs.

**Person-to-machine.** The person-to-machine channel serves the customer using an automated system, such as a website or a telephone-driven menu. The speed of response is comparable to the face-to-face remote channel. Service capacity is automated and centralized, making variable costs fairly low. Variable costs are further reduced because of the high degree of customer participation when making a service request through this channel. For example, an individual making trades or transfers between accounts using the Internet or an automated phone system does all the work, while the bank simply provides access to a server with the appropriate content. This channel has a very low variable cost, but the channel is not effective at providing a high degree of customization or a high level of interaction. It is best suited for standardized services where a high level of service inventory is carried. Delta is planning to provide self-service check-in machines at airports, which are ideal for travelers with routine check-in, such as those not checking in any bags. Such a channel would not be appropriate for customers with ticketing changes, given the need for customization and interaction.

**Machine-to-machine.**    The machine-to-machine channel is an interaction between two automated systems. An example of a machine-to-machine interaction is an investor's computer updating her portfolio status by downloading information from Schwab's systems. As the interaction is automated, no personal intervention is required and all service capacity can be centralized. Thus, this channel has a very low variable cost of providing service. This channel is best suited for standardized, regular transactions. This channel, however, is not very effective if a high level of customization or interaction is required. For example, standard monthly payments such as mortgages and telephone bills can automatically be transferred from a customer's account without any personal intervention. Any nonstandard payment, however, will require a person to write a check or convey the payment details to the bank.

**Hybrid models.**    In most cases, it is not appropriate for a firm to use a single service delivery channel for all its customers and services. As discussed earlier, service delivery channels have different costs and capabilities. The face-to-face direct channel is expensive, but allows the service chain to provide a high level of customization and interaction. The person-to-machine channel, in contrast, cannot handle customization well, but has a much lower cost structure. Thus, it is best for a service chain to use an appropriate hybrid set of channels when serving a large number of customers with a variety of needs.

For example, answers to a set of standard questions from a person applying for a passport are available in a set of frequently asked questions from the U.S. Department of State website for free. For more complex questions, people can call a 900 number and speak to an agent. For even more complex services, they must visit a passport office. The Department of State has used a hybrid set of channels that are targeted appropriately. The face-to-face direct channel is used for the most customized transactions requiring a high level of interaction, whereas the person-to-machine channel is used to answer standard questions. Similarly, Delta is offering the person-to-person, person-to-person remote, and person-to-machine channels for check-in at airports. The person-to-machine channel is targeted at travelers with the most routine check-in; the person-to-person channel is targeted at travelers requiring some interaction, but not having too much service time variability; the person-to-person remote channel is targeted at travelers with high service time variability. This hybrid allows Delta to provide appropriate service at low cost.

**Customer participation.**   It is important to observe that different service delivery channels lend themselves to varying degrees of customer participation in performing the service itself. Increasing customer participation can decrease the need for service capacity, but care should be taken to ensure that customer errors do not lead to more work for the service chain. The person-to-person channel typically has less customer participation, whereas the person-to-machine channel usually has a high degree of customer participation. For example, transferring funds between accounts through a bank teller requires a limited amount of customer participation, whereas the same transaction on the web uses the customer to execute the transaction. An excellent example of a service chain using extensive customer participation is eBay. Customers can buy and sell items on the web using services provided by eBay. eBay simply provides the web tools that allow customers to execute these transactions. Clients do all the work, both on the selling and buying side. The use of extensive customer participation allows eBay to lower the cost of this transaction relative to companies that run live auctions. This has made it practical for clients to sell relatively low-cost products on eBay.

**Information.**   The information driver is extremely important because it facilitates the use of other drivers to lower cost or improve responsiveness in the supply chain. Information allows service capacity to become flexible and be allocated appropriately across customers and over time. For example, service capacity at banks has become flexible primarily because of the availability of information with each teller that allows him or her to perform all transactions. Most service inventory is carried in the form of information. For example, service inventory in the form of information allows a customer service representative at McMaster Carr to view all relevant information on a customer as soon as he or she calls in. Finally, information requirements vary significantly with the choice of service delivery channels. We now discuss how information may be used along with other drivers to support different customer needs.

**Service capacity network.**    Information plays a key role in decreasing the level of service capacity required and better matching available capacity with demand. Information can be used to decrease the level of capacity required by lowering the service time and also the variability of service time. For example, call centers are able to pull up all information on repeat customers from their telephone number. This cuts the service time and allows customer service representatives to process orders much faster. Information can also be used to decrease the variability of demand by analyzing the available demand data. For example, McDonald's analyzes demand data during peak periods in 15-minute increments to be able to predict capacity requirements very accurately. This decreases the need for unnecessary additional capacity.

Information is key to increasing flexibility in the available service capacity. For example, at a call center, if service representatives are to be able to go from answering e-mails to answering the phone as inbound calls increase, they must have information that triggers this change and also information technology that allows them to perform both services from their station. The result is a better matching of available capacity with demand, better customer service, and lower overall cost of capacity. Information is also used to vary prices and thus demand for services. A classic example of this approach is the airlines, since use differential pricing and seat availability over time to increase revenues while better matching supply and demand.

**Service inventory.**    Most service inventory is carried in the form of information. Examples include information in the form of answers to frequently asked questions, calculators that allow standard analysis of a client's investment portfolio, and information in a database of potential candidates carried by the HR division in a firm. In each case, service inventory in the form of information allows the service chain to decrease the amount of work that needs to be performed after the service request arrives. eBay uses information to build service inventory that facilitates web transactions for its clients. It provides search engines that allow buyers to look for and find products. It provides reviews on both buyers and sellers from previous customers, and engines that run auctions for each product. The company uses information technology to increase customer participation in the process and thus significantly lower the cost of each transaction.

Information technology also helps customers to easily access the service inventory. For example, customers gain access to both frequently asked questions and the calculators described earlier using the Internet. Schwab's customers conduct trades on their own over the company website. Finally, eBay allows customers to directly conduct all transactions using the Internet.

**Service delivery channel.**   All service delivery channels other than the person-to-person direct channel rely on information and information technology to operate. The person-to-person remote channel uses the phone or web to create the remote contact. It is important for the remote service provider to have access to both customer- and service-related information to be effective at providing remote service. For example, an insurance company like GEICO that only provides remote service has outstanding information systems to support the customer service representatives. The person-to-machine channel relies on good organization of information to function and be effective. For example, Delta's use of self-service check-in machines will only be effective if travelers can easily complete their transaction. If most travelers are unable to check-in using a machine and must then go to an agent, the total cost of the service increases and customer experience worsens relative to not using the person-to-machine channel. Machine-to-machine transactions are based on good organization and standardization of information. For example, electronic data interchange (EDI) transactions that automate ordering at firms only works when the buyer and seller firm have an agreed upon standard for information transfer.

**Service recovery.**   Service recovery design is an essential element of overall service chain design. Service recovery design choices for the four drivers are influenced by the desired responsiveness and efficiency. In general, responsiveness for a successful service recovery should be higher when compared to the responsiveness of the service chain itself. Service errors result in a loss of reputation that hurts future demand. Therefore, the cost of a service error is often higher than the lost margin from not providing service. This justifies the higher responsiveness for service recovery.

The implication is that service recovery systems should have a higher level of service capacity, should carry a high level of service inventory, use a more responsive delivery channel, and have suitable information support. For example, consider an airline with a delayed flight. Several passengers on this flight are likely to miss connections. For service recovery, an airline should try to provide the following capabilities:

- Service capacity in the form of an agent at the gate and more aggregated service capacity in the form of agents at a service desk.
- Service inventory in the form of information on other connections that may be suitable for each passenger. Passengers may automatically be booked on these flights to save time if they agree to the new connection.
- Service inventory in the form of hotel bookings, calls to people the passenger wants to inform, and possibly an apology or bonus air miles.
- Initial service recovery is preferably delivered through the person-to-person direct channel even though it increases costs, because service recovery requests are likely to be different for each customer and a high level of interaction can improve the customer's perception of service recovery.
- Information systems must be structured to support each of the other choices. For example, information could be used to identify passengers who will miss their connections and arrange for alternative flights.

# TOOLS FOR DEFINE AND DEVELOP

## PROCESS ANALYSIS

A process flow diagram is a way to present a process. An example of a flow chart is shown in Figure 6.3. The flow chart for a stay at a motel is presented for illustrative purposes. A customer decides to stay at a motel conveniently located near a highway exit. The customer parks his car and approaches the clerk at the front desk and checks in.

**FIGURE 6.3.** Stay at Motel (Used with permission from Lovelock and Wirtz, 2004, 31).

Process flow concepts that are applicable to manufacturing could be applied to services as well (Anupindi et al. 1997, Chapter 3). A few definitions and core concepts are presented here (Anupindi et al., 1997).

*Average flow rate* or *throughput* of the process is the average number of flow units that pass through the process per unit of time.

*Average flow time* for a flow unit through the process is the average time that a typical flow unit spends within process boundaries.

*Average inventory* is the average number of flow units within the process boundaries.

The preceding three parameters interact through a relationship called Little's Law, as follows:

Inventory (I) = Flow rate R/Flow time (T)

Managers could focus on two measures, and the third one would be automatically fixed. For example, managers might want to focus on the process time and inventory (proxy for wait time).

The law is valid for a process within steady state or stable processes. For example, a branch office for insurance claims;

this office processes 20,000 claims per year, and assuming a typical claim takes on average two weeks:

Flow rate, R = 20,000 per year or 20,000/50 (assuming 50 weeks per year) = 400 per week

Flow time, T = 2 weeks.

Average claims in the processing system = I = R x T = 400/week x 2 weeks = 800 claims.

In general, managers want to control the following process improvement:

· Increase throughput.
· Decrease process cost.
· Reduce inventory (in process wait) because total time taken = wait time + process time.

## BLUEPRINTING

Lyn Shostack (1982, 1984, and 1987) suggested blueprinting as a technique for extending the process flow diagram. A blueprint is a modification of a flowcharting process and involves drawing a customer visibility line in the middle. Separating what a customer sees or experiences at the front stage is the distinguishing feature of a blueprint. Often, companies do not have processes and procedures in place to deal with front line steps. Backstage processes or steps often are well documented. Blueprinting has the following purposes and advantages:

· Clarifies the interaction between customer and employee. Employees see the effect of their actions in the overall scheme of things.
· Helps facilitate the integration of functional boundaries between marketing, operations, and HR management. The map removes any bias from any one functional area and works as a common document.

- To assign tasks by HR manager and could be used for recruitment.
- To analyze potential fail points in the system and develops an appropriate strategy. Foolproofing could be used to address these issues. Companies could set standards against the performance measures. Service systems include employees' errors and customer errors. Designing foolproofing (or as often called poke-yoke) requires minimizing both types of errors. Customers often need to be prepared for the transaction by providing information, checking ahead a transaction (getting a call from dentist's office one day before the visit), and ensuring that customers are ready for the transaction.
- To design and redesign the service delivery system.

Blueprint has the following challenges and issues, no standard symbols are used, and users must decide who should draw the blueprint (generally a team of cross functional team should draw the map), and if a time dimension should be placed on the horizontal axis.

Two examples of blueprints showing a bike repair shop and a bank are presented in Figures 6.4 and 6.5.

Key components of the blueprint are as shown in Figures 6.4 and 6.5:

- Definition of standards for each step in front stage activities
- Customer actions at front stage
- Line of interaction
- Front stage actions by customer-contact personnel
- Line of visibility
- Backstage actions by customers contact personnel
- Support processes involving other service personnel
- Support processes involving information or supporting technologies

**Self-service in service production.** The self-service aspect of a service is described in a separate section due to its impor-

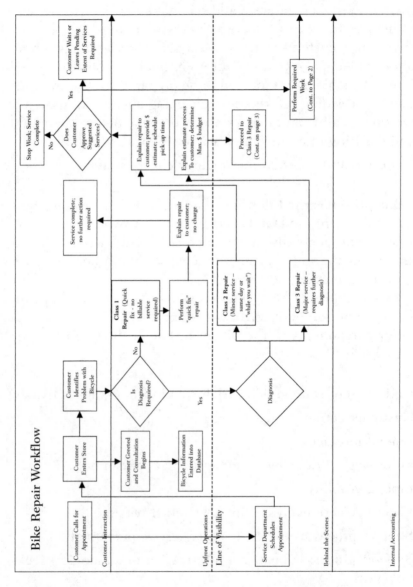

**FIGURE 6.4.** Blueprint for a Bike Repair Shop

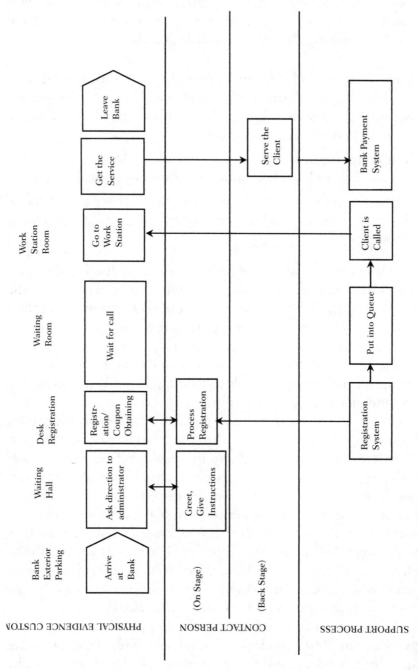

**FIGURE 6.5.** Blueprint for a Bank Transaction

tance. Some customer participation is inevitable in any service process. How much customer participation is ideal? The level of participation varies greatly and has often been categorized into the following three levels (Bitner et al., 1997):

- Low: Systems and employees do most of the work. Offerings tend to be standard and generally in the push mode. An example is a movie theatre where customers only pay at the counter and experience the movie.
- Moderate: Input from the customer is required to provide the service. Examples: Hair salon and annual physical exam.
- High: Customers actively participate in the production of these services. Examples: personal training and management consulting.

*Self-service.*    Customer involvement is at the highest level here. Coproduction started from retail stores where, for the first time in the 1930s, customers were required to select their own groceries from the selves, put them in the cart, and transport the cart to the checkout counter (some stores presently even provide self-checkout counters). eBay and Priceline.com provide all the necessary tools to customers, and customers can pick and customize the offerings.

Companies need to be very careful in deciding self-service channels, as the decision has a long-term strategic impact. First, a firm needs to know what drives its customer from one channel to the other channel. Self-service channels provide flexibility, cost savings, convenience, and greater control to customers. Customers love the convenience part of this channel; however, customers get frustrated when the technology fails or somehow the channel doesn't function properly. The biggest issue with the self-service channel is the lack of inclusion of a service recovery system with the channel. Often, there is no easy way to solve the problem on the spot (Lovelock and Wirtz, 2004).

At the end of the *define and develop* phase, a firm should agree on the issue (includes finding out what customers are expecting), understand the alignment of the project to the operations strategy (and eventually to the corporate strategy),

and agree on indicators and evaluation metrics. The blueprinting process helps everyone understand the context and help
draw boundaries for the project. A service chain design framework is helpful in understanding the relationship of the project
to corporate strategy.

## KEY TAKEAWAYS

· No matter which combination of methodologies a firm
  chooses to adopt, an alignment with the corporate strategy
  and goals is a necessary and required condition for the initiative to succeed.

· Service chain design is the missing piece of the puzzle that
  links customer requirements with a firm's corporate strategy
  and goals. No comprehensive methodology is provided in literature on how to design a service chain.

· Service chain design drivers include service chain capacity,
  service delivery channel, information, and service inventory.
  Service inventory is the extent of degree of completion analogous to work-in-progress inventory in manufacturing setting.

· A few tools for define and develop phase include process
  analysis and service blueprinting.

# MEASURE AND TRENDS

This chapter describes the second phase of Six Sigma methodology. The phase includes deciding the measurement strategy, measuring the appropriate indicators, and setting the baseline and goals based on these measurements.

The previous chapter defined and developed the methodology. The next step in the methodology is to collect data, perform measurements, and identify trends. We will discuss a few things that could go wrong while determining what to measure. Next, we will discuss a few tools. Some of these tools are relatively new to Six Sigma methodology. However, we strongly feel the need to introduce new tools to Six Sigma methodology.

In general, two types of issues could occur while determining a measurement strategy for a service company:

1. The measurement system is not aligned with the service delivery system. The alignment is based on company-specific strategies, business sector context, and the life cycle of the offering and the firm. We refer to the Haywood-Farmer model for this strategic alignment. Hope and Muhlemann (1997) presented a table based on the Haywood-Farmer cube, as shown in Table 6.3. According the concept, service companies are categorized based on three characteristics—labor intensity, customization, and

customer contact/interaction. Measurement strategy will heavily depend on the position of a company. Companies in one category will have very similar measurement issues to deal with. Roger Schmenner (2004) recently presented another classification scheme where "degree of variation: customization for and interaction with customers" is used on the x-axis and "relative throughput time: measured for a service transaction as compared to others in the industry" is used as the y-axis.

2. The second issue could arise due to the rigidity of the measures. Measures need to keep up with the life cycle of the company or offering. A company in its startup phase will require a different set of measures compared with the same company in its maturity phase.

Table 7.1 shows performance metrics for typical companies. These metrics are outcomes for a particular business context and could be used to measure and track the performance of a particular company.

TABLE 7.1    Typical Performance Metrics for Service Companies

|  | HOTEL | RESTAURANT | THEATER | CAR RENTAL | AIRLINE |
|---|---|---|---|---|---|
| Units of inventory | Room | Seat | Seat | Car | Seat |
| Unit of time | Night | Hour | Hour or performance | Day | Hour or seat-mile |
| Revenue per available time-based inventory unit | Revenue per available room night | Revenue per available seat hour | Revenue per available seat per performance | Revenue per available car day | Revenue per available seat miles |

Fitzgerald et al. (1991) have provided some guidelines for types of measures that could be used in service businesses. Tables 7.2 and 7.3 present such a list of measures.

TABLE 7.2    Types of Generic Measures in Service Businesses

| DIMENSIONS OF PERFORMANCE | TYPES OF MEASURES |
|---|---|
| Competitiveness | Market share and position, sales growth |
| | Measures of customer base |
| Financial Performance | Profitability, liquidity, capital structure, market ratio |
| Quality of Service | Reliability, responsiveness, comfort, friendliness, access, availability, and other dimensions |
| Flexibility | Volume flexibility, delivery speed flexibility, and specification flexibility |
| Resource Utilization | Productivity, efficiency |
| Innovation | Performance of the innovation process |
| | Performance of individual innovation |

Source: Fitzgerald et al. (1991). Performance measurement in service businesses, CIMA, 8

TABLE 7.3    Types of Specific Measures in Service Businesses

| DIMENSIONS | TYPES OF MEASURES (PROFESSIONAL SERVICES) | TYPES OF MEASURES (MASS SERVICES) |
|---|---|---|
| Competitiveness | Percent success in tendering | Number of customers |
| | Percent repeat business | Market share |
| | Relative market share | Relative prices and product range/variety |
| Financial Performance | Staff costs, profit per service | Return on net assets |
| | Debtors and creditors days | Working capital |
| | Value of work in progress | Profit per market segment |
| Quality of Service | Investment in training | Equipment availability |
| | Percent nonchargeable: chargeable hours | Product range |
| | Percent on time project delivery | Customer processing time |
| | Customer satisfaction | Customer satisfaction |

TABLE 7.4    Types of Specific Measures in Service Businesses
(*Continued*)

| DIMENSIONS | TYPES OF MEASURES (PROFESSIONAL SERVICES) | TYPES OF MEASURES (MASS SERVICES) |
|---|---|---|
| Flexibility | Percent orders lost due to late delivery | Monitoring of queue length |
| | Staff skill mix | Number of part time and floating staff |
| | Percent hours bought in from other offices | Customer satisfaction with service availability |
| | Customer satisfaction with delivery speed | |
| Resource Utilization | Ratio to supervisors to staff | Costs per customer |
| | Ratio of hours chargeable to clients and nonchargeable hours | Revenue per customer |
| | | Percent utilization (percent seat filled on a plane) |
| Innovation | Number of new services | Percent new: existing offerings |
| | New service introduction lead times | Research and development costs |
| | Percent training spend invested in new services | |

Fitzgerald et al. (1991). Performance measurement in service businesses, CIMA, 121–122.

Finding the relationship between outcome measures and causal factors is a challenging task for managers. A few measures that affect the outcome measures are customer-related and some are internal factors. First we will describe customer-related measurement schemes. The zone of tolerance model is one such model. The zone of tolerance model and performance measures were presented by Fitzgerald et al. (1991).

# CUSTOMER PERSPECTIVE: TOLERANCE ZONES ARE DYNAMIC

Zeithmal et al. (1993) first used the term "zone of tolerance." Johnston (1995) further examined these zones of tolerance, with each zone having an upper and lower boundary. Broadly speaking, there are three zones of tolerance, and each zone has three categories:

1. Preperformance expectation: Unacceptable, acceptable, and more than acceptable
2. Service processes: Less than adequate performance, adequate performance, and more than adequate performance
3. Outcome state: Dissatisfaction, satisfaction, and delight

Outcome becomes much more complicated for a multistage service. Let us take hair cutting as an example. The process involves reception, waiting, consultation, hair washing, hair cutting, hair drying, and payment. According to Johnston (1995), if sufficient numbers of the stages are perceived to exceed the upper boundary of the zones of tolerance, then customers will be delighted. Johnston also suggests that the size of the zone of tolerance may be dynamic rather than static and influenced by excellent or poor service in the early stages of the process. For example, a customer arrives at a hotel to find that his/her reservation has been lost and the front-line staff is rude and unhelpful. For this customer, the zones of tolerance for the following stages will be narrower compared with the zones of tolerance if those mis-steps did not happen. Similarly, with a very good start of a service, zones of tolerance could be wider at the later stages.

Quality/operations managers need to understand the dynamic nature of this process. The process needs to be managed in a manner that will eventually delight the customer.

## Customer Is Unhappy, What Should We Do?

While deciding on a particular measurement system, managers also need to take the customer complaint management system into account. Managers need to understand why customers complain, why some unhappy customers do not complain, and what customers' expectations are about their complaints. Customers usually complain for the following reasons: to obtain compensation from the firm, to vent their anger, to help improve the service, and for altruistic reasons.

Research shows that about 5 to 10% of dissatisfied customers approach a firm to formally complain (Tax and Brown, 1998). Often customers are unhappy and just leave the firm. Customers do not complain for multiple reasons. Tax and Brown (1998) found that satisfaction for the majority of customers that complain is determined by the following three dimensions of fairness:

1. Procedural justice: Customers expect the firm to apply fair procedures that include flexibility of systems and considerations for customers' input into the system.

2. Interactional justice: Customers expect an explanation for the failure, and employees should make a genuine effort to resolve the issue. Customer perception of the effort must be honest, genuine, and polite.

3. Outcome justice: Outcome justice relates to the compensation part of the effort. Customers expect to be compensated for the inconvenience and economic losses incurred.

Recovery efforts by a firm should be looked at as opportunities for improvement, customer input, and to win back customer loyalty. The following principles should be followed for an effective service recovery measurement system:

- Principle 1: Make it easy for customers to give feedback—Available data will be much more useful. Firms need to make it easier for customers to complain by providing various channels and means. Some companies even have separate toll-free numbers for this purpose and have links prominently displayed on the company website.

- Principle 2: Provide proactive and effective service recovery—The procedures need to be planned, and employees should be appropriately trained and empowered to make decisions.

- Principle 3: Learn from customers' feedback—Feedback should be appropriately documented and acted upon.

- Principle 4: If appropriate, provide unconditional service guarantees—Unconditional service guarantees provide the following advantages to a service firm (Hart, 1990): Set clear standards at the firm level, provide meaningful feedback from customer, and force a firm to understand why it fails, and encourage the firm to identify potential fail points.

- Principle 5: The service recovery process should be well documented and blueprinted. One example is shown in Figure 7.1.

# TOOLS FOR SERVICE QUALITY MEASUREMENT

A long list of tools could be used for measurement strategy. We will selectively describe a few tools. Some tools have been covered in more detail in other Six Sigma references.

The following tools will be discussed here:

- Statistical sampling
- Process flow chart
- Data envelopment analysis
- Regression analysis
- Hypothesis testing

Tools like histograms and Pareto charts are not discussed at length. Control charts will be described in Chapter 9.

## DATA COLLECTION: STATISTICAL SAMPLING AND SAMPLE DESIGN

Statistical sampling is used for measuring customer satisfaction and for conducting similar analyses. The most common issue with a customer satisfaction survey is that results are often not managerially relevant. Before designing a particular survey or collecting a particular data, ask the following questions- Is this relevant? How are we going to use the data? How often should this survey be conducted? Who is answering these questions?

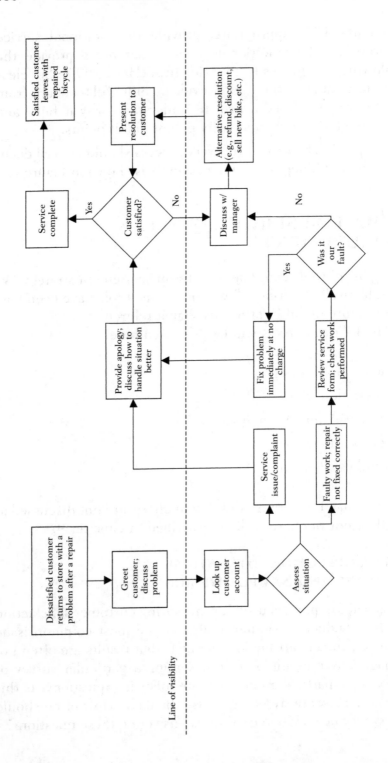

**FIGURE 7.1.** Service recovery blueprint.

Research design could consist one or more of the following steps:

· Exploratory research: Secondary sources and primary sources
· Descriptive or quantitative research

Exploratory research is often necessary at the beginning to design the survey in customer's language. Exploratory phase provides customer input in form of customer vocabulary and range of customer perceptions. Exploratory analysis often involves a focus group of customers. Some experts have suggested designing surveys around business processes. A survey designed around business processes should have direct managerial implications and will have an owner. We should also start the process where it usually ends and then we should work backward. As a first step we need to determine how the research results will be implemented. As a second step we need to ensure the implementation of the results, determine what the final report should contain and how it should look.

Data collection during descriptive research phase takes two forms- data collected over time and cross-sectional data or a data being collected at a point in time. Descriptive phase involves a questionnaire containing set of questions. A good questionnaire matches the survey objectives with the respondent information. Various types of measurement scales are used to capture information. Following steps are  followed to draw a sample:

· Define the population
· Identify the sampling procedure: Probability samples (simple random, stratified) or non-probability samples (judgmental or convenience)
· Select the sampling plan
· Determine sample size
· Select the sampling units

Determination of the appropriate sample size is a crucial element during the process. Statistical theories provide an answer to this question. Appropriate sample size is a function of specified confidence level, acceptable magnitude of error, and sample standard deviation. Another important aspect of the process is measurement accuracy in terms of reliability and validity of data. Total error will consist of sampling error, respondent error and administrative error.

Detailed description of calculations is beyond the scope of this book. Readers are requested to refer to appropriate business statistics textbook for further in-depth background on this topic.

The previous chapter also described SERVQUAL methodology to measure the gaps between customer expectations and perceptions of the real service. Managers need to keep in mind the following issues of the SERVQUAL instrument (Hope and Muhlemann, 1997):

· Timing and order of the questionnaire. If the expectations part of the questionnaire is administered first, one distinct issue arises- expectations may be raised making it more difficult to meet those raised expectations.

· The questionnaire is not a short one. Sample questionnaire is presented in Figure 7.2. Some customers might feel that answering 49 (22 questions for each section and 5 questions distributing 100 points among service quality dimensions) questions is a laborious and time-consuming task.

· The tendency in the expectations section to circle "very important" for all aspects. Answers could be double checked with the third section about relative important of each dimension.

## PROCESS FLOW CHART

Process flow charts give a visual presentation of the process. The process of flowcharting is familiar to most readers of this textbook. The textbook by Schmenner (1998) is an excellent reference for this purpose. The author has presented following detailed process diagrams:

## EXPECTATION SECTION

Directions for the survey are provided here. Each of the statements is accompanied by a 7-point scale anchored at the end by the labels "strongly diagree" (=1) and "strongly agree" (=7). Service quality dimension heading shown here are not included in the questionnaire. Headings are shown here for simplicity only.

*Reliability:*

1. When excellent telephone companies promise to do something by a certain time, they will do it.

2. When customers have a problem, excellent telephone companies will show a sincere interest in solving it.

3. Excellent telephone companies will perform the service right the first time.

4. Excellent telephone companies will provide their services at the time they promise to do so.

5. Excellent telephone companies will insist on error-free records.

*Responsiveness*

6. Employees of excellent telephone companies will tell customers exactly when services will be performed.

7. Employees of excellent telephone companies will give prompt service to customers.

8. Employees of excellent telephone companies will always be willing to help customers.

9. Employees of excellent telephone companies will never be too busy to respond to customer requests.

*Assurance*

10. The behavior of employees of excellent telephone companies will instill confidence in customers.

11. Customers of excellent telephone companies will feel safe in their transactions.

12. Employees excellent telephone companies will be consistently courteous with customers.

13. Employees of excellent telephone companies will have the knowledge to answer customer questions.

*Empathy*

14. Excellent telephone companies will give customers individual attention.

15. Excellent telephone companies will have operating hours convenient to all their customers.

*(continued on next page)*

**FIGURE 7.2.** Sample SERVQUAL Questionnaire (Adapted from Parasuraman, et al. 1991, pp 420–50).

16. Excellent telephone companies will have employees who give customers personal attention.

17. Excellent telephone companies will have the customers' best interest at heart.

18. The employees of excellent telephone companies will understand the specific needs of their customers.

*Tangibles*

19. Excellent telephone companies will have modern-looking equipment.

20. The physical facilities at excellent telephone companies will be visually appealing.

21. Employees of excellent telephone companies will be neat-appearing.

22. Materials associated with the service (such as pamphlets or statements) will be visually appealing in an excellent telephone company.

## PERCEPTION SECTION

*Reliability*

1. When ABC promises to do something by a certain time, it does so.

2. When you have a problem, ABC shows a sincere interest in solving it.

3. ABC performs the service right the first time.

4. ABC provides its service at the time it promises to do so.

5. ABC insists on error-free records.

*Responsiveness*

6. Employees of ABC tell you exactly when services will be performed.

7. Employees of ABC give you prompt service.

8. Employees of ABC are always willing to help you.

9. Employees of ABC are never too busy to respond to your requests.

*Assurance*

10. The behavior of employees of ABC instills confidence in customers.

11. You feel safe in transactions with ABC.

12. Employees at ABC are consistently courteous with you.

13. Employees of ABC have the knowledge to answer your questions.

*Empathy*

14. ABC gives you individual attention.

15. ABC has operating hours convenient to all its customers.

16. ABC has employees who give you  personal attention.

17. ABC has your best interest at heart.

**FIGURE 7.2.** Sample SERVQUAL Questionnaire *(Continued)* (Adapted from Parasuraman, et al. 1991, pp 420–50).

18. The employees of ABC understand your specific needs.

*Tangibles*

19. ABC has modern-looking equipment.

20. ABC's physical facilities are visually appealing.

21. ABC's employees are neat-appearing.

22. Materials associated with the service (such as pamphlets or statements) are visually appealing at ABC.

## POINT ALLOCATION QUESTIONS

Please allocate a total of 100 points among the five features according to how important each feature is to you- the more important a feature is to you, the more point you should allocate to it.

1. The ability of the telephone company to perform the promised service dependably and accurately.

2. The willingness of the telephone company to help customers and provide prompt service.

3. The knowledge and courtesy of the telephone company's employees and their ability to convey trust and confidence.

4. The caring, individualized attention the telephone company provides its customers.

5. The appearance of the telephone company's physical facilities, equipment, personnel and communications materials.

**TOTAL POINTS ALLOCATED    100**

**FIGURE 7.2.** Sample SERVQUAL Questionnaire (*Continued*) (Adapted from Parasuraman, et al. 1991, pp 420–50).

- A worker-placed line flow process and a service factory (Burger King, Noblesville, IN): Process description, capacity planning and workforce management are highlighted.

- A service shop (Ogle, Tucker Buick): Process description (writing the service order, dispatching, quality control, parts department, and body shop, etc.), flow of the process and of information, and demands of the process on workforce and management are presented.

- Mass service (Lazarus, Castleton Square Mall): Layout, workforce, transactions, and manager's job, flow of process and of information, and demand of the process on the workforce/management are presented.

- A professional service (Arthur Anderson, Charlotte): Managing the practice, flow of the process and of information, and demand of the process on the workforce/management are presented.

- A project (Geupel DeMars, Indianapolis, a construction management firm): Project organization and timetable and planning and control are presented.

*Managing Business Process Flow* by Anupindi et al. (1997) is another good text for quantifying the process analysis. The text is used at the graduate level to introduce operations management to students. Various examples are presented.

## Productivity Measurement

Productivity is measured as a ratio of outputs to inputs. The overall objective is to gain more output from a given level of input or to obtain the same level of output from a lower level of inputs. The financial measure of interest to the investor is return on investment or return on capital employed. Another measure could be just the ratio of output to input in terms of operational input.

This measurement technique is introduced as a benchmarking tool. The technique could also help in identifying best and poor practices, target setting for the units, resource allocation, and monitoring the efficiency change over time.

# DATA ENVELOPMENT ANALYSIS AS A BENCHMARKING TECHNIQUE

Data envelopment analysis (DEA) is also called frontier analysis. Metters and King-Metters (2003, Chapter 15) provide an overview of this technique. Service businesses often have multiples sites, as compared to a few sites for manufacturing. Services often include customer participation and hence should be closer to the customer. The following table shows the number of sites for different services.

Table 7.4 was adapted from Metters and King-Metters (2003) and is based on *The Franchise Handbook* (Summer 2001) data.

TABLE 7.4    Number of Sites for Typical Franchise Operation

| SERVICES | NUMBER OF SITES |
| --- | --- |
| Budget Rent a Car | 3,240 |
| Bank of America | 4,490 |
| Jani-King | 7,475 |
| Dunkin' Donuts | 4,736 |
| Burger King | 8,246 |
| Subway | 14,700 |
| Holiday Inn | 1,758 |
| Radio Shack | 7,070 |

The sheer number of locations makes it impossible for a manager to have a hands-on approach.

## WHAT IS DEA?

DEA is a linear programming technique; however, knowledge of DEA is not a prerequisite to understand the technique. Common performance measures used at a business are profitability, revenue, and market share, to name a few. Profitability and revenue measurement are problematic for many businesses having multiple sites. A bank customer might open an account close to his/her workplace, but probably uses the branch close to his/her home. How should a bank allocate cost and revenue to each location? Profitability is often a misleading number for a company in its startup phase. Therefore, a firm might want to include other measurements like market share, sales volume, customer service, and contribution margin.

## DEA CONCEPT

DEA defines the performance by the following equation:

$$\text{Performance} = \text{Results obtained (output)/resources used (input)} = \text{efficiency}$$

DEA provides the most optimal conditions for each unit (or decision making unit, DMU; here we will use units to signify DMU). Therefore, the performance of a unit will be the maximum possible under a given situation. DEA assigns weights for output and input (and weights are not preassigned) and solves (maximizes) the following equations (functions):

$$\text{Output (results)} \times \text{a weight for each result for a specific unit}$$

Given that (for all the units)

$$(\text{output} \times \text{weight})/(\text{input} \times \text{weight}) = 1.0$$

and for a specific unit

$$\text{input} \times \text{weight} = 1$$

The final result of DEA output is a relative efficiency number assigned to each unit. This efficiency number is not a theoretical efficiency number. For example, a unit with an efficiency number of 1.0 is not a theoretically perfect unit. However, this unit could be used as a benchmark for other units having score lower than 1.0.

Two popular software packages are available for performing DEA analysis. Frontier analysis, marketed by Banxia Software, and Warwick DEA software.

## ADVANTAGES AND LIMITATIONS OF DEA

Advantages of DEA over other traditional and more subjective methods are (Metters and King-Metters (2003):

- Data reduction: Multiple outputs are reduced to one single number. Multiple performance measures are included. Units with similar goals should be compared using this technique.
- Objectivity: Exercise provides an objective measure for performance outcome. Appropriate weight is provided for doing the calculations. These weights could vary unit-to-unit depending upon the situation.
- Neutralizes the effect of environmental changes: External environment might affect all the units and the results will still be valid.

Users requiring further information and free software download are directed to go to www.deazone .com. The website is an excellent resource and was created in honor of A. Charnes, W. W. Cooper, E. Rhodes, and R. D. Banker, who originally developed DEA. Other introductory DEA material is available at http://mscmga.ms.ic.ac.uk/jeb/or/dea.html. The site is intended for new users of the technique. DEA has been applied to the following business contexts: banks, retail stores, management evaluations, fast food restaurants, benchmarking, manufacturing, police stations, hospitals, tax offices, prisons, defense bases (army, navy, air force), schools, and university departments.

The following limitations should be kept in mind while applying this technique:

- The technique implicitly assumes that input and output are correlated.
- The technique is a point technique, and small measurement error could cause a big difference.
- Hypothesis testing or a similar statistical analysis is not possible after performing DEA analysis—DEA is a nonparametric technique.
- The technique can be computational-intensive at times.

## Step-by-Step DEA Analysis

The following step-by-step procedure should be followed while performing DEA analysis:

· Specify decision making unit. We need to define the boundary for the unit where we will measure input and output. We should be comparing similar decision making units.
· Define input to the decision making unit.
· Define output to the decision making unit.
· Reevaluate the correlation between output and input to make sure that we have the right variables.
· As a rule of thumb, number of sites > 2 * (number of input + number of output). If we have five inputs and four outputs, we need to have data from at least 18 sites for our analysis to have a reasonable validity.

Eventually, DEA could be used to make a strategic decision by identifying weak players, benchmark groups, and groups needing improvement.

## PREDICTING PERFORMANCE: REGRESSION ANALYSIS

The purpose of a regression analysis is to build a model describing the relationship between a dependent variable and several independent variables. In general, we have one dependent variable and multiple independent variables. Regression involving multiple independent variables is called a "multiple regression." For example, attendance at a stadium will depend on the following parameters: day of the week, as we could expect more people over the weekend; day or evening time of the game; weather; and other games in the town.

Sometimes, independent variables could be "dummy variables" with a value of zero or one, showing the presence or absence of certain characteristics; for example, presence or absence of toher game in town could be a dummy variable in the preceding case. A dummy or qualitative variable is one that only takes on the values 0 and 1—the value of 0 shows the absence and a value of 1 shows the presence of this effect.

The regression technique is a valuable tool after identifying *root causes*. The regression model quantifies the cause and effect relationship. For example, a hospital concerned about the length of time required to get a patient from the emergency department to a patient bed may identify several potential causes, such as number of patients in the emergency room, availability of a nurse, availability of medical/surgery units, or readiness of the bed.

An automotive insurance company may want to predict the risk factor for drivers before deciding on the premium. This risk factor (dependent variable) will likely depend on the age of the driver, marital status, dependent driving children, distance to work place, type of car, claims filed in the past, incidents report during the last three years, and age of the car, to name a few. Once information about the independent variables for a driver has been compiled, the company looks at historical data and determines the risk factor and associated premium. Based on the regression analysis, the insurance companies found that the age of drivers makes a difference in their driving style or pattern.

People deploy regression analysis continually, whether using the formal or informal methods of calculation. People estimate correlations between events they observe around them. In a Six Sigma project, the regression analysis can be used to prioritize independent variables, or to establish a causative relationship between output and input. Once the relationship or a model has been established based on historical data, one could predict the dependent variable for a given independent variable. Regression analysis can be seen as analysis of scatter plots by adding best-fit lines and quantifying the relationship between the dependent and independent variables. The regression analysis includes the regression equation, residual variance, and R-square. A regression equation defines the best-fit line. The best-fit line is the line that relates the dependent variables to the independent variables. It produces the smallest difference between the actual and predicted values.

The following aspects of regression analysis must be understood for its effective use:

- Assumption of linearity: Before going ahead with the analysis, it's a good idea to look at the scatter plot of two variables to confirm this assumption.

- Variability in the independent variables: Not all the independent variables have equal variation.

- Spurious correlation: An implicit assumption one makes while using the regression analysis is that the dependent variable is caused by independent variables. However, the stronger statistical correlation does not represent a causative relationship between the dependent and independent variables. For example, one could correlate rain fall in a particular geographical location to unemployment and try to predict employment level based on the amount of rain. This relationship would be a spurious one.

- Choices of number of variables in building the process model: It is very tempting to include more variables to make the model look more complete. We need to test the significance of each variable and ask the question whether the model will be worse off if we drop this variable. The balance is between parsimony and completeness. As a rule of thumb, we need to have at least 20 times as many observations as one has variables.

- Multicollinearity: Two independent variables are either completely or very strongly correlated with each other. For example, assume we have length in meters and width in inches as two variables. These two variables happen to be completely correlated. One must recognize such collinearities.

- Residuals say it all: Residuals are normally distributed and independent. We should produce histograms for the residuals, as well as normal probability plots, to validate the assumptions. The residual plot will also indicate the presence of "outliers."

- Error-related assumptions:
  - Expected error is a random variable with a mean of zero.

– Homoskedasticity: Error variable has a constant variance.

– Serial independence: Error terms are independently distributed.

– Normality of error: Errors are normally distributed.

The purpose of regression analysis must be clearly stated and understood. To make the decision to use regression analysis, one must consider the following questions:

· Is regression the right tool for analysis?
· What is the dependent variable?
· What are independent variables?
· What tool should be used to perform regression analysis?

The following model shows a typical regression model equation.

$$y = \beta_0 + \beta_1 x + \epsilon$$

The model predicts a dependent variable, $y$, based on the values of independent variable, $x$. Error term, $\epsilon$ (epsilon), is often assumed to be zero and not written as part of the model. Beta-zero is the constant or y-intercept and will correspond to a value if values of all independent variables were zero. Beta-one here is the coefficient for the independent variable.

Each estimate of coefficient is subject to sampling error and has a distribution. Regression output provides estimates of standard deviation. R-squared values and adjusted R-squared values, not shown with the equation, tell the percent variance in $y$ that is explained by knowing $x$. Adjusted R-squared values will be equal to or lower than R-squared values. Adjusted R-squared values will be lower and lower as the number of independent variables increases.

A natural extension of the analysis we have considered so far is to have more than one predictor variable ($x$).

$$Y = \beta_0 + \beta_1 \times X_1 + \beta_2 \times X_2 + \beta_3 \times X_3 \ldots + error$$

Essentially everything stated so far about simple linear regression carries through for multiple regressions. The only technical problem is that formulas for the betas and associated variances involve matrix inverses and thus are impractical to apply without software.

## SELECTING VARIABLES

Two methods can be used to determine how many and which variables to include in a model. The methods to build the regression model could be the: 1) statistical method using t-ratio and p-value, or 2) model builder judgment. Statistical method alone to eliminate a variable is not a good process to determine or exclude a variable and should not be the basis recommended for the selection of the model's coefficients.

The term *parsimonious* is used to indicate that the model should contain only "necessary" variables. While one can always identify other factors that can influence sales at a retail store as an example and might be included in the model, the best models are ones that contain as few variables as possible. Including too many variables in a model results in both "over-fitting" and imprecise estimates of the model's coefficients. Therefore, it is important to limit the model to critical variables. For example to predict sales at a retail store as a function of various marketing and sales variables, one would like to include the type of advertising into the model rather than simply using the presence of advertising. If the type of feature advertising is not going to be used to design the firm's strategies, and the difference in the effects of the different types is small, then there is no reason to include several feature advertising variables (one for each type) in the model. Deciding when a model is not parsimonious is difficult to determine, but the rule should be: "Do not include a variable unless it is absolutely necessary."

In the real world, it is commonplace for data to not been collected for key causal variables. This is because the retailer or the data supplier did not have reason to collect it or it is too expensive to collect. Therefore, while a variable may be important, data are not available for it. The rule is to do the best one can with the data available. A model will never be perfect, and some analysis is better than no analysis.

An example of regression output is to determine the effect of customer income and the presence of competition on the sales of an offering. Here we shall consider dummy and slope dummy variables. In this case, competition would be a dummy variable, having a value of 0 or 1, indicating absence or presence of competition. However, the difference in sales based on competition alone will not be a fixed value. The regression model can be written as follows:

$$\text{Sales} = \text{constant} + \text{constant} * \text{income} + \text{constant} * \text{competition (0 or 1)}$$

However, one could argue that the effect of competition will not be a fixed value, but will depend on the income group of our customers. We need to use a slope dummy variable for this purpose. Slope dummy is a product of a dummy variable (in this case, competition) and another variable (income, in this case). The revised model can be written as follows:

$$\text{Sales} = \text{constant} + \text{constant} * \text{income} + \text{constant} * \text{competition (0 or 1)} + \text{constant} * \text{competition (0 or 1)} * \text{income}$$

One needs to check for p-values for the significance of these variables. R-squared value will indicate the variation that can be explained with this model.

## PERFORMING REGRESSION ANALYSIS

The following are the steps for performing a regression analysis:

1. Decide whether regression is the right tool to quantify a cause and effect relationship.
2. Construct a scatter plot for various variable data to check for the assumption of linearity.
3. Check for any missing or incomplete data.
4. Identify the dependent variable and select dependent variables.

5. Run regression using the appropriate software tool.

6. Check for R-squared, p-values (significance), and signs of different coefficients. These three numbers together will indicate the validity and usefulness of our model.

7. Check the plot for residuals and test for normality, independence, and constant variance assumptions.

8. Drop any insignificant variable or transform variable if certain assumptions are violated. Rerun regression.

9. Once we are satisfied with our model, write down the model.

10. Use the above model for predictions.

11. Check the model for common sense, statistical, and economic significance.

12. Check for validity in terms of signs of the coefficients of independent variables. Ascertain whether the predicted sign of the model makes sense.

For simple analysis between the two variables, one can draw the scatter plot and evaluate the relationship in terms of positive or negative correlations. Calculating the correlation coefficient, R squared ($R^2$) will establish the strength of the correlation between two variables.

## POWER OF TESTING: TESTING HYPOTHESES

Hypothesis testing is a statistical technique that is used to support an experimental statement. The primary purpose of this technique is to convince stakeholders that the statement made is likely true or not true. The technique involves setting up two hypotheses, null and alternative statements, that include all the possibilities and do not overlap with each other. $H_a$ denotes the alternative hypothesis, and $H_0$ denotes the null hypothesis. The outcome of any hypothesis testing includes the two possibilities as follows:

1. Rejecting the null hypothesis

2. Not rejecting the null hypothesis

The outcome depends on the statistical significance of evidence. The statistical significance of evidence is compared with a threshold value based on the required "level of significance" or confidence. The statistical significance of evidence is called the p-value, expressed in probability terms. As a rule of thumb, low p-values will correspond to strong evidence against the null hypothesis and hence support the alternative hypothesis. A high p-value corresponds to weaker evidence.

A criminal trial in a court of law provides a good analogy to understand this concept. A prosecuting attorney in a trial case wants to prove to the jury that the defendant is guilty of a crime. Whatever hypothesis we are trying to prove is the alternative hypothesis. In this case, the defendant being guilty of a crime is an alternative hypothesis. All other possibilities not included in an alternative hypothesis are included in the null hypothesis. Here, the null hypothesis is that the defendant is not guilty of a crime. These two hypotheses should cover all the possibilities and should not overlap at all. Two possible outcomes of this trial are "not guilty" and "guilty."

The test can never result in rejecting the alternative hypothesis or equivalently accepting the null hypothesis. If a jury does not reject the null hypothesis that the defendant is not guilty, it simply means that the evidence was not strong enough to prove the defendant guilty. It does not mean that the defendant was innocent. It is very important to make sure when setting up a hypothesis test that what we hope to prove is stated in the alternative hypothesis.

A hypothesis test could lead to wrong conclusions though, with predefined, low probabilities. There are two types of errors—type I and type II error. Setting the level of significance is one major step, and the value will depend on the context. In some contexts where we are looking for very strong evidence, we might set up type error at 1% or 0.01. If we are not that concerned about the error, we could likely set this value at 10% or 0.10. We should always make sure that we set up an alternative hypothesis first.

For typical process improvement, a threshold p-value of 5% or 0.05 is used for type I error. This threshold value is also called a standard of proof required or level of significance denoted by the letter alpha, α—a phrase similar to "beyond any reasonable doubt." A lower level of significance will make it harder to prove our point, and a higher level of significance will allow us to prove our point more often. If significance level is set at 0.05 or 5% level, 5% of the time, we will conclude a significant difference by chance; 95% of the time, the difference will be real.

The probability of making type II errors is denoted by beta. The main tool to improve the reduction of type II errors is to gather more data. Type II is the error of missing an improvement while some exists. Because the cost of missing an improvement is not as significant as the cost of type I error—that is the risk of saying there is an improvement when none exists, the beta is set at 10%, or 0.10.

The sampling distribution is dependent on the null hypothesis. We assume that this distribution is a normal distribution. For a large sample, the assumption is justified by the central limit theorem. Normal distribution is characterized, as usual, by its mean and its standard deviation.

## MECHANICS OF HYPOTHESIS TESTING CALCULATIONS

After computing a value for t-statistics (depending on the sample size), we need to determine a p-value or theoretical t-values. These values can be calculated either by using t-tables or using MS Excel. The TDIST function in MS Excel does p-value calculations for us. However, the TDIST function in Excel does not allow us to enter negative values. For example, if we wish to find the p-value corresponding to −1.7, or our estimator was 1.7 standard deviations below the value in the null hypothesis, we are not able to enter −1.7. Because the t-distribution is symmetric, the area above −1.7 is the same as the area below 1.7, which equals 1-TDIST (1.7, degree of freedom, 1). In this case, test statistics have an approximate t-distribution with a mean of 0 and n−1 degree of freedom.

So far, we have used t-distribution. However, normal distribution is used for testing the hypothesis for population proportions. We follow the same procedure as described earlier to formulate test statistics for population proportions. However, we will be using the different formula. Population proportion will follow normal distributions and not a t-distribution. The Excel function used in this case is NORMSDIST.

## STEPS FOR TESTING A HYPOTHESIS

The theory of hypothesis testing is simply a statistical formalization of common sense procedures. It makes the following assumptions:

1. Make a statement about the desired change ($H_a$).
2. Make a statement of no change ($H_0$).

The following types of hypotheses are commonly formulated:

*One-sided, greater-than hypothesis:*

$$H_0: \mu \leq 5 \qquad H_a: \mu > 5$$

*One-sided, less-than hypothesis:*

$$H_0: \mu \geq 5 \qquad H_a: \mu < 5$$

*Two-sided, not equal to hypothesis:*

$$H_0: \mu = 5 \qquad H_a: \mu \neq 5$$

The type of hypothesis will determine the type of test (i.e., the right tail, the left tail, or the two tail). When the two tail test is used, the $\alpha$-risk (e.g., 0.05) is distributed on both sides of the distribution.

Regardless of population parameters, the testing of a hypothesis includes the following steps:

1. Identify what we would like to prove and make that statement as our alternative hypothesis.

2. Set up a null hypothesis.
3. Determine level of significance ($\alpha$).
4. Gather data and calculate the t- or Z-value for sample distribution.
5. Find the p-value using the appropriate function—TDIST or NORMSDIST.
6. Compare the statistic to the theoretical values, or p-value to the level of significance.
7. Draw a conclusion. Reject the null hypothesis if the p-value is smaller than the significance level; otherwise null is not rejected.

To evaluate the data, begin by assuming that the null hypothesis is correct. The null hypothesis determines the sampling distribution of our estimator. Assume that this distribution is a normal distribution and the assumption is justified by the central limit theorem. Next, calculate t-statistics for the hypothesis by using the following relationship:

$$\text{t-statistics} = \text{(estimator-equal value in null hypothesis)/} \\ \text{standard deviation of estimator}$$

In essence, t-statistics measures how far away our observed estimator is from the value we would expect if the null hypothesis were true. This distance is measured in terms of standard deviation (similar to Z-value concept). The interpretation of t-statistics, therefore, could be expressed as follows. Our estimate is "value of statistic" standard deviation away from the value in the null hypothesis. Test statistics will have a t-distribution. Next, we calculate p-value using the t-statistics values. We can find p-value using MS Excel or t-tables. The TDIST function will be used if we have t-distribution.

The next step is to compare the p-value with the level of significance. If p-value is lower than the level of significance, accept the alternative hypothesis. If p-value is higher than the level of significance, do not accept the alternative hypothesis or choose the null hypothesis.

This chapter is only an overview of hypothesis testing. The testing is often done after identifying the causes for particular events. We confirm certain hypotheses using these techniques, and at times, we will require more data to make a better decision.

This section concludes our discussion of measures and trends. We provide a suggested sequence of tools:

- Identify data collection plan and obtain data: Use statistical sampling and process flow analysis.
- Benchmark performance against an internal or an external peer group: Data envelopment analysis
- Quantify cause and effect relationships: Regression analysis
- Confirm relationships: Hypothesis testing

## KEY TAKEAWAYS

- Dynamic nature of customer perspective before and during service transaction or service delivery should be factored into the design of an effective measurement system.
- Service recovery efforts should be looked at as opportunities for improvements.
- Multisite service locations can be benchmarked using quantitative techniques like Data Envelopment Analysis, DEA.
- Trends and cause and effect could be quantified and tested using techniques like regression and hypothesis testing.

# ANALYZE AND INNOVATE

This chapter provides a look at the analyze and innovate phase of the methodology. The analyze phase is the next obvious step after the measure step. The step includes analyzing available information for further actions, identifying failure modes, and identifying root causes leading to the best course of further actions.

The intangible world of services needs objectivity to measure and improve the performance of associated business processes. The innovate phase provides a process to make breakthrough improvements. The crucial phase of Innovate in Six Sigma methodology is relatively new. This chapter will discuss the role of creativity and mind mapping as one technique. An innovation technique that is relatively new to the scene called TRIZ (acronyms in Russian) will also be discussed. The technique was developed for products; however, the technique has applications in service world.

Firms often have an abundance of data. However, identifying actionable and usable information leading to breakthrough improvement is the key to success to any improvement methodology. Therefore, the role of information and knowledge management will be discussed at the end of this chapter.

In summary, this chapter will describe the tools that are useful for the analyze and innovate phases: cause and effect identification, failure mode and effects analysis, mind mapping, TRIZ, and information and knowledge management.

## ANALYZE

Objectively defining and establishing a baseline of perform-ance are the initial steps to understand what and how great a problem exists. Knowing the current state of a process is a pre-requisite to establishing the goal or the future state. Many problems are solved by the time a problem is clearly defined and measured. As someone has said, a well defined problem is half solved. For those problems that do not reveal themselves after the initial definition and quantification, the analyze phase is pursued to solve the problem. To solve the problem, one needs to perform root cause analysis to anticipate any problems that could prevent effective implementation, regression to establish the relationship between inputs and causes and effects, and hypothesis testing for evaluating the new process-es. Many tools can be used for root cause analysis and devel-oping a solution. The following set of tools represents the most used set of tools to solve the problems:

1. Fishing for the causes: Cause and effect analysis, also known as fishbone or Ishikawa diagram. This tool is an excellent tool to explore potential causes that could impact the problem. For further convergence or potential solution the team then prioritizes the causes.

2. Looking for trouble: Failure modes and effects analysis (FMEA) is a great tool to anticipate potential problems to prevent some from occurring.

3. Predicting performance: Regression analysis is used to build process models, or to quantify or prioritize a rela-tionship between various causes and effects.

4. Evaluating change through power of testing: Process improvement requires process changes. Hypothesis testing or comparative tests are used to evaluate the process changes.

Regression and hypothesis testing tools have already been discussed in Chapter six. In this chapter, we will only discuss cause and effect analysis and failure modes and effects analysis.

## FISHING FOR THE CAUSES:
## CAUSE AND EFFECT ANALYSIS

Those who succeed in problem solving in quality control usually perform a useful cause and effect analysis. Cause and effect analysis is a tool to identify potential causes of a given problem. It is an easy tool to use. If implemented correctly, cause and effect analysis can save significant resources by collectively identifying potential causes. A well done cause and effect analysis can solve the problem during the analyze phase, thus saving further effort in the improvement phase.

Dr. Kaoru Ishikawa, President of the Musashi Institute of Technology in Tokyo, first utilized a diagram similar to the cause and effect diagram. Therefore, the tool is sometimes called an Ishikawa diagram or a fishbone diagram because of its resemblance to the skeleton of a fish.

The benefits of conducting cause and effect analysis include identifying potential causes, increasing understanding of process issues, recognizing the importance of various process variables, and heightening the awareness to the process yields. The graphic nature of the diagram allows teams to organize large amounts of information about a problem and to pinpoint possible causes. Specifically, the cause and effect diagram has the following advantages:

1. Lists potential causes of a problem.
2. Shows a relationship between causes and the problem.
3. Encourages cross-functional teamwork, improving understanding of the process.
4. Sometimes identifies the root cause of a problem.
5. Simple to use and easy to understand.

Cause and effect analysis or the fishbone diagram segregates potential causes in 4 to 6 major categories. The four main categories are material, method, machine, and manpower (people power, or mental power). The other two categories some users like to add to a cause and effect diagram are measurements and environment. However, if the team feels a need for broader categorization, more categories are added. Employees representing various departments such as production, maintenance, quality, purchasing, and calibration identify the potential causes in a brainstorming session. The team members can identify the causes for each branch either separately or together.

Successful analysis requires a group leader facilitating the brainstorming session. The method essentially identifies all potential causes of variations or effects that may contribute to a problem or a cause. Taking a process view of work performed, we classify causes of variations into the following categories:

1. Input variations: Materials, tools, information, requirements, procedure, equipment condition, set-up conditions, and people skills.

2. Process variations: Machine maintenance, equipment mounted or stand-alone monitoring devices, compliance to procedures or methods, material consistency, operator discipline, consistency, and data collection, software version or errors, and environmental conditions.

3. Output variations: Verification method, inspector skills and consistency, calibration of measuring or verification devices, product design quality, product performance parameters, workmanship standards, and methods to handle nonconforming material.

A sample cause and effect diagram is shown in Figure 8.1. The effect of the problem statement is written in the head of fish, or the box. Major branches or bones are attached to the head. Twigs or subbranches are attached to the major branches as a potential cause is further explored for an actionable, verifiable, and measurable cause.

Once the potential causes are listed, the team prioritizes the causes, either by a branch or overall for the problem. When

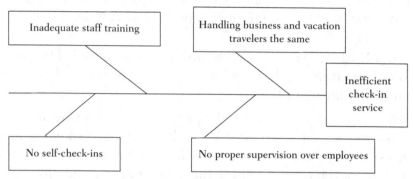

**FIGURE 8.1.** Check-in Services at a Hotel Chain: Cause and Effect Diagram

team members struggle to select important causes, each member must be required to select at least one cause on each branch that is considered the most critical. Order ranks the causes and identifies the root cause or the major causes. While listing causes at a lower level, each potential cause must be questioned with at least five "whys" to get to the actionable, quantifiable, and verifiable root cause. While searching for a root cause of a complex and chronic problem, one may need to develop a process flow chart and then conduct the cause and effect analysis at each process to identify various causes. In this case, there may be multiple fishbone diagrams. After identifying one or more potential root causes, an action plan is developed to alleviate the root cause of the problem. If there is a consensus about one root cause, then a corrective action is developed to remedy the root cause. When there are multiple potential causes, the next level of techniques is deployed to learn more about the process.

## CONSTRUCTING A CAUSE AND EFFECT DIAGRAM

To construct a cause and effect diagram, the problem must be clearly defined and the team members must be identified. The team leader is designated to facilitate the team exercise and members willing to participate. Then a team meeting is set up where basic rules of equal participation apply—each member gets a chance to identify cause(s) without being questioned or criticized. The following lists various steps in constructing the cause and effect diagram, or fishing for the causes:

- Step 1: Identify and state the problem to be solved. The problem is written in a box, and a long arrow, or the backbone, is drawn pointing to the box.
- Step 2: Ensure that the right people are present during the construction of the cause and effect chart. Typically, one must have representatives from the area management, material or purchasing, maintenance, engineering, quality, production, and other departments as necessary. Make the seating arrangement in a U-shape or oval shape to increase interaction among participants.
- Step 3: Construct main branches, and label them People (manpower or mental power), Material, Method, and Machine. Other branches (measurements and environment) can be added if the team deems necessary.
- Step 4: Conduct a brainstorming session to identify potential causes. During the brainstorming session, any potential cause is listed without questioning. To explore further, the team asks at least "five whys" in trying to get to the deeper understanding of the relationship between the cause and the effect. Each "why" peels away a layer of ambiguity and gets closer to the root cause of a problem. As causes are identified, subbranches are attached to the main branches. Each cause can be questioned for availability of data as well. The following is a set of guidelines that can be followed during the brainstorming session:
  - Team members do not evaluate or criticize others' ideas during the process. Criticism of others' ideas will only shut off the flow of ideas.
  - Everybody, irrespective of rank and position, should be given equal air time during the brainstorming session. Senior or dominating members of the team should not monopolize or dominate the discussion.

- – Get as many ideas as possible. At the beginning, we don't need to worry about the quality of ideas. Encourage quantity over quality at the early stage of the process. Each member will first be asked to contribute one idea before the second round of ideas, and so on.

- – Encourage piggybacking, in which one member of the teams get ideas from other team members and extends those ideas to new depths. Generally, one member's idea becomes a launching pad for the other ideas.

- – Record every idea.

- Step 5: Ask why this problem occurs and write the answer below the problem. Have team members write their ideas on a 3 × 5 cards or loose sheets of paper before starting the discussion. Once all the causes are listed, the team identifies the most likely causes. At this stage, the team discusses various causes and evaluates their significance based on their process of knowledge. Sometimes teams cannot make a decision soon enough; they end up looking at higher level causes, which dilutes the team activity. The best way to quickly decide which are most likely causes is to ask the team to start removing less likely causes. This narrows down the list of potential causes to more likely causes. The following are probing questions about key factors to gain insight into them:

  1. Is this cause a variable or an attribute?

  2. Has the cause been operationally defined?

  3. Is there a control chart or are other data available?

  4. How does this cause interact with other causes?

- Step 6: Prioritize the most probable cause for corrective action. Prioritization is done based on the importance of most likely causes, available resources, and ease of verification of causes. The objective is to identify at least one cause on each major branch with team consensus. By selecting one critical cause per branch, the team reduces the number of variables from approximately 25 to 4.

- Step 7: After the most probable causes are identified, the team must decide which one cause is most likely the root cause. If the team members can agree on the root cause(s), the exercise moves into the solution development phase.

After the root cause has been identified, an appropriate corrective action is identified to remedy the problem. The remedial actions must be validated to ensure the effectiveness of the corrective action.

During the brainstorming session, the team leader plays an important role in constructing the cause and effect diagram. The success of the cause and effect analysis depends on experience and participation of team members. A well constructed cause and effect diagram and well run brainstorming session makes team members feel happy and creates a sense of breakthrough or turning on their light bulbs.

Figure 8.2 shows a cause and effect diagram for employee turnover. The problem or the effect is "high turnover rates." We start the process by drawing the fishbone and drawing six major branches—recognition, career planning, selection, understanding the task, employee involement, and environment. Structured brainstorming sessions will bring out likely causes and help the team complete all four branches of the fishbone diagram. After identifying all possible causes, the team needs to rank the causes and identify the most likely cause.

An example of a generic cause and effect diagram is shown in Figure 8.3. The effect or the problem statement is written in the head of fish. We classify major causes of the problem and draw bones to the horizontal lines. The most common bones are materials, method, machines, and people. These bones will

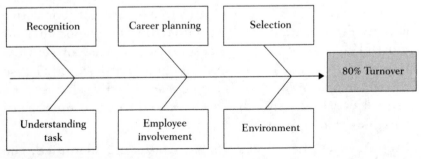

**FIGURE 8.2.** Employee Turnover: Cause and Effect Diagram

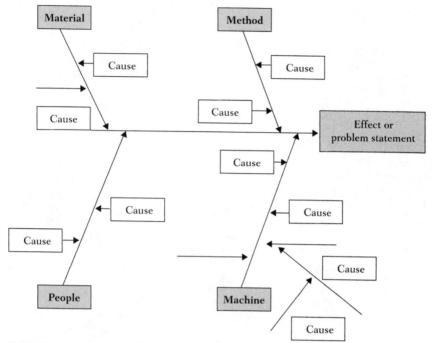

**FIGURE 8.3.** Generic Fishbone Diagram Template

be different and will change depending on the context of the problem. Further brainstorming sessions add to the causes on each bone. As stated in the next section, the brainstorming needs to follow certain guidelines to bring out all possible causes to the problem.

The example in Figure 8.4 identifies root causes of the "server is down" problem. The problem or the effect is "Network is down." We start the process by drawing the fishbone. At the end of the horizontal line or in the head of the fish, we write down the problem. Next, we draw four bones—hardware problems, people, software problems, and outside problems. Note, we don't always need to have the same four bones. Successful identification of bones is the first step in drawing the fishbone diagram. Structured brainstorming sessions will bring out likely causes and help the team complete all branches or bones of the fish. After identifying all possible causes, the team needs to order rank all likely causes and iden-

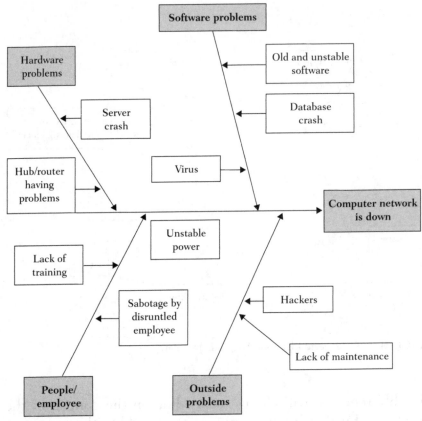

**FIGURE 8.4.** Example of a Fishbone Diagram

tify the most likely cause. Often, the mostly likely cause might not be the root cause and could be an effect. In this situation, the team will need to construct another fishbone diagram by putting the most likely cause as the "effect." As an example, sabotage by a disgruntled employee could be the likely cause of the problem. However, the team needs to further identify the root cause of this effect.

## EXAMPLE: HOTEL BUSINESS PROCESS

As part of a classroom project at DePaul University, students analyzed business processes at a hotel chain. At this hotel chain catering to business travelers and vacation travelers, the results of a recent survey suggested inefficiencies in the check-in time of customers, which causes delays in getting customers to their rooms. Researchers analyzed the following four areas of interest from a service quality perspective:

1. Check-in process
2. Restaurant services
3. Leisure activities
4. Employee attrition

Researchers felt that this inefficiency might be strongly linked to hotel staff not recognizing the different customers they are servicing or not segmenting the customer base properly. They tended to treat both segments in the same manner.

To collect our data, researchers visited the hotel and collected primary data. Check-in processing time was noted for each customer. The location on a typical day has three customer service representatives. Checkout process time was also observed and noted. Evidently, there are some obvious differences across service representatives for the same service. Service time also depended on whether customer service representatives were very busy.

The majority of customers are vacation travelers, often requiring special accommodations for their stay, which might cause delays in their check-in time, also delaying check-ins for other customers. However, much of the hotel's customer base is business travelers, whose requests are often very general and mostly stay the same across all customers. Recognizing and further researching this difference might help the hotel solve its check-in problems.

There are three types of options for service providers, to increase customer satisfaction:

1. Provide empowerment to customers in the form of self-check-in counters or provide web-based check-in.
2. Reduce variability in the process of check-in by making it a standard process with very well spelled-out steps.

3. Provide additional buffer capacity to deal with additional demand situations.

**Check-in services.**    Business travelers are used to using self-check-in machines. In this case, installing a self-check-in at some locations makes sense. Business travelers using the machine would decrease the traffic at the customer service desk. Service delivery channels are part of the whole service design. Often, wait time depends on the service capacity available and how the service capacity is distributed across different service delivery channels.

If the management would like to reduce the overall variability in the Customer Service Department, management should focus on motivating workers to work faster. In this case, the task of check-in has to be standardized to a greater degree. A standard task will reduce the variability of time employees spend with each customer. Making the service task standard could also involve an information support system.

Reduced variability will create more predictable service. More predictability will generate greater satisfaction for customers looking for faster service. Providing additional capacity has its own disadvantages. Buffer capacity is an expensive option and employee training and expertise could be compromised if a firm decides to rely too much on a temporary workforce.

**Restaurant services.**    The next service analyzed at the hotel chain was the restaurant. The customer survey and house of quality analysis revealed that hotel customers are not satisfied with the restaurant service they are offered. One hotel location has three restaurants. Restaurant food quality is average and customers often are drawn outside the hotel area to dine.

In contrast, competing high-end hotels offer an upscale restaurant, providing a unique dining experience. By opening an upscale restaurant, this particular hotel chain would be reaching out to the wants of both business and vacation travelers' needs. This would also increase the hotel's profits from the restaurant, as more customers would be willing to stay in the hotel to dine instead of searching for a restaurant in the city, or worst yet, picking another hotel as their place to stay on the basis of a nicer restaurant.

**Leisure activities.**    The third item analyzed at the hotel chain was leisure activities. Our findings show that the hotel provides its customers with an insufficient leisure experience. While spa packages and a gym are available on site, these are low in quality. Options are also limited. We could argue that this hotel chain has made a strategic decision not to provide these choices. However, the hotel could consider partnering with an upscale spa service provider. In other words, they could outsource some services that are noncore, according to the vision of the hotel chain. Similar services could be provided for a gymnasium and other related leisure activities. The customers staying in the hotel would be able to take advantage of the club's facility and equipment at no extra charge.

**Employee turnover.**    Using data from J.D. Power's past surveys of turnover ratios, we see that over the past 10 years, turnover has decreased drastically. According to the general manager's qualitative feedback, these decreasing turnover rates are due to the amount of quality training and employee empowerment. A fishbone or a cause-and-effect diagram was used to identify the problem by the company to target and solve a $4 million yearly expense!

## LOOKING FOR TROUBLES—FMEA

Failure modes and effects analysis (FMEA) is a systematic analysis for identifying potential failures with the objective of preventing their occurrence so that the probability of the customer seeing a failure is minimized. The potential failures are those that customers perceive as a failure. Here the customers are both internal as well as external. In performing FMEA, one attempts to identify each potential mode of failure, its effect, and its severity, and address its causes. After the causes have been identified, the failure mode and its impact are mitigated through corrective actions. FMEA is normally conducted while developing a solution for the problem, thus minimizing the surprises during the implementation. Normally, FMEA is deployed in the product or process design phase.

In essence, FMEA is a risk minimization technique that helps to identify a risk with a proposed solution in a Six Sigma project, investigate its root cause, and initiate efforts to reduce the risk. The technique was first used in the aerospace industry in the mid-1960s to find problems with an aircraft before it ever left the ground. Typically, the technique has been used in many industries such as aerospace, automotive, and pharmaceutical. Applying FMEA during the application of DMAIC to improve the profitability of a company is as critical an application as it is in the human-safety-sensitive industries. FMEA offers the following advantages:

1. Identifies potential failure modes that could have been overlooked and makes the solution more effective.

2. Ranks potential failure modes based on severity, occurrence, and detection. The ranking allows the prioritization of failure modes and monitors the effectiveness of the solution while recalculating the ranking.

3. Continual application of the FMEA process during the product or process life cycle leads to better quality and reliability of products and higher customer satisfaction.

4. FMEA creates a document that contains a significant amount of knowledge about a product and process. FMEAs can contribute to the knowledge management effort for developing innovative solutions.

An important factor for the successful implementation of a FMEA program is timeliness. FMEA is more effective when it is proactively applied in the design phase of a process, product, or service. Services involve intangibles and perceptions more than products. Customer participation in the service process makes it even more challenging, considering customers also include internal functions such as design, engineering, marketing, finance, and sales. The FMEA technique can help in anticipating and reducing potential errors passed on to the internal customers.

FMEA can be used in evaluating of design concepts, process selection and improvement, design process analysis, software development, and services. FMEA methodology consists of grading failure modes for severity, potential causes for occurrence, and controls for detection. The definition of these terms may vary across companies and products. Therefore, one must standardize the definitions before applying the FMEA for consistency and accuracy.

- Severity: The seriousness of the impact of the potential failure mode on the functionality of the product or on customer applications. It gives an idea of the resulting loss generated directly or by adverse impact on the subsequent operational steps. The severity is graded on a scale of 1 to 10, where 10 relates to life-threatening situations and 1 implies minimal impact.

- Occurrence: The frequency with which a cause of potential failure may occur. The frequency of occurrence is estimated based on process knowledge and historical performance. In the absence of historical data or past knowledge, the frequency of occurrence may be estimated based on similar processes, or as determined by the cross-functional team. The occurrence is graded on a scale of 1 to 10, where the grade of 10 implies certainty of an event and the grade of 1 implies a hypothetical situation.

- Detection: The relative probability with which the impact of a cause can be detected through appropriate controls (i.e., inspection, test, or process control). Detection is also graded on a scale of 1 to 10, where the 10 implies difficult to detect, and a grade of 1 implies a certain containment of adversely affected material.

## PERFORMING FMEA

To perform an FMEA while working on a Six Sigma project, the proposed solution or the process change is evaluated for potential problems or challenges that could prevent the desired results from occurrence. The following are steps taken to perform an FMEA:

1. Identify the project and understand all components and their functions.

2. Select a cross-functional team from all affected work groups who could contribute toward completing the FMEA analysis.

3. Gain consensus on the ranking criteria for severity, occurrence, and detection. Tables 8.1 and 8.2 show the typically used guidelines.

4. Draw the process flow charts and describe the functions of each component.

5. Identify necessary inputs to the process such as material, method, machine, and people actions. Anticipate and list all potential failure modes for each input at various steps in the process. Team members can brainstorm for ideas and potential failure modes based on their experience and process knowledge.

**TABLE 8.1.** Generic Procedures for Performing FMEA Analysis

| ITEMS AND FUNCTION | POTENTIAL FAILURE MODE | POTENTIAL EFFECTS OF FAILURE | POTENTIAL CAUSES OF FAILURE | DETECTION METHODS AND QUALITY CONTROL | DET | RPN= SEV* PROB* DET | RECOM- MENDATION ACTION |
|---|---|---|---|---|---|---|---|
| Part Name and function | Possible modes of failure | Consequences of failure on part function and on the next assembly | Such as inadequate design and improper materials | Measures available to detect failure before they reach the customers | | Total RPN | List actions for each failure mode identified as significant by the RPN rating |

TABLE 8.2. Generic Scoring Guideline for Assigning Different Values to Severity, Probability and Detection

| SEVERITY | SEVERITY RANKING | PROBABILITY OF FAILURE | RANKING | DETECTION DETECTION | RANKING |
|---|---|---|---|---|---|
| Hazardous without warning | 10 | Very High: Failure is almost inevitable ( >1 in 2) | 10 | Absolute Uncertainty | 10 |
| Hazardous with warning | 9 | 1 in 3 | 9 | Very Remote | 9 |
| Very High | 8 | High: Repeated failures (1 in 8) | 8 | Remote | 8 |
| High | 7 | 1 in 20 | 7 | Very Low | 7 |
| Moderate | 6 | Moderate: Occasional failures (1 in 80) | 6 | Low | 6 |
| Low | 5 | 1 in 400 | 5 | Moderate | 5 |
| Very Low | 4 | 1 in 2,000 | 4 | Moderately High | 4 |
| Minor | 3 | Low: Relatively few failures (1 in 15,000) | 3 | High | 3 |
| Very Minor | 2 | 1 in 150,000 | 2 | Very High | 2 |
| None | 1 | Remote: Failure is unlikely (< 1 in 1,500,000) | 1 | Almost Certain | 1 |

6. Evaluate the severity of the impact of potential failure mode. Consider the degree of cosmetic, function, and safety aspects in estimating the severity of the impact. The impact can be bucketed into catastrophic, critical, major, minor, and negligible categories. Accordingly, assign the severity ranking on a 1 to 10 scale.

7. Investigate potential causes of the failure mode. It is difficult to identify causes, construct a cause and effect diagram. Evaluate the frequency of occurrence and assign the ranking on a scale of 1 to 10.

8. Calculate risk priority number (RPN) by multiplying severity, occurrence, and detection as follows:

$$RPN = Severity \times Occurrence \times Detection$$

9. The maximum possible number for RPN is 1,000. RPN provides a relative ranking of causes of the failure modes. One can sort them by RPN in descending order. Based on the severity, RPN, and available resources, one can establish an action plan to reduce RPN. An action can be triggered for an RPN as long as it makes economic sense.

10. The team performing FMEA can establish a threshold value of RPN for taking action. If severity is greater than 7, some action must be initiated to address the impact.

11. Identify ways to reduce the risk of the highest priority failure mode by using poke-yoke and error-proofing methods.

12. A plan to reduce RPN is developed to address necessary failure modes by addressing the following actions:

    a.  Assign responsibilities for further action

    b.  Outline an action plan to reduce or eliminate failure mode

    c.  Implement corrective action and reassess RPN

    d.  Continue until all issues are addressed

FMEA is increasingly being used in the service sector. For example, a physical therapy clinic is interested in performing

FMEA. A patient makes an appointment to meet a physical therapist. In Table 8.3 shows a list of various failure modes or errors that could occur during the therapy process. Once failure modes are identified, severity and detection grading are assessed on a scale of 1 to 10.

It appears that improper treatment given to a patient has the highest RPN and is therefore a likely candidate for process improvement.

TABLE **8.3.** An Example of Physical Therapy Treatment at a Clinic

FAILURE MODE AND EFFECTS ANALYSIS

Date: _____

Process Name: Patient receiving physical therapy     Process Number: _____

| FAILURE MODE | A) SEVERITY | B) PROBABILITY OF OCCURRENCE | C) PROBABILITY OF DETECTION | RISK PREFERENCE NUMBER, RPN = AXBX C |
|---|---|---|---|---|
| Patient missing appointment | 2 | 3 | 7 | 42 |
| Patient providing confusing information about the symptoms | 5 | 2 | 6 | 60 |
| Improper treatment given to a patient | 9 | 2 | 5 | 90 |
| Patient fails to understand physical therapy benefits | 4 | 2 | 4 | 32 |

# INNOVATE

## Nonlinear World of Creativity: Mind Mapping

Often, we think in linear terms and try to project trends in linear forms. For example, we might want to project a certain trend. Business situations involve nonlinear thinking, and we need to become familiar with this process. For example, if we wish to project business performance in the year 2010, we could project the current trend into the future in a linear fashion or we could prepare a multiscenario road map.

Peter Russell joined with Tony Buzan in the mid 1970s, and together they taught mind-mapping skills to a variety of international corporations and educational institutions (http://www.peterussell.com/mindmaps/mindmap.html). We could also prepare a mind map to find the factors affecting business performance. In this section, we will describe the nature of mind mapping and its use for services.

The process to draw a mind map includes:

- Step 1: Focus on the center and draw the center. Depending on our language ability and cultural background, as we read from left to right and top to bottom, our left-brain education system has taught us to start in the upper left-hand corner of a page. However, our mind focuses on the center. Therefore, mind mapping begins in the center of a page with a word or image. These words or images symbolize what you want to think about, placed in the middle of the page.
- Step 2: Brainstorming and creativity part. Let your brain go wild! This step involves brainstorming and registering all "brain dump" on the paper. Brainstorming provides an outlet to the creative mind; therefore, we cannot be judgmental in this part of the process. The important part is also to have fun during this step while jotting down new and creative thoughts. We could remove some material if we think it is not relevant for our purpose.
- Step 3: Free associate. As ideas emerge, print one- or two-word descriptions of the ideas on lines branching from the central

focus. Allow the ideas to expand outward into branches and subbranches. List all ideas without judgment or evaluation.

- Step 4: Think fast. Your brain works best in 5 to 7 minute bursts, so capture that explosion of ideas as rapidly as possible. Key words, symbols, and images provide mental shorthand to help you record ideas as quickly as possible.

- Step 5: Break boundaries. Break through the "8-1/2 × 11 mentality" that says you have to write on white, letter-size paper with black ink or pencil. Use ledger paper or easel paper or cover an entire wall with butcher paper.... the bigger the paper, the more ideas you'll have. Use wild colors, fat colored markers, crayons, or skinny felt-tipped pens. You haven't lived until you've mind mapped a business report with hot pink and day-glow orange crayons.

- Step 6: Judge not. Put everything down that comes to mind, even if it is completely unrelated. If you're brainstorming ideas for a report on the status of carrots in Texas and you suddenly remember you need to pick-up your cleaning, put down "cleaning." Otherwise your mind will get stuck like a record in that "cleaning" groove and you'll never generate those great ideas.

- Step 7: Keep moving. Keep your hand moving. If ideas slow down, draw empty lines, and watch your brain automatically find ideas to put on them. Or change colors to re-energize your mind. Stand up and mindmap on an easel pad to generate even more energy.

- Step 8: Allow organization. Sometimes you see relationships and connections immediately, and you can add subbranches to a main idea. Sometimes you don't, so you just connect the ideas to the central focus. Organization can always come later; the first requirement is to get the ideas out of your head and onto the paper.

Some tips while making a mind map include:

- Use only key words or images, wherever possible.

- Make the center a clear and strong visual image that depicts the general theme of the map.
- Create subcenters for subthemes.
- Put key words *on* lines. This reinforces the structure of notes.
- Print, rather than write in script. It makes them more readable and memorable. Lowercase is more visually distinctive (and better remembered) than uppercase.
- Use color to depict themes and associations and to make things stand out.
- Anything that stands out on the page will stand out in your mind.
- Think three-dimensionally.
- Use arrows, icons, or other visual aids to show links between different elements.
- Don't get stuck in one area. If you dry up in one area, go to another branch.
- Put ideas down as they occur, wherever they fit. Don't judge or hold back.
- Break boundaries. If you run out of space, don't start a new sheet; paste more paper onto the map. (Break the 8-1/2 × 11 mentality.)
- Be creative. Creativity aids memory.
- Get involved. Have fun.

**Advantages of a mind map.**    Mind maps follow the brain's activities. The brain doesn't work in nice neat lines. Ideas are natural to the brain. The organization of a mind map reflects the way our own brain organizes ideas.

Mind maps allow the brain to see through the links and associations. In our brain, one activity might have multiple associations and links. Again, these associations are not linear, but do have some pattern. Mind maps allow the brain to see this pattern in association and links.

Mind maps use key words and images, and we don't have space for including full phrases. Because we are using only key words, we could likely fit a lot more information on one page. We could call this the big picture advantage.

Mind maps are easier to recall. Words and images have different retention impact on individuals. Visual things are easier to recall in the future, as compared with recalling linear notes or recalling a text.

As stated in step 1, we start from the center. Starting from the center has the advantage of working out and filling out the page in all possible directions.

Mind maps are easy to review, and a group could have a starting discussion point regarding an issue using a mind map. We could also remember key points from the mind map, as highlighted in the mind map with color or in bold.

Little software is available to draw a mind map. However, the important part is the process of mind mapping, as compared to the mind map itself. We want to emphasize this point regarding the process because many users get lost and forget the value of the process. We review two software packages here for the reader's benefit. Some of the language is taken from the vendor's material.

MindManager is easy to use and can be formatted to produce clear, easy-to-read Mind Maps, making it a very powerful communication tool. MindManager caters to everyone, from individuals to large organizations.

ConceptDraw MINDMAP is recommended for Mac users. This cross-platform software helps you generate, organize, discuss, and present ideas in a clear, visual form.

MindGenius introduces an alternative way of working that challenges existing techniques and is guaranteed to develop the productivity, agility, and self-belief of individuals, teams, and organizations.

**Uses of mind mapping.**    During the innovate phase, and to some extent in the analyze phase, mind mapping could be used for following purposes:

*Notes.*  Whenever information is being taken in, mind maps help organize it into a form that is easily assimilated by the brain and easily remembered. Mind maps can be used for noting anything—books, lectures, meetings, interviews, phone conversations, etc. In service settings, mind maps could be used in creativity exercises.

*Recall.* Whenever information is being retrieved from memory, mind maps allow ideas to be quickly noted as they occur, in an organized manner. There's no need to form sentences and write them out in full. They serve as a quick and efficient means of review and therefore keep recall at a high level.

*Creativity.* Whenever you want to encourage creativity, mind maps liberate the mind from linear thinking, allowing new ideas to flow more rapidly. Think of every item in a mind map as the center of another mind map.

*Problem solving.* Whenever you are confronted by a problem—professional or personal—mind maps help you see all the issues and how they relate to each other. They also help others quickly get an overview of how you see different aspects of the situation and their relative importance.

*Planning.* Whenever you are planning something, mind maps help you get all the relevant information down in one place and organize it easily. They can be used for planning any piece of writing—from a letter to a screenplay to a book (I use a master map for the whole book, and a detailed submap for each chapter)—or for planning a meeting, a day, or a vacation.

*Presentations.* Whenever I speak, I prepare for myself a mind map of the topic and its flow. This helps me organize the ideas coherently, and the visual nature of the map means that I can read the whole thing in my head as I talk, without ever having to look at a sheet of paper.

## TRIZ

The systematic methodology of inventive problem solving known as TRIZ (from the Russian acronym for theory of inventive problem solving, or TIPS) was developed in the former

Soviet Union during the 1950s by Genrich Saulovich Altshuller and his colleagues. Today, TRIZ is in use throughout the world, and there are many variants of this method (e.g., SIT [structured inventive thinking]).

One of the core ideas of TRIZ is that the relevant inventions in one field may not be known to workers in another, with the result that much effort may be spent in reinventing an existing solution. Fifty years ago, Altshuller began a quest to overcome this impediment both by formulating a methodical approach to creating an invention, and by providing a list of known methods for solving problems that would have relevance in many different fields (Smith and Sudjianto, 1997).

Altshuller examined a large number of patents, looking for the hallmarks of truly creative inventions. He found that often the same problems had been solved in various technical fields using one of only about 40 fundamental inventive principles. As a result, he and his colleagues made a new classification of these patents without regard to their industry basis. By removing the subject matter, Altshuller was able to elucidate the problem solving process. He categorized the patent solutions into the following five levels (Savransky and Stephan, 1993):

- Level 1: Routine design problems solved by methods well known within the specialty or the company. (About 32% of solutions were at this level.)
- Level 2: Minor corrections to an existing system, by methods known within the industry (about 45%).
- Level 3: Fundamental improvements to an existing system, by methods known outside the industry (about 18 %).
- Level 4: New generations using a new scientific (rather than technological) principle to perform the primary functions of the system (4%).
- Level 5: Rare scientific discoveries or pioneering inventions of essentially a new system (less than 1%).

TRIZ is a combination of algorithms and principles. Its key concepts are:

- Ideality: All systems evolve toward increasing system ideality. Ideality is defined as the ratio of the sum of a system's useful functions over the sum of the system's undesired effects (Savransky and Stephan, 1993).

- ARIZ: An acronym for algorithm for inventive problem solving, ARIZ is a structured method of thinking so that a higher level problem is transformed into a lower level of difficulty, making the solution more apparent.

- Contradictions: Contradictions are both technical and physical. Technical contradictions are situations where the improvement of one characteristic of the system results in the deterioration of another. Physical contradictions result when a physical attribute should be increased to improve one function of the system, and decreased to improve another. Traditionally, trade-off or compromise is used to handle contradictions. TRIZ always seeks a solution without compromise.

- Substance-field analysis: Sufield or substance-field analysis is a systems modeling language developed by Altshuller. It consists of combinations of three standard elements (field, substance object, and substance tool) and is used to describe the simplest workable engineering system.

- Laws of system evolution: Based on a study of the development of thousands of products and technologies, Altshuller discovered a number of patterns or trends associated with system evolution. S-curve analysis is used to reveal the current level of maturity for the technology of interest.

- Knowledge base of inventive principles and effects: Derived from a study of 40,000 patents, TRIZ contains a data-base of inventive principles that can be used to resolve contradictions.

Algorithm of solving inventive problems (ARIZ, the Russian acronym) is a series of actions for the innovation process. These steps are like a bridge between a problem and the solution. Steps are:

1. Formulation of the task: We use standards or universal steps. An example could be given for a professional service

firm, as provided in the book, in a different context. All projects taken up by a professional service firm can be described as brain, gray hair, and procedure projects (Meister, 1993). *Brain projects* are highly customized client projects and are at the cutting edge of knowledge and innovation. Each brain project is unique in nature, and activities necessary to complete this type of project cannot be completed in advance. *Gray hair projects* are somewhat more familiar and some tasks are known in advance. Gray hair projects have a medium degree of customization. *Procedure projects* usually have a well-defined problem solving exercise. The client might want a second opinion or just may not have the personnel to perform procedure projects. The first step therefore is to know whether the project is of procedure, gray hair, or brain nature. The second step is to identify whether the problem can be solved using one of the standards. In other words, is this problem a typical problem, a new application to a new business context, or a completely new problem in a new business context?

2. Model of the problem: A model will usually have an offering (or a product) and a tool. Other extraneous factors that could affect the outcome are either controlled or ignored. We are using a system approach here to solve the problem and to build our model (Sterman, 2000). System thinking helps us understand complex systems and helps us formulate policies and guidelines. The long-term side effects of decisions are taken into account.

3. Analyze the model: Determine what elements of the model should be changed. We should focus our attention on changing tools and, if and only if it's impossible to change the tools, we should focus on the external environment.

4. Ideal final result formulation: We need to make a specification list abiding by ideal final result formulation. A top-down approach with each step makes the area of search narrower and narrower. Diagnosis will determine where we need to focus most. We should end up with a list of diag-

noses at this step. Contradictions are analyzed to find the ideal solution here.

5. Develop a list of found solutions. This is a reserve of typical methods and principles or standards. As we find more and more solutions, we need to keep adding those solutions to the list. Authors of TRIZ analyzed over 40,000 patents to develop such a table.

6. Use existing methods and solve problems in other areas.

7. Self-check: We could find flaws in ARIZ methodology and make corrections accordingly. An example could be a pattern of mistakes made by the customers and keeping a table on such a pattern. Such a table will directly help in service recovery efforts.

All of the preceding material is related to creativity and innovation and can be very helpful for idea generation. The technique has a heavy slant towards manufacturing. However, the technique can be equally applied to service sector.

## INFORMATION AND KNOWLEDGE MANAGEMENT

Information and knowledge management has increasingly become a strategic competitive advantage for many companies. Information is used in the form of prepackaged, unfinished service acts that will be acted upon later. These unfinished information actions are waiting for the customer to arrive into the system. A service firm has to grapple with the following issues: What type of information should be collected from customers when they arrive? Where should this information reside in the system? How should this information be disseminated and used within the firm? Importance of information as a service delivery system design process is discussed in this section.

Service companies are working harder to provide knowledge and information at their finger tips. A recent article in *Harvard Business Review* (Davenport and Glaser, 2002) describes such examples. According to the authors, "The key to success to take specialized knowledge into the jobs of highly skilled workers is to make the knowledge so readily accessible that it can't be avoided." A health care example is discussed in the article to illustrate the usefulness of such an approach. Professionals have instant access to online journals and databases, care protocols or guidelines and particular diseases, and interpretive digests prepared by the company. The partner handbook is accessed about 3,000 times a day. Professionals could make more informed decisions faster. Professional services will greatly benefit from such an approach.

The key to success is to develop such a customized system. Off-the-shelf software packages will not be useful for such applications. The company therefore must develop the majority of these systems from scratch. An organization also needs to have a "measurement-based" culture.

There could be misuse of technology as well, and customers feel frustrated. An article in *The Wall Street Journal* by Jane Spencer (May 8, 2002) describes such a problem. The problem posed is about the need for call centers to deal with customer complaints in spite of spending billions on automation. We view it as a strategic decision. The firm needed to find out what type of activities should be automated and what type of processes should be dealt with interactively.

This section concludes our discussion of this chapter. The next chapter will describe the embed phase of the Six Sigma methodology.

## KEY TAKEAWAYS

- The analyze and innovate phases of Six Sigma methodology helps strengthen and sustain long term competitive advantage of a service firm.
- Failure Mode Effects Analysis (FMEA) could be applied to transactions and services in an effective manner.
- The nonlinear world of brainstorming and problem solving requires techniques like Mind Maps to capture and communicate information and ideas.

- TRIZ, a technique originally developed for innovation in the manufacturing world, has a direct application in transaction and services.

# EMBED

How is Six Sigma different from a total quality management toolset? Six Sigma efforts are aligned with the strategic objectives of the firm. These objectives include cost reduction, quality improvement, and enhanced flexibility and speed. In this final chapter of this section, we will describe the implementation aspects of this methodology. The last step, called "embed," is divided into three steps. The first step is training employees on the methodology and its alignment with corporate goals. Training will take multiple forms and will be context dependent. We will describe project management aspects of this methodology in the second section of this chapter. The third section includes case studies and a template using current Six Sigma methodology.

## TRAINING

As noted earlier, training is context dependent, and within one company, the training will depend on the level of engagement with the methodology. The bottom line is that everybody understands the two key premises of this methodology. The method is heavily data driven, and our objective is to reduce variation within the bounds of corporate objectives. The data-driven approach forces one to look into measurement approaches and analytical tools.

## DIFFERENT PERSONALITY FOR DIFFERENT ROLES

The basic idea in a service is a decoupling of high contact and low contact service activities. High contact and low contact activities should be separated. Employees will require a different skill set for these two sets of activities, and consequently, training will be different. At a hotel, an employee working at the front desk will have different training and skills as compared with an employee in the order processing room. Broadly speaking, public relations and interpersonal skills are appropriate for high-contact activities, and technical and analytical skills are more appropriate for low-contact activities (Metters et al., 2003). Companies generally will buffer the low-contact activities from customers. The strategy prevents these employees from random customer arrivals and nonstandard customer demands. Efficiency improvement is critical for these services.

Job functions need to be broken down into smaller tasks, and tasks are standardized for low-contact activities. Employees will perform a few select tasks, reducing the variance. Focused factory strategy is a corporate strategy in this direction. Overall impact is reduced inherent variance.

Low- and high-contact split have various marketing and strategic implications. Managers need to balance operations cost saving against other losses from such a change.

Service organizations are complex systems, and diagnosis will not provide one single answer to the issue. Employee empowerment directly affects customer satisfaction. However, firms need to keep a system perspective and should diagnose problems in terms of vicious cycles (Norman, 2000). The following elements of a system should be analyzed together:

- Degree of empowerment.
- Measurement system and relevance of measurement system.
- Alignment of compensation and reward system with corporate strategy.
- Availability of information and professional development (including training) to employees.

## WHY, WHEN, AND HOW TO EMPOWER EMPLOYEES?

Employee empowerment will depend on the business context, cost, quality, and flexibility. Broadly speaking, firms obtain the following benefits from empowerment:

· Quicker response to customer needs, leading to higher customer satisfaction and loyalty.
· Quicker resolution to service recovery, hence leading to higher customer satisfaction.
· Employees feel happier. Employee satisfaction has been related to customer satisfaction in service firms (Heskett et al., 1997).
· Innovation and new idea generation. Customers and employees are a great source of new service ideas.
· Higher employee retention and hiring costs.

However, on the cost side, firms need to watch out for the following costs of empowerment:

· Higher investment in training and selection of employees.
· High labor cost likely due to higher proportion of permanent staff, as compared to temporary staff.
· Inconsistent service delivery likely due to customized service delivery and customized service recovery, leading to possible violation of fair play.

Organizations choosing a higher degree of empowerment have the majority of knowledge and training at the employee level. If these employees leave the firm, the firm loses corporate knowledge. These firms must learn to convert gains in individual learning into enhanced organizational learning. Professional service firms are a prime example of these type of firms with a higher degree of empowerment. How does one convert individual learning into corporate learning? A few tools are rotating teams, idea exchanges, and having a good knowledge management system in place.

Similarly, firms with a production-line approach will have a majority of learning at a higher management level. Employees have very little opportunities for growth, education, and professional development. In these situations, firms need to provide opportunities for growth and professional education to employees. Again, why should a firm worry about these issues? The answer lies in employee satisfaction and the firm's knowledge management system.

## MEASUREMENT SYSTEM

The previous chapter discussed the need for an effective measurement system. We wish to add to that discussion and describe an effective measurement at the macro level and its relationship with compensation and rewards system.

A manager at a service firm has to deal with the following questions while designing an appropriate measurement system for quality management:

- How many total measurements are appropriate?
- How often should these measurements be taken?
- What is the right balance between the objective or quantifiable measurements and subjective (nonquantifiable) measurements?
- What is the right balance between internal and external measurements?
- How often should the measurement system be changed or updated?

There is no straight answer to the preceding questions. Before deciding on any measurement, we should answer the questions, Why do we need this measurement? and How are we going to use these data? As a first step, managers need to find customers' need attributes. The system should be outward looking, and internal measures should have some relationship with external measures. Measurement systems need periodic revisiting, especially when firms enter from one phase to another (for example, from being a start-up company to a mature company).

**Hard versus soft measurements.**  Service firms can easily measure hard data. For example, an airline can measure number of late flights, and a package carrier can measure number of missed/lost packages. How should one measure friendliness, professionalism, courtesy, and helpfulness? Companies do have some discrete idea about these so-called soft variables. Some companies have attempted to measure friendliness by using minimum number of eye contacts and smiles. An airline has been known to specify the minimum number of passengers a cabin attendant should talk to before and after the take off of a flight. Companies have attempted to put boundaries on professionalism and courtesy by having experienced managers specify the parameters during a phone conversation at a call center.

**Measurement should be based on customer needs.**  Airlines only tell us the departure and arrival time of the flight. Why not have an airline provide an estimate of arrival time at the airport (or estimated departure time from the office) and estimated departure time from the airport after arrival? Companies need to keep customer needs in mind while establishing a measurement system. For a business-class traveler, the need is to attend the meeting and not to sit in a seat in a plane. Some airlines have even started to provide check-in at a downtown hotel in a major metropolitan city to take care of business-class travelers.

## COMPENSATION AND REWARD SYSTEM

The compensation and reward system should be in alignment with a firm's empowerment approach. Employees should be rewarded, and the reward system should be consistent with the overall firm strategy. If employees are doing a very good job at keeping customer complaints low, these employees should be rewarded.

## INFORMATION AND PROFESSIONAL DEVELOPMENT

Employees need necessary, actionable information handy and appropriate training for using these tools. If a firm wishes for its call center employees to solve customer complaints, the firm needs to provide access to appropriate information to these employees. These employees should also be trained in dealing with unhappy customers. The bottom line is that employees like to see the economic benefit of their job, and they will be able to contribute only if appropriate information and tools are provided. Cross-training is required if employees are responsible for multiple tasks.

Neglecting these elements, together with organization structure, leads to a vicious cycle or a cycle of failure. How should a firm turn a cycle of failure into a cycle of success? A firm should follow these strategies to break the cycle of failure (Schlesinger and Heskett, 1991):

- Careful selection of employees: Employees should be carefully screened. Many firms require multiple interviews before hiring an employee. Employee expectations should be aligned with the reward and compensation system at the company. These firms take extra care in selecting and developing its employees.

- Provide accurate information about the job and organization: Such information helps set the expectation of new employees. For example, a company could provide a paid two- to three-day work experience at the work site before final selection.

- Focus on quality at the core: A firm should set a minimum standard. Offerings falling below this quality standard will have consequences.

- Focus on job ladder and professional development: Employees should have a very clear idea about their career path and job ladders leading to senior positions. Very often, companies fail to provide even basic career counseling and advice.

- Scorekeeping and continuous feedback: Employees would like to see the progress of various initiatives. The company, even at a business unit level should keep track of its performance and, the performance score should be easily available to all employees. A call center could have a score posted

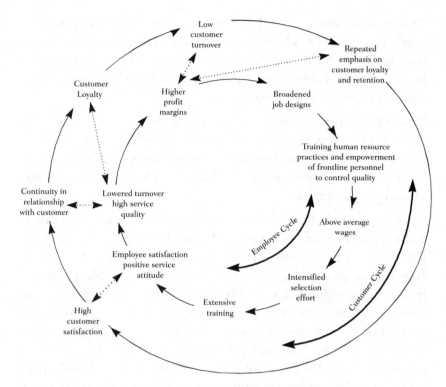

**FIGURE 9.1.** Cycle of Success (from Schlesing and Heskett, 1991)

on a whiteboard for easy view. The score could be any measure that a firm has decided to adopt.

## PROJECTS

Six Sigma projects should be managed just like any other projects. Project selection should take place on a portfolio basis to maximize the impact. The portfolio approach maximizes the return on investment. Two overlooked areas in the life cycle of a project are project definition and project termination. The project definition has long-term strategic impact on the outcome of the project. Project termination similarly should include a debriefing about learning from the project, stakeholder's feedback, and participant's feedback. Phases of Six Sigma methodology in a service firm will include the following steps:

- Project selection and definition: The organization should be ready to take on such projects. Readiness will include establishing the right infrastructure and choosing the right project leaders. Project selection will be based on setting the baseline and appropriate long-term goals.

- Demonstration of early wins: Managers need to show the impact of projects. Impacts likely should be measured in terms of corporate performance: for example, customer satisfaction, cost, response time, added flexibility, market share, or increased revenue.

- Manage knowledge/information: Knowledge management is a strategic weapon for a service firm. The weapon has a direct influence on sharing best practices within the firm. Knowledge should be converted into an actionable form and hence be readily usable by the firm's employees. The just-in-time knowledge concept is becoming increasingly common in many service business contexts. A doctor would like to have ready access to customized research while diagnosing a patient at her clinic. A call center employee at a hotel chain taking a call from an existing customer to make a reservation would like to have access to customer preferences. Very often, sharing of best practices is a function of organization structure and reporting structures within an organization. In general, if so-called "silos" are deep, or in other words reporting only occurs at a very high level, the more difficult it will be to share best practices. The organization needs to bring down common reporting to a lower level for easy communication and better information flow.

- Innovation and continuous improvement: So far, the emphasis of Six Sigma methodologies has been on variability reduction and continuous improvement. We feel the need to include innovation as a separate step in Six Sigma methodology. Innovation could be in different areas: process innovation, offering innovation, business model innovation, and others. The innovation step does not preclude statistical process control, or SPC. SPC is a powerful tool to measure and control variability in any process. SPC has been used as early as 1986 for services (Mundy et al., 1986). Limits of the control chart in a service environment are dynamic, as noted earlier. The dynamic nature becomes much more visible in a multistep service offering. Control limits for later stages will depend heavily on customer experience in previous stages.

· Identify service operation drivers: Managers need to understand the critical driver(s) of operations. Drivers are service capacity, service delivery channels, information, and service inventory, as defined in a previous chapter. Understanding these drivers will lead to an answer to the following questions—What is an appropriate degree of pooling for this system? Should employees be cross-trained? What should be buffer service capacity? What channels are appropriate for this design?

## ILLUSTRATIVE CASE

The following case is used for illustrative purposes only. The case is based on a real business situation, but the company identities have been hidden due to competitive reasons.

### CALL CENTER OPERATIONS

The call center is part of the Department of Internal Medicine. Internal medicine is one business unit of this health care facility. The call center gets calls for appointment scheduling, emergency medical advice, and various other queries. The introduction of the call center operations at the Department of Internal Medicine helped the department in several ways. First, it streamlined the process of appointment scheduling and easier tracking. Second, it took scheduling responsibility away from the ward assistants and secretaries, thereby enabling them to be available for other work. Finally, the call center employees were able to take on certain support functions including translation, administrative duties, query handling, emergency medical advice for first aid, and the like.

The call center unit functions five days a week from Monday to Friday between 8 am and 5 pm. (Refer Table 9.1 for daily call volume.) When a caller dials the unit's number, the customer care representative (CCR or "agent") attends the call. Nine customer care representatives process the calls. The call goes to the agent waiting the longest. Occasionally, calls are missed if the CCR does not attend a call forwarded to him.

TABLE 9.1    Average Call Volume at the Call Center

| DAY | TOTAL NUMBER OF CALLS |
| --- | --- |
| Monday | 660 |
| Tuesday | 663 |
| Wednesday | 625 |
| Thursday | 514 |
| Friday | 478 |
| Average | 588 |

If all the CCRs are busy, then the caller is kept on hold and hears recorded music interspersed periodically with the message, "You are in a queue. Please bear with us and our staff will attend to you shortly."

If after 240 seconds (4 minutes) of holding, none of the customer care representatives are able to attend to the caller, she is automatically directed to a message recording facility and hears the message, "Sorry for keeping you waiting. Unfortunately, none of our staff has been available to attend to your call. Please leave your message after the tone, and your request will be attended to." At this point, the caller has the option to abandon the call and opt out of queue or to leave a message in the mailbox. Agents retrieve the voice mail messages and perform the necessary actions based on the message. Part of an agent's time is spent making these outgoing calls.

Voice mail performs the following functionality:

· Records the message left by the customer
· Plays back the message to the agent when accessed
· Transfers a voice mail from one agent to another

Some patients are not comfortable conversing in English. In this case, CCRs are often called upon to assist doctors and nurses in translation and interpretation. The CCRs are also trained by the nurses to handle certain emergency queries. In case they cannot respond to a query, the caller is put on hold and the nurses or doctors are consulted. The time spent consulting a nurse or a physician, or the time during which the agent is away from the terminal and unavailable due to other responsibilities, is clocked as "walk away time" for the employees. The other duties of the call center employees include arranging for special patient requests such as wheelchairs, and paperwork pertaining to appointments (called "postmessage" activities). The manager also gives them noncall-center–related work to be completed when they are not attending calls or doing postcall-related activities. The time spent on rest breaks is clocked as "not ready time." The CCRs are allowed a half hour lunch break staggered between 1 pm and 2 pm. They are also allowed a 15-minute break during each pre- and post-lunch time period. They are required to punch in the "not ready" button on their consoles before going on their lunch break or any other scheduled or unscheduled breaks.

Some of the CCRs are called more frequently than others for translation assistance for their knowledge of Italian or Spanish, and one of them is called for her knowledge of Chinese. These CCRs have clocked a very high proportion of walk-away time. The other performance parameters also show a wide variation. (Refer to Tables 9.2 and 9.3 for CCR performance ratings found in a study by a Customer Care Department.) These exhibits show that the processing time varies not only by the type of calls, but also depends on the CCR taking the call.

**TABLE 9.2**    Average Time Required to Service Each Call Type (in seconds)

| | AGENT 1 | AGENT 2 | AGENT 3 | AGENT 4 | AGENT 5 | AGENT 6 | AGENT 7 | AGENT 8 | AGENT 9 | AVERAGE |
|---|---|---|---|---|---|---|---|---|---|---|
| Pres Refill | 191 | 186 | 199 | 196 | 199 | 188 | 201 | 192 | 189 | 193.4 |
| Future Appointments/ confirm | 230 | 226 | 267 | 222 | 220 | 228 | 199 | 219 | 230 | 226.8 |
| Managed care | 222 | 199 | 265 | 231 | 206 | 225 | 189 | 222 | 251 | 223.3 |
| Acute care/ phone advice | 117 | 111 | 142 | 114 | 109 | 114 | 91 | 103 | 107 | 112.0 |
| Billing & admin issues | 82 | 80 | 88 | 84 | 83 | 77 | 111 | 79 | 81 | 85.0 |
| Repeat calls (return) | 129 | 122 | 146 | 178 | 170 | 135 | 212 | 176 | 166 | 159.3 |
| Lab results | 182 | 190 | 196 | 180 | 181 | 165 | 190 | 181 | 170 | 181.7 |
| Misc | 39 | 33 | 52 | 38 | 31 | 28 | 50 | 35 | 32 | 37.6 |

**TABLE 9.3**    Average Time/Call (sec) by Agent

| | AGENT 1 | AGENT 2 | AGENT 3 | AGENT 4 | AGENT 5 | AGENT 6 | AGENT 7 | AGENT 8 | AGENT 9 |
|---|---|---|---|---|---|---|---|---|---|
| Monday | 106 | 125 | 115 | 119 | 139 | 113 | 112 | 92 | 125 |
| Tuesday | 135 | 137 | 115 | 130 | 133 | 129 | 111 | 89 | 164 |
| Wednesday | 92 | 84 | 91 | 151 | 158 | 105 | 119 | 122 | 141 |
| Thursday | 116 | 54 | 134 | 129 | 120 | 129 | 125 | 106 | 142 |
| Friday | 97 | 102 | 105 | 110 | 145 | 104 | 124 | 116 | 153 |

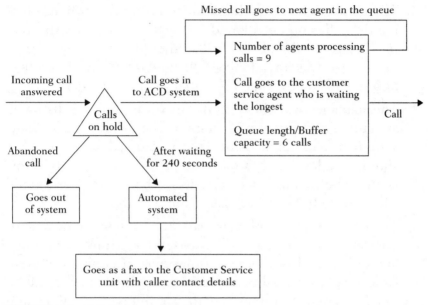

**FIGURE 9.2.** Call Center System Flow

The quality of customer interaction with the CCRs is measured by periodic customer satisfaction surveys, and monitoring is done by random checks and a "listening in" facility. No standard methodology is used for measuring customer satisfaction. Figure 9.2 shows the call center system flow.

## CHALLENGES

The following questions emerge as part of this case:

1. The unit does not seem to have a very good idea about customer perception of quality. The call center is still viewed as a cost center rather than as a revenue center. Well-defined quality dimensions need to be spelled out.

2. Agents are the first point of contact and most likely were responsible for creating lasting impressions about the efficiency and quality of care of the department. Revenue lost will be due to patients abandoning the calls. What is the amount of revenue lost due to these lost calls?

3. The manager noted with satisfaction that the proportion of missed calls and abandoned calls was lower than expected, roughly 4% to 8% of total calls. Should the manager strive to improve this abandoned call standard? With an average of 588 calls a day, that amounted to about 35 callers!

4. The manager was thinking of answering a few calls using an automated answering system, but what type of calls should be answered using this automated system? What should the level of flexibility be for service capacity? How should she manage her capacity—what incentives should she provide to keep her agents motivated?

5. The service chain of this call center has to deal with greater demand as well as service time variability. Variability is also affected by the type of services offered: for example, variability for acute care/phone advice will be much higher, as compared to variability for billing and administration issues. The service chain for this call center needs to exploit the low variability to lower costs and be efficient for low variability services.

6. Additional capacity is needed due to the presence of the variability—difference in the types of calls, difference in processing time of different types of calls. Safety service capacity must be higher than the average capacity needed to serve all customers. How should this variability be reduced and how much additional capacity should be made available? Is this a problem of insufficient capacity or poor scheduling? Key service capacity–related decisions at the call center are as follows:

   · Level: How much capacity to have? This decision will be driven by total buffer capacity and total variability faced by the service chain.

   · Degree of centralization: Capacity is centralized in this case.

   · Flexibility: How flexible should the capacity be? Time flexibility in terms of scheduling needs to be improved to better match the available service capacity to the demand.

- Ownership: Should the department think about outsourcing the calling center part of the business?
- For low variability and predictable services, we could likely automate those services so the degree of completion could be much higher.
- Where should the inventory reside (distance from customer)? At this point, we are only thinking about the call center, but there is a possibility of making other channels available and bringing service inventory closer to customers.

## SIX SIGMA TEMPLATE

We developed a template to manage Six Sigma projects in a service environment. The template is shown in Figure 9.3.

The template is divided into five phases, namely define, measure, analyze, innovate, and embed. Phases in real life do not have such a distinct start and finish, as is presented here.

## DEFINE

First, we need to define the objectives of our Six Sigma project. Objectives could be a combination of any of the following:

- Who are our customers? Patients, insurance companies, employees, or all of them.
- Reduce blocked and abandoned calls to zero.
- Improve customer satisfaction.
- Improve top line and/or reduce cost.
- Improve employees' retention.

There could be various likely approaches to achieve these improvement goals. Suppose our goal is to reduce blocked and abandoned calls to zero. We could think of the following different ways to achieve this goal:

- Improve variability of the process so that scheduling and capacity allocation will become easier to implement.

| Project Title: | Project Leader: |
|---|---|
| Team Members: | Project Start: |

| Estimated Project Selection Parameters:<br>Probability of Success (P) ❑    Cost (C) ❑    Time (T) ❑<br>Savings (S) ❑    Project Index (PI) ❑ |
|---|
| Project Description: |
| Project Goal and Objectives: |
| **Define Phase** |
| Customer(s):<br><br>Customer Critical Requirement (based on house of quality of similar analysis):<br><br>Any critical requirement from employees: |
| Project Scope |
| Resources Required and Their Source |

**FIGURE 9.3.** Six Sigma Methodology Template (Define Phase)

| Stakeholder Analysis: |
|---|

STAKE HOLDER

| SUPPORT | CUSTOMER | MGMT. | EMPLOYEES | QUALITY | SUPPLIER |
|---|---|---|---|---|---|
| Passionately Committed | | | | | |
| Supportive | | | | | |
| Compliant | | | | | |
| Neutral | | | | | |
| Opposed | | | | | |
| Hostile | | | | | |
| Not Needed | | | | | |

Legend: X—Present level of commitment; O—Required level of commitment

Customer Service Quality Dimensions (Use SERVQUAL or Similar Instrument):

1.

2.

3.

4.

5.

Critical to Quality (Operation) Requirements:

FIGURE 9.3.  Six Sigma Methodology Template (Define Phase)

**Force Field Analysis:**

| Drivers: | Distracters: |
|---|---|
|  |  |

Problem Attributes in the order of significance:

Other observations:

**Measurements:**

Customer Related (Primary/External):

Operations Related (Internal):

Employee Related (Internal):

**FIGURE 9.3.** Six Sigma Methodology Template (Define Phase)

**Process Map:**

| # | ACTIVITY | PROCESS TIME | AVERAGE WAIT TIME | TOTAL TIME | CAPACITY OF THE STEP (PER UNIT TIME) | BOTTLENECK (SLOWEST RESOURCE) |
|---|---|---|---|---|---|---|
| | | | | | | |
| | | | | | | |
| | | | | | | |
| | | | | | | |
| | | | | | | |
| | | | | | | |
| | | | | | | |
| | | | | | | |
| | | | | | | |

All times are in minutes

Summary of process and blueprinting analysis:

Front room bottleneck activity: _____

Backroom bottleneck activity: _____

Critical backroom operations
- 1
- 2
- 3
- 4

Critical front-line operations
- 1
- 2
- 3
- 4

**FIGURE 9.3.** Six Sigma Methodology Template (Define Phase)

**Process Flow Chart:**

Basic Symbols (Additional symbols can be used consistently):

| Start/End | Predefined | Wait/Delay |

| Activity | Decision/Test | Storage |

Process flow diagram

**FIGURE 9.3.** Six Sigma Methodology Template (Define Phase)

**Service Blueprint:**

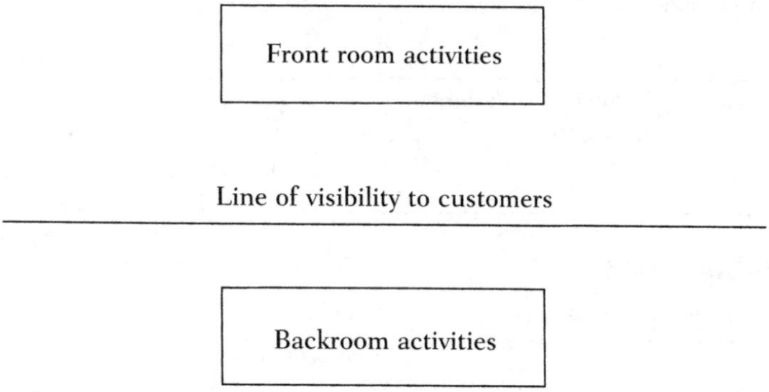

Front room activities

Line of visibility to customers

Backroom activities

Steps:

- Identify the process to be blueprinted.
- Identify the customer of customer segment.
- Map the process from customer's point of view.
- Draw the line of visibility.
- Map contact employee actions.
- Link customer activities and contact employee activities to back room (support activities).
- Draw any time scale, if required, horizontally.

FIGURE 9.3. Six Sigma Methodology Template (Define Phase)

| Project Title: | | | Project Leader: | | |
| Team Members: | | | Project Start: | | |

Estimated Project Selection Parameters:
Probability of Success (P) ❑          Cost (C) ❑          Time (T) ❑
Savings (S) ❑                    Project Index (PI) ❑

**MEASURE PHASE**

Project Descriptive Statistics:

| MEASUREMENTS | AVERAGE | STANDARD DEVIATION | COMMENTS |
| --- | --- | --- | --- |
| | | | |
| | | | |
| | | | |
| | | | |
| | | | |

**Cost of Quality**

| BACKROOM FAILURES ITEMS | FRONT LINE FAILURE ITEMS | DETECTION ITEMS | PREVENTION ITEMS | SUMMARY | COST |
| --- | --- | --- | --- | --- | --- |
| Rework | Customer complaints | Process control | Quality planning | | |
| Waste | Warranty charges | Peer review | Training program | Internal | |
| Recovery cost | Liability insurance | Supervision | Quality audits | External | |
| | Legal judgments | Customer comments | Data analysis | Appraisal | |
| | | Inspection | Supplier selection for quality | Prevention | |
| | | | | | Total |

**FIGURE 9.3.** Six Sigma Methodology Template (Measure Phase)

| Performance Measures (Use columns as appropriate). | | | | | |
|---|---|---|---|---|---|
| MEASUREMENT | QUANTIFIABLE? | ACCEPTABLE UPPER RANGE | ACCEPTABLE LOWER RANGE | DPMO | SIGMA |
| | | | | | |
| | | | | | |
| | | | | | |
| | | | | | |
| | | | | | |
| | | | | | |
| | | | | | |
| | | | | | |
| | | | | | |
| | | | | | |
| | | | | | |
| | | | | | |
| | | | | | |
| | | | | | |
| | | | | | |
| | | | | | |
| | | | | | |
| | | | | | |
| | | | | | |
| | | | | | |
| | | | | | |
| | | | | | |
| | | | | | |
| | | | | | |
| | | | | | |

**FIGURE 9.3.** Six Sigma Methodology Template (Measure Phase)

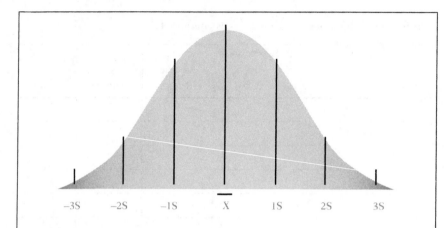

-3S      -2S      -1S      $\overline{X}$      1S      2S      3S

**Statistical Depiction of a Characteristic**

(Write actual values for a process characteristic).

List of Unacceptable Key Characteristics and Their Variation:

| CHARACTERISTICS | MEASURE OF VARIATION |
|---|---|
|  |  |
|  |  |
|  |  |
|  |  |
|  |  |
|  |  |
|  |  |
|  |  |
|  |  |
|  |  |

Comments:

Keep in mind, acceptable limits by customers are dynamic in nature and will often depend on number of stages in a process.

| Project Title: | Project Leader: |
|---|---|
| Team Members: | Project Start: |

**FIGURE 9.3.** Six Sigma Methodology Template (Measure Phase)

| Project Title: | Project Leader: |
|---|---|
| Team Members: | Project Start: |

Estimated Project Selection Parameters:
Probability of Success (P) ❏          Cost (C) ❏                    Time (T) ❏
Savings (S) ❏                         Project Index (PI) ❏

## FAILURE MODE AND EFFECTS ANALYSIS FOR
## IDENTIFYING POTENTIAL CAUSES

| PROCESS | POTENTIAL FAILURE MODE | POTENTIAL EFFECTS OF FAILURE MODE | SEV. | POTENTIAL CAUSES OF FAILURE MODE | OCC. | CURRENT PROCESS CONTROLS | DET. | RPN |
|---|---|---|---|---|---|---|---|---|
|  |  |  |  |  |  |  |  |  |
|  |  |  |  |  |  |  |  |  |
|  |  |  |  |  |  |  |  |  |
|  |  |  |  |  |  |  |  |  |
|  |  |  |  |  |  |  |  |  |
|  |  |  |  |  |  |  |  |  |
|  |  |  |  |  |  |  |  |  |
|  |  |  |  |  |  |  |  |  |
|  |  |  |  |  |  |  |  |  |
|  |  |  |  |  |  |  |  |  |
|  |  |  |  |  |  |  |  |  |
|  |  |  |  |  |  |  |  |  |
|  |  |  |  |  |  |  |  |  |
|  |  |  |  |  |  |  |  |  |
|  |  |  |  |  |  |  |  |  |
|  |  |  |  |  |  |  |  |  |
|  |  |  |  |  |  |  |  |  |
|  |  |  |  |  |  |  |  |  |

**FIGURE 9.3.** Six Sigma Methodology Template (Analyze Phase)

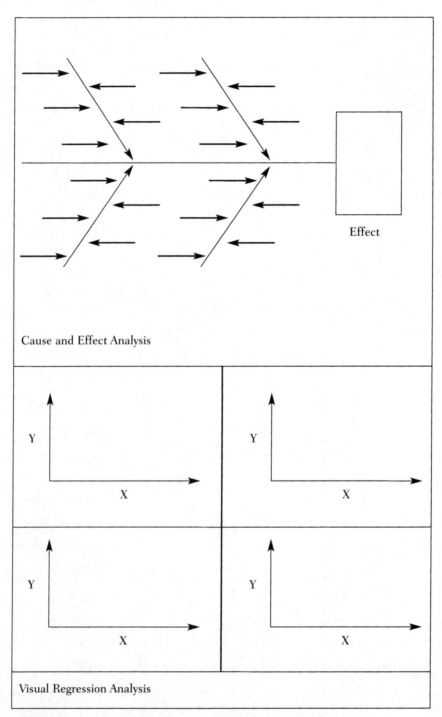

FIGURE 9.3.  Six Sigma Methodology Template (Analyze Phase)

**Hypothesis Testing**

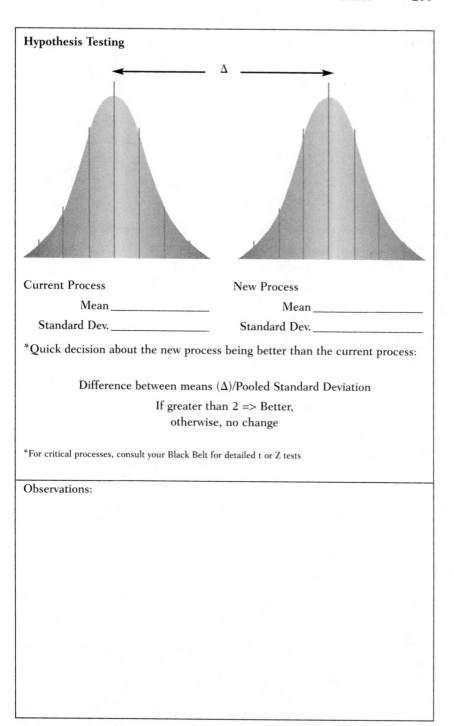

Current Process

      Mean _____

    Standard Dev. _____

New Process

      Mean _____

    Standard Dev. _____

*Quick decision about the new process being better than the current process:

Difference between means ($\Delta$)/Pooled Standard Deviation

If greater than 2 => Better,
otherwise, no change

*For critical processes, consult your Black Belt for detailed t or Z tests

Observations:

**FIGURE 9.3.** Six Sigma Methodology Template (Analyze Phase)

| Project Title: | Project Leader: |
|---|---|
| Team Members: | Project Start: |

Estimated Project Selection Parameters:
Probability of Success (P) ❑          Cost (C) ❑          Time (T) ❑
Savings (S) ❑          Project Index (PI) ❑

**Innovation type: Process or business model or offering**

Key innovation opportunities:

1.

2.

3.

Innovation for Six Sigma Project
• Budget
• Key leaders: development and execution phase
• Facilities/infrastructure
• Metrics to measure the performance of innovation projects

| STEPS | COMMENTS |
|---|---|
| Define the issue. | |
| Transition to an innovation team. | |
| Look for best practice internally: | |
| Benchmarking internal units and identify best practice units and practices. | |
| Look for best practices externally (same sectors and other sectors) and identify best practices. | |
| Develop and analyze solutions. | |
| Pilot test and implement. | |

**FIGURE 9.3.** Six Sigma Methodology Template (Innovate Phase)

| A.I.# | IMPROVEMENT ACTION ITEM | RESPONSIBILITY | COMMITTED DATE OF COMPLETION |
|-------|-------------------------|----------------|------------------------------|
| 1. |  |  |  |
| 2. |  |  |  |
| 3. |  |  |  |
| 4. |  |  |  |
| 5. |  |  |  |
| 6. |  |  |  |
| 7. |  |  |  |
| 8. |  |  |  |

**Design of Experiment—Full Factorial:**

| FACTORS | LEVEL 1 (CURRENT) | LEVEL 2 (NEW) |
|---------|-------------------|---------------|
| X1. |  |  |
| X2. |  |  |

**Expected Response:**

Experiment Design:

|  | Factor 2 Level 1 | Factor 2 Level 2 |
|--|------------------|------------------|
| **Level 2** Factor 1 | $F1_2, F2_1$ | $F1_2, F2_2$ |
| **Level 1** | $F1_1, F2_1$ | $F1_1, F2_2$ |

Factor 2
Level 1    Level 2

| Project Title: | Project Leader: |
|----------------|-----------------|
| Team Members: | Project Start: |

**FIGURE 9.3.** Six Sigma Methodology Template (Innovate Phase)

| Project Title: | Project Leader: |
|---|---|
| Team Members: | Project Start: |

Estimated Project Selection Parameters:
Probability of Success (P) ❑     Cost (C) ❑          Time (T) ❑
Savings (S) ❑                    Project Index (PI) ❑

Project Risk Management: Process and People Management

| CRITICAL INPUT PROCESS PARAMETERS FOR CONTROL | CRITICAL IN-PROCESS PROCESS PARAMETERS FOR CONTROL | CRITICAL OUTPUT PROCESS PARAMETERS FOR CONTROL |
|---|---|---|
|  |  |  |
|  |  |  |
|  |  |  |
|  |  |  |
|  |  |  |

| CRITICAL RISK FACTORS | PLAN FOR MANAGING CRITICAL RISK FACTORS |
|---|---|
| 1. |  |
| 2. |  |
| 3. |  |
| 4. |  |

FIGURE 9.3. Six Sigma Methodology Template (Embed Phase)

| X-Bar chart, stage 1 (Dynamic control limits and dependent on stage 1) |
| --- |
| Mean |
| 3s |
| 2s |
| 1s |
| Mean |
| −1s |
| −2s |
| −3s |

| X-Bar chart, stage 2 (Dynamic control limits and dependent on stage 2) |
| --- |
| Mean |
| 3s |
| 2s |
| 1s |
| Mean |
| −1s |
| −2s |
| −3s |

**FIGURE 9.3.** Six Sigma Methodology Template (Embed Phase)

**Commitment to Six Sigma Initiative**

| COMMITMENT | RESULTS ACHIEVED | COMMENTS |
|---|---|---|
|  |  |  |
|  |  |  |
|  |  |  |
|  |  |  |

Corrective Actions for Six Sigma Initiative

| CONCERN WITH SIX SIGMA | ROOT CAUSE | CORRECTIVE ACTION |
|---|---|---|
|  |  |  |
|  |  |  |
|  |  |  |
|  |  |  |

Management Review Actions

|  |  |  |
|---|---|---|
|  |  |  |
|  |  |  |
|  |  |  |
|  |  |  |
|  |  |  |
|  |  |  |

**Communication plan**

| COMMUNICATION TYPE | FREQUENCY | COMMENTS |
|---|---|---|
|  |  |  |
|  |  |  |
|  |  |  |
|  |  |  |
|  |  |  |
|  |  |  |
|  |  |  |

**FIGURE 9.3.** Six Sigma Methodology Template (Embed Phase)

· Provide additional capacity and or cross-train staff.
· Provide additional channels so that some capacity will be directed to these additional channels.

Next, we need to define customer and employee perspectives. What do customers really care about? What is important to employees? These steps will involve carrying out surveys and finding out priorities from these two stakeholders. Stakeholder analysis is a next obvious step to understand the whole system and to get an idea about the likely success of this project.

The last step is to understand the process and the service blueprint. Measurement strategy will heavily depend on the process and service blueprint. The measurement system will classify critical indicators.

## MEASURE

Once goals are well defined for the team, and appropriate measurement variables have been defined, we move to the measurement phase. Critical measurements have already been identified and need to be documented. Some measurement is already provided in exhibits for this case; for example, processing time for each activity, demand pattern, processing time by agent, and type of services requested by customers. Sufficient data about abandoned calls, blocked calls, customer perception of quality, and employees' perception of the service should be collected.

Some variables are easily measured. Other variables will likely have discrete distribution, and these variables present standardization issues in services. A number of these variables will vary depending on the context. We could think of courtesy in this context, as a variable. We could likely give some guidelines to provide a consistent experience. Some likely ways to measure courtesy could be—how often did the agent mention the patient by his name? Did the agent greet and follow a standard script and ask what additional services that clients might require?

The call center unit should be benchmarked with similar units available within the company.

## ANALYZE

Data available need to be analyzed in this phase of Six Sigma methodology. We will use tools like cause and effect, regression analysis, benchmarking analysis, and hypothesis testing in this part.

Failure mode and effects analysis will provide a ranking of top likely failure, and the fishbone diagram will provide the root cause. The next step is to quantify our cause and effect relationship using regression analysis and test the hypothesis using hypothesis testing.

At the end of the analyze phase, we should have a pretty good idea about the existing service delivery system.

## INNOVATE

Improvement alone is not sufficient in this competitive environment. Often, service concepts cannot be patented and firms have to rely on proprietary knowledge. Firms wanting to stay ahead in the game need to continuously innovate at the corporate level and at the operational level. The innovate phase of Six Sigma includes innovative methods and innovative processes that will help a firm improve by order of magnitude.

Here, a call center needs to look for innovating ways to improve performance. TRIZ has provided some guidelines in the area of innovation for products, and we recommend similar steps for services. Firms need to look for best practices in other business contexts, and need to look for so-called "contradictions."

## EMBED

The project needs to be executed well. The embed part includes all the execution elements, including documenting the learning from the project after the project is completed.

## CHALLENGES TO IMPLEMENTATION

Ghobadian et al. (1994) identified the following four barriers to the implementation of service quality improvements:

1. Lack of visibility. Problems are often not visible to the service provider. To create visibility, we often need to collect preliminary data.
2. Difficulty in assigning specific accountability: How do we attribute a certain quality problem to a certain stage? Accountability management is crucial to the success of any project.
3. Time to improve service quality. The service business is a people business, and it takes longer to change attitudes and beliefs.
4. Delivery uncertainty. The unpredictable nature of people (customers and employees together) makes it harder to design and deliver.

All these problems could be overcome by using the right toolbox, and with top management commitment. Techniques like blueprinting and continuous customer feedback are good ways to monitor and create visibility. The concept of creating waste visibility in a process proposed under the Toyota production system in manufacturing, should be applied. The invisible problem could create more damage. The service chain design will help us design a system that is aligned with the corporate strategy.

In a service context, a few reasons for failure are presented here:

- Lack of commitment from top management: Quality should be used as a strategic tool, and senior management support is mandatory for any quality program to succeed.
- The measurement system does not keep up with evolving customer needs: For example, security is quickly becoming an important quality dimension in transaction services. Firms need to update their measurement system and include new customer needs.
- Roles are unclear: Role ambiguity is another reason for failure. Quite often, in the innovation phase of new offerings development, authority and responsibilities do not match.

Managers responsible for the innovation process lack authority to manage team members from a cross-functional background.

- Losing sight of the whole system approach: Changing one part affects other parts of the system.

The DMAIE methodology tries to overcome these deficiencies and has the following advantages:

- It is a system approach, and the corporate alignment is considered while executing the methodology.
- The system goes through continuous innovation, and innovation is an intricate part of the methodology.
- The method is data driven and objective in nature.
- It considers the interpersonal aspects of the service delivery system.

## KEY TAKEAWAYS

- The degree of training and development of employees at a firm depends upon the degree of empowerment provided to employees.
- Corporate level measurement system need to have an appropriate balance between hard versus soft measurements. The ratio of hard and soft measurements will depend upon the lifecycle of a product and maturity of a firm.
- Top management commitment is crucial in addition to rigorous toolbox for success of any effective Six Sigma methodology.

# DESIGNING FOR SERVICE

# CHAPTER
# TEN

# AXIOMS OF
# SERVICE DESIGN

The term "service" often means different things to different individuals; in this book, we are using it for business activities that are performed to satisfy customer needs, where the major end product is customer experience and not a physical article. Our focus in designing services is for high-volume, low-cost activities. However, most of the discussion and tools are also relevant to low-volume services. Sometimes high-volume services are referred to as transactional services. Examples of transactional services are stock trading, insurance, banking, retailers, and Internet auctions.

Every service has two types of activities: information generation and information processing. We have called these activities interactions and transactions, respectively. Services are provided by the combination and interdependency of four elements: people, processes, infrastructure, and goods. In most cases, people are key to interactions. Processes support transactions, and infrastructure provides environmental and supporting technology to make interactions and transactions possible. Finally, goods may be a part of or outcome of the service.

Interaction in service means a communicative action or activity involving at least two parties that reciprocally affect or influence each other. Every service activity starts with an objective, and is completed by the exchange of information and/or goods. For example, the transaction of an insurance claim has two parties: the insured person and the insurance company. It starts with an objective of payment for the insured event, as per an agreement between two parties, and will be completed on the acceptance of the claim and payment of the insured amount or rejection of the claim with reasons communicated and accepted by both parties.

It is important to understand that education of service design by this book can at best only provide seed; it needs the fertile soil of experience, environmental conditions of organizational culture, and nutrients of management support to grow a tree that will bear the fruits of flawless service delivery.

# DEFINITION OF SERVICE PROCESS

A service process is a set of activities that is performed to transform given input(s) into a useful output(s). Every process has input of information—activities that consume resources to generate outputs. Some of the outputs are useful, and some are waste; some are expected, and some are unexpected. A good process has minimal variation in the expected activities and outputs and is robust against variations in inputs. Processes are generally identified in terms of events that define start and finish points; at both of these points, information and/or goods are exchanged. Once a process starts, a chain of predefined events or subprocesses begins, which culminates at the predefined end point. For the insurance claim transaction, the starting point is the accident and the next step is the communication of accident information by the customer to the insurance company representative. This triggers a chain of events designed for the insurance claim process.

Most of us have experience with service processes and may even have designed one. Every service transaction follows a process; it may be very simple and quick, so you may not pay attention to the process and its elements. Examples of simple transactions are buying groceries at the supermarket, withdrawing money at the bank ATM, and going on a family vacation. Examples where more complex transactional processes may be followed are buying a house, trade union agreements, and head-of-state election.

If a business organization is structured around functions, process activities will be performed across functional departments. It is important to understand that the customer wants to receive service and is not concerned about the internal working of an organization. It is the organization's responsibility to prevent the customer request from falling through the cracks of functional boundaries. Many companies are moving toward a process-oriented organizational structure or a matrix of functions and process.

As defined by Shostack and discussed in Chapter 6, blueprint is a map of the process that constitutes service. It divides the service process activities in two parts: one that is visible to the customer and others that are not visible. In most cases, the interaction part is visible to the customer and the transaction is hidden. Karl Albrecht, in his book *At America's Service*, calls interactions or those activities visible to the customer as part of the service process as a "Moment of Truth." Figure 10.1 shows a blueprint of the restaurant service process.

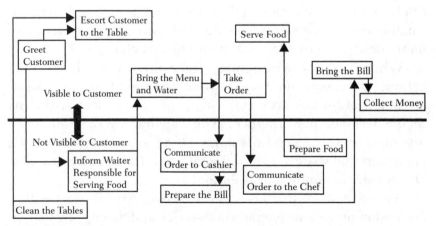

**FIGURE 10.1.** Blueprint of Restaurant Service Process

# DFSS FOR SERVICES

Let us first understand what we mean by design. It can be explained in simple terms as a set of definitions that describe either physical objects or instruction for activities, or a combination. Every design has input that gets transformed into desired and undesired output. The better the design, the less undesired output created, and a minimum amount of resources are consumed in performing activities. For the insurance claim process, input is the accident information, and output is the decision on the claim, which may be accepted as claimed, accepted with modifications, rejected, or require more information. Process design will describe how the accident information will be collected and what activities will be performed, as well as how the decision will be made, communicated, and executed. A good insurance claim process design will ensure that true claims are accepted and false claims are rejected, with a minimum amount of information exchange in the least amount of time.

Design for Six Sigma is a combination of axioms, methodology, and tools that, if applied properly, will produce designs capable of Six Sigma performance level. Methodology of design for Six Sigma (DFSS) for transactional processes, as described in detail later in the chapter, is: define, measure, analyze, design, and optimize (DMADO). Examples of tools are: TRIZ, Pugh's concept selection, QFD, Robust Design, and Systems Engineering. A list of tools associated with different phases of methodology is provided at the end of this chapter.

When it comes to designing either physical goods or transactional processes, no easy steps or tricks exist that will allow an organization to start generating goods or services at the Six Sigma level of performance. The capabilities that allow an organization to successfully apply DFSS are rooted in people, organizational structure, and operational strategies and tactics; DFSS can't be applied in isolation.

As the focus of this book is on service, we will start with a discussion on various elements of service and, later in this chapter, continue with axioms, DFSS methodology, and tools for transactional service processes. In later chapters, we will also extensively discuss implementation aspects related to people, organizational structure, operational strategies, and tactics.

## BASIC ELEMENTS OF SERVICE

Every service involves two types of activities: one where the customer interacts with the provider and the second when the information generated is processed through transactions. It is important to understand that interactions are not limited to human contacts; service interaction is related to customer experience. Any moment that leaves an impression on the customer about the service is an interaction. A service need not be rendered by a person; even when people are the chosen means of execution, they are only part of the process. For example, for a restaurant interaction, it may start from an advertisement in the Yellow Pages. For a complete transaction of a restaurant's service, the process has activities such as taking the customer's order, which the customer sees, and activities such as preparation of a meal in the kitchen, which the customer doesn't see.

Most of the service's activities involve a process and a physical product/enabler combination. The process enables the transaction to be completed by exchanging or processing information. In our restaurant example, it is part of the process to motivate the customer to choose the restaurant; however, the restaurant's furniture and the food itself are the physical objects necessary to provide the services—but these are not part of the process. In this book, we focus more on process design as the enabler of the service activities and not on the physical product and environmental design, though many of the ideas, methods, and tools are equally applicable to physical product design. From an information perspective, in every service, there are two types of activities: one is related to the generation of information and the other is related to processing of the information. Information is generated during the interaction, and it is processed during the transaction. Both of these types of activities are supported by processes and physical enablers.

The tree diagram in Figure 10.2 graphically depicts the service components and their relationship in a simple-to-understand flow. If you move toward the left, it provides the reason for that specific element; if you move toward the right, it provides more specifics about that element.

In summary, if we have efficient and effective communication and processes supported by proper goods, environment, and equipment, we have an excellent service in place. Our focus in this book is on information management and less on the physical enabler, as physical enablers tend to be very specific to the service. For example, a restaurant is a service business, and this book will help in designing and improving the Six Sigma quality level of interactions and transactions. For interactions, this book provides axioms and methodologies to develop communication strategies, guidelines, and work instructions. For transactions, this book provides axioms and methodologies to develop processes, procedures, and work instructions.

## AXIOMATIC APPROACH TO DESIGN

Axioms are self-evident truths, which can't be proved, but have no counterexamples. These are abstractions of existing knowledge embedded in many things that people use or observe rou-

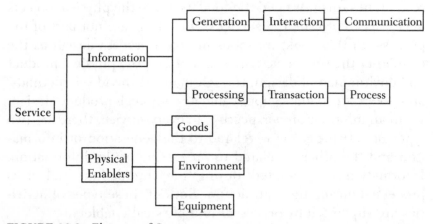

**FIGURE 10.2.** Elements of Service

tinely. The axiomatic approach to service design is based on the assumption that there exists a fundamental set of principles that determines good practices. Suh (1990) has identified two design axioms, that govern all types of designs ranging from physical products to processes for services. First, axioms deal with the relationship between customer needs and requirements to process, or physical variables that will define a design. Second, axioms deal with the complexity of design.

## SUH'S FIRST AXIOM OF INDEPENDENCE

Suh's first aximiom says maintain the independence of functional requirements (FRs). This means that when making a decision on a design concept and details, we need to ensure a one-to-one relationship between FRs and design parameters (DPs) at any given level of hierarchy. For a best design concept, if we change any design parameter, it will only impact one functional requirement.

To understand Suh's independence axiom we need to first understand the concepts of FR, DP, and hierarchy of both. FRs are defined such that all the customer requirements that a design must satisfy are captured without redundancy or overlap. They describe the objective of a design for a specific need in the simplest possible form. Design parameters are those variables that define a solution for meeting the functional requirements. Both functional requirements and design parameters exist at several levels; the lower the level, the more detailed it is.

Shannon and Weaver (1949) explained the mathematical theory of communication. The key idea is that as long as the communication rate stays below channel capacity, the chances of error are minimal. Through their study of the random processes, they defined a level of complexity, called "the entropy," below which the signal cannot be compressed. The principle of entropy was generalized and is used in many disciplines as a measure of uncertainty or complexity. In the design context, Suh (1990) equated entropy with required information content for a design and proposed it as a measure of complexity for his second axiom.

## SUH'S SECOND AXIOM OF MINIMUM INFORMATION

Suh's second axiom says minimize the information content of a design. The design that satisfies the first axiom of independence, and the design that has the maximum chances of success, is the best design.

## AXIOM-BASED TRANSACTIONAL SERVICE MODEL

We propose an axiom-based transactional service model, which is not in conflict with Suh's generalized design axiom model although, it is specific to services. The axioms of transactional service are based on maximizing customer value and meeting business objectives. These axioms are abstract and universally true for all types of transactional services. The axiom-based transactional service model will provide a framework for our DFSS approach, as well as criteria for evaluation.

**First axiom.**    Service interactions should be designed to generate or exchange quality information.

The ultimate objective of service interaction is to ensure customer satisfaction and achieve the business objectives. Interactions take place at all stages of services; at every stage, both customers and providers may have different reasons for interacting. For example, in the beginning, the provider is trying to sell the service, and the customer is trying to assess the capability of the provider to make the buying decision. A successful interaction will be that which will meet the objective of both. The first axiom states that we need to design service interaction so that quality information gets generated and communicated to both parties at all stages. Table 10.1 summarizes interaction objectives of customers and providers at the beginning, during, at completion, and after the service is rendered.

For further understanding of interactions, we need to define the attributes that define the quality of communication for transactional services. Attributes related to communication or interactions may have different interpretation for human and nonhuman contacts, and providers and customers will value different attributes. No doubt attributes that are important to customers will also matter to providers, as customer sat-

TABLE 10.1    Interaction Objectives

| | STAGES OF THE SERVICE | | | |
|---|---|---|---|---|
| | BEGINNING | DURING | COMPLETION | AFTER |
| Customer | Communicate the need and assess the provider's capability | Know the status and communicate any change in requirements | Verify that service meets the expectations | Feedback on things gone wrong and right |
| Provider | Sell the service, understand customer needs, and set the right level of expectations | Input for internal processes and any change in the original expectations | Communicate the final outcome | Continuous improvement and repeat business |

isfaction is one of the most important drivers for a successful business. The following attributes define the quality of interaction from a customer's perspective:

· Responsiveness
· Empathy/understanding
· Assurance

In a human interaction, responsiveness is related to a customer's perception of how much the service provider is willing to help. Empathy is related to whether individual attention is given; in other words, the service provider is able to understand the customer's unique requirement and is not trying to provide a "one size fits all" solution. Assurance is related to how confident the customer feels about the knowledge and ability of the provider to meet the customer's requirements.

When it comes to a customer interacting with nonhumans, for example, in an automated answering service for customer questions, these attributes have a slightly different meaning. Responsiveness is related to how easy or difficult it was to interact and get the information. Empathy is a human interaction attribute; however, the related attribute is understanding, which means that a customer's requirements were properly understood during the interaction, assurance means how the customer feels after the interaction is over about the ability of system.

If interaction is human to human, it can't be standardized. However, our first axiom tells us that it needs to be designed to provide value to both parties. Earlier, we discussed attributes from the customer's perspective; now let us look from the provider's point of view. The following attributes will define quality from the provider's perspective:

- Speed
- Usability
- Comprehensiveness

Speed is related to how much time the provider spends in generating or exchanging required information, and how accurately it's done. Usability means how much of the collected or generated information will be utilized in providing the service. Comprehensiveness tells if the interaction completed its designed purpose or if one of the two parties will have to go back and ask for more information.

For example, if you go to a garage to get your car repaired, the clerk at the desk will collect information from you about the symptoms of the problem, which will help in diagnosing the cause and giving you an estimate of repair. The clerk will also collect other information that will be required for billing and feed the data for the provider's internal processes. During this interaction, if you look from the customer's perspective, he will be satisfied with the interaction if it gives him the feeling that the clerk was very responsive, willing to listen his problem, seemed to understand the severity or urgency (empathy), and was able to provide satisfactory answers. Finally, the customer must feel assured about the capability of this garage being able to repair his car based on this interaction. From the garage owner's point of view, he would like to minimize the clerk's time spent in collecting and recording the information correctly (speed), collect only the information that will be useful for repairing this car or for his business in some other way (usability), and make sure that all the required information is collected (comprehensiveness) at the very first time.

In the example of the insurance claim process, the first stage is the information exchange between the customer and insurance company representative about the accident. Here, speed is related to recording facts correctly with less time, and usability is related to collecting more information than needed for processing the claim. Comprehensiveness is related to the completeness of the information collected to process the insurance claim. The customer's perception of representatives who are willing to help, pay attention to the customer's situation, and are knowledgeable about claims processing will decide the responsiveness, empathy, and assurance attribute for this interaction.

**Second axiom.**     Transactions should process customer interactions efficiently and securely. The second axiom implies that if a transaction in business is directly or indirectly not processing output from the customer interaction, it should be eliminated. All transactions need to be performed with the required level of security and efficiency. The second axiom assumes that the first axiom is valid and interaction has resulted in generating quality information. Efficiency in a transaction has the following three attributes:

- Accuracy
- Resources spent
- Relevancy

Accuracy is the attribute for the facts contained in the information; it can be measured based on errors in interpreting the data generated from the customer interaction. Resources spent on processing information are important, as they will define the cost and time it takes to process information. Any transactional output that is generated without any usability is irrelevant and a waste of resources. Relevancy is also related to the scope.

Once a service provider collects information from the customer, the provider has a moral and/or legal obligation to ensure that it is not accessed by any unauthorized person and security is not compromised.

In our car repair garage example, let us assume the first transaction will be the estimate for repair generated and given to the customer for acceptance. Accuracy in this example is how close the estimate is to the actual cost of repair. How long it took to generate the estimate and other inputs it required to generate the estimate (e.g., mechanic's time, etc.) will define the resources spent. Finally, the estimate should have everything, but nothing more than that which the customer needed to know to approve the repair; this defines relevancy.

## USING AXIOMS FOR DESIGNING SERVICES

At every stage of service, either all or a subset of these attributes will define the quality of an interaction or transaction. For all the service elements, there should be a defined minimum standard for each of the relevant attributes, and performance below that level will be considered a defect. Attribute-related to information processing or transactions and the provider's perspective in the interaction are objective, while attributes related to interaction from the customer's perspective are subjective and based on customer perception. The objective of design for Six Sigma is to design service transaction processes and communication methods that will minimize the likelihood of defects per million opportunities.

Design the service interaction and transaction elements such that they have a maximum likelihood of achieving a required level of performance during the execution of service. This will ensure that the probability of providing the best customer experience and achieving business objectives are maximized during the actual delivery of service. To use the axiomatic design model, we need to define the concept of value so that we can establish a common scale of measurement.

**Measurement of value.** To have customer satisfaction and business objective fulfillment, there needs to be compatibility between customers' expected values and designed value in service process elements. Every element of the service process needs to be designed so that it adds perceived value by a target group of customers. To understand this, we need to define value in service. A typical definition of value is the ratio of functionality and cost. For defining service value, we need to understand the value of an attribute, as explained by Noriaki Kano, a renowned Japanese expert in total quality management:

· Must be: The customer expects these attributes in service as a minimum. If they are unfulfilled, the customer will perceive very limited or even zero value, but even if they were completely fulfilled, the customer would not perceive a very high value. Must attributes don't increase the value, but they do decrease it to almost zero or even negative; for example, if you go to a barber shop for a haircut, and the barber doesn't know how to cut hair, nothing else really matters. The value of service is zero.

· Satisfiers: These have a linear effect on value—the more these attributes are met, the higher the value perceived by the customer. Satisfiers may also have a lower or upper limit below or above which these will become "Must Be" type attributes.

· Delighters: These do not cause dissatisfaction when not present, but increase the value when present.

All the preceding attributes can again be divided into two categories from the measurement point of view:

1. Higher the better (HTB)
2. Lower the better (LTB)

Occasionally, you may find a "nominal the better" type attribute for service; however, it can also be considered as LTB by taking its deviation from the target as a measurable quantity.

To calculate service value, we need to perform following steps:

- Identify all the "must be" attributes.
- Divide them into HTBs and LTBs and establish lower and upper limits, respectively. Ensure that service has all "must be" attributes above the lower limit or below the upper limit.
- Identify satisfiers and delighters.
- Develop a common measurement scale for all the attributes.
- Divide all satisfiers and delighters in two groups: HTB and LTB.
- Calculate the ratio of HTB to LTB.

To increase the value of service, we need to increase the ratio of HTBs to LTBs while ensuring that all the must-be attributes are present at least at their minimum level. Value for transactional services is the ratio, where the numerator measures the level of fulfillment of those attributes that customers want more and more, such as accuracy in transactions and responsiveness in interactions. The denominator measures things like resources, which you would like to keep at a minimum level. Any organization that strives to stay ahead of competition has no choice but to grow the numerator. The real trick is to continuously reduce the denominator at the same time.

# DFSS METHODOLOGY

DFSS methodology provides the "how to" for our axiomatic model of service process design and execution. To satisfy both axioms, we need a design methodology supported by tools.

DFSS methodology is a five-phase process: Define, Measure, Analyze, Design, and Optimize. We refer to this methodology as DMADO. These phases are interdependent on each other, and some of the activities are sequential in these phases. However, there is lot of overlap and back and forth movement between phases. It is a structured methodology and expectation is that before starting any measurement activities, one will complete certain tasks in the define phase. However, you will not wait to complete the define phase before starting measurements. Similarly, analysis can't be started unless some define and measurement activities are completed. However, you will not wait to complete either of these two phases before starting analysis.

## DEFINE

The first phase of the DFSS methodology for transactional services is to define the project. The objective is to develop an understanding of potential business opportunities and create a vision to be followed throughout the process of design. This is the most important phase of all the five phases, as it will establish the goal and set the direction for all future activities. The following are the specific activities for the define phase of project:

1. Develop a project charter that will include the following:
   a. Broad service definition
   b. Business case
   c. Vision and milestones
2. Develop initial project plan

**Project charter.**   Create a project charter that will include the definition of the opportunity, the statement of vision for this opportunity, scope, business case, champion and other stakeholders, team, potential risks, and roadblocks.

*Broad definition of service.*   Decide the generic area of service. As a first step, this will help in focusing the efforts at subsequent stages. The broad definition can be created by mapping the strengths of the organization with the basic needs that will be satisfied by this service and later converting needs into high-level customer attributes.

*Business case.* Identify who the target group of customers is and the needs of these customers, which you will satisfy by providing the proposed service. The business case for the service should identify what will make the service compelling and what are potential associated risks. It should link service with the overall strategy of the organization. At the define phase, the business case usually focuses on connection to business strategies, market opportunity, technical feasibility, and initial project cost. A more detailed financial business case will be developed during the analysis phase. Even at this early stage in our axiomatic DFSS approach, we emphasize preparing a value statement, which will define what target value you are aiming at and how it differs from other competitive services in the market.

*Vision and milestones.* Vision captures what gives the most value to customers; it should also bind the people in the organization toward a common goal. A real vision is concrete, specific, guiding, and engaging, and it focuses peoples' efforts. Once a vision is established, the next step is to develop intermediate goals or milestones.

**Initial project plan.** Create detailed plans for work to be done by defining tasks that need to be accomplished, along with a timeline. A tool that helps you develop the task structure is called a *work breakdown structure*. Tools that can be used for understanding interdependencies of tasks and timelines are Gantt chart, network diagrams, PERT, and CPM.

The output of the define phase will be captured in two documents: project charter and project plan. Both of these documents are living in nature, and they will change as we will progress. However, it is important that we create a reasonably good focus and plan at the very start of a service design project.

## MEASURE

Six Sigma is a data-driven approach. Whether we are improving or designing a product or service, data are at the heart of the project. After creating a focused definition and assessing the initial feasibility based on the available data, we need to collect additional information to proceed with the project. Without a proper framework, the data collection will become a very complicated task. For service design, we need to understand the concept of domains (see Figure 10.3). It will help tremendously in developing a measurement plan with the proper framework. For service, three domains are:

· Customer
· Service process
· Service delivery

The objective of the measurement phase is to collect information regarding customer attributes, process elements, and service delivery details. Initial activities in the measurement phase are geared toward generating additional data and information on customer attributes, and later, information on process elements and execution variables will be collected. The following are some of the initial tasks of the measurement phase:

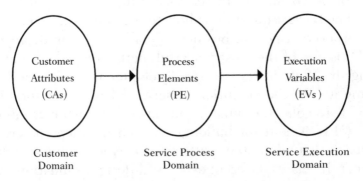

**FIGURE 10.3.** Service Domains

· Capture voice of customer
· Value-based measurement system
· Benchmarking
· Plan and collect the data

**Voice of the customer.**    The voice of the customer (VOC) needs to be heard in order to provide quality services because it is the customer who defines quality, not the corporation. The companies that are leading today and are poised toward shaping tomorrow in the service industry have learned to provide service to customers as they want it. These successful companies are never tied to conventional wisdom or industry norms. These far-sighted corporations identify opportunities, create a vision, and listen to the customer to design and deliver services that no one else can match easily. Examples of such companies are Wal-Mart, Microsoft Corporation, Southwest Airlines, and eBay. In each case, a company is providing a service in a very competitive market. However, they stay ahead of the game by listening to the VOC and letting the competition play catch up. In Internet auctioning, eBay leads by continuously listening to the customer and creating services that the competition can't match. Only companies with a commitment to listen and to serve customers can consistently produce delighted customers. And only by delighting customers can companies produce growing profits decade after decade.

Understanding that customers are the most important element of any business is necessary, but not sufficient for success. The biggest question is how to listen and incorporate the VOC in every aspect of business. Quality function deployment (QFD) is one of the most powerful methodologies for incorporating the VOC in the service development process and integrating it with the execution of service. Up-front efforts must be made to collect, evaluate, and categorize customer information. QFD is a team building, consensus-oriented, creative approach that allows the synthesis of new ideas in a structured manner. It relates different aspects of design, validation, delivery, cost, reliability, and technology while preserving the customer's voice as the driving force for the service. However,

QFD is not the tool to collect the VOC. QFD translates VOC into detailed customer attributes and standards that can be deployed throughout the process. For collecting VOC, the following tools may be used:

· Surveys
· Service review
· Customer advisory panels
· Focus group
· Employee field reports

*Measurement system development.*   Define the unit of measurement for each attribute and create design standards by specifying the desired level of performance. Determining targets and specification limits is not always easy. Target setting is both an art and a science. Targets are often set arbitrarily because of the lack of information on benchmarks and satisfaction/performance relationships. This ultimately impacts the quality of the design. Even if formal mathematical methods cannot be used, some thought and analysis is critical to setting good targets.

The general concept behind setting targets is the need to understand what the customer values, the strengths and weaknesses of the organization versus competition, and finally, what is the market trend. Due to limited resources, the highest performance targets should be set for those attributes that have the potential of providing the highest value to the customer and most benefit to the business. Therefore, in general, targets must be set to exceed competitor benchmarks for the most important attributes. This must be balanced against the resources required for meeting all the targets with the resources available to the organization. If the organization is not capable of performing at the target levels without a significant investment, then a business case needs to be developed to acquire more resources or modify targets. There is no magic formula for generating targets and measures—they are based on the team's knowledge, the data obtained from the VOC, benchmarking, internal surveys, and discussions between team members.

*Benchmarking.* For conducting benchmarking, first we need to identify the benchmarks. We need to identify market leaders for the process elements and delivery methodologies. Some of the information on process elements may be available from the published studies, reports, and articles. Companies that excel in particular areas are often described in trade journals, or their results are published as part of conference proceedings or reports on awards such as the Malcolm Baldrige. Benchmark databases are also commercially available. Benchmarking databases are also a source of performance information. The following are the generic categories of benchmarking opportunities:

- Customer satisfaction
- Customer loyalty/defection
- Core service process performance
- Support process performance
- Employee performance
- Supplier performance
- Technology performance
- Innovation performance
- Financial performance

*Data collection plan.* It may require a significant amount of resources to collect and analyze data. Before developing a plan to collect data, the following activities need to be performed:

- Review the focused opportunity statement
- Decide what data are needed from each domain (i.e., customer, process, and delivery)
- Collect and review existing internal data first
- Identify the gap

- Decide what type of data is required to be generated
- Decide the sources and methods for collecting data

Develop a data collection plan with the following details:

- Objective for each type of data
- Who will collect the data
- Timeline
- Budget

## ANALYZE

The third phase of DFSS methodology is to analyze the existing and newly created information, generating as well as evaluating concepts, and developing a basic process flow. The measure phase established the performance requirements that the service to be designed should meet or exceed to provide expected value to the customers. In the analyze phase, we will determine the alternative concepts for the service process and delivery that will perform to these requirements. We will analyze these concepts and select the best ones for the next stage of the design process. For information analysis, the choice of method depends on the nature of the data and objective of the analysis. For quantitative data, statistical analysis may be suitable; for qualitative data, logical analysis such as cluster analysis, pattern recognition, affinity diagram, and prioritization techniques may provide the answers. For concept generation and evaluation, creativity techniques should be used. Examples of creativity techniques are brainstorming, Pugh's Concept Selection Matrix, TRIZ, and mind mapping. For developing process flow and logical relationships, techniques such as supplier, input, process, output and customer (SIPOC), workflow model, and flow-charting are used.

**Analyze data.**    The collected or generated data need to be analyzed to better understand customer requirements and priorities. Analysis of data should establish the primary and secondary functions of the service being designed. Functions are the customer expectations that a service should satisfy, no matter what technology or method is used for the process and delivery. The first step in QFD is to map customer requirements to the specific critical to quality (CTQs). It is accomplished in the measure phase.

The second step in QFD is to take the CTQs identified in the measure phase and relate them to the functions identified for the new designs in the analyze phase. This will provide focus on where the design effort should be spent to have the maximum value for the customer. Also, with an understanding of the relationship between CTQs and functions, clear requirements can be established for the critical functions. This will help in developing new concepts capable of meeting functional requirements. Data analysis should clearly indicate following:

· Priority of functions based on customer value.
· Which functions need new concepts and for which existing processes are they sufficient?
· Which functions can be reverse-engineered using competition's design?

For all the high-priority functions that need innovative ideas, new concepts need to be generated and evaluated as a next step in the analyze phase.

**Concept generation and evaluation.**   Concept generation is an activity in which creativity is crucial for the success. In a group situation, the team can only—by following the principle of suspending judgment—generate "out of box" solutions. The team must have sufficient knowledge about the process; however, an open mind and a guided creative process are essential. Brainstorming is the most commonly used method for generating ideas. A more effective and structured brainstorming team can also use some methods for stimulating ideas. Alex Osborn's (1953) pioneering book, *Applied Imagination*, talks about "questions as spurs to ideation," and outlines about 75 idea-spurring questions. Mind maps, developed by Tony Buzan (1994), are an effective means of note taking and are useful for the generation of ideas by associations. To make a mind map, one starts in the center of the page with the main idea and works outward in all directions, producing a growing and organized structure composed of key words and key images. The systematic methodology of inventive problem solving known as TRIZ can also be very useful in developing new concepts. One of the core ideas of TRIZ is that people in one field may not know the relevant inventions in another, and as a result, the team may spend a lot of time reinventing the wheel.

During the concept evaluation, one must continually ascertain that the proposed concept is feasible. This feasibility can often be determined using the approach attributed to Enrico Fermi. A Fermi problem has a characteristic profile. Upon first hearing it, one finds it very complex, and feels certain that not enough information has been provided to find a solution. Yet, when the complex situation is divided into smaller elements, each one may be simple enough to find the answer without the help of expert knowledge or complex calculations. In most of the cases, an estimate can be made that comes remarkably close to the exact solution (Magrab, 1997). Fermi's intent was to show that, even in very complex and uncertain situations, one can proceed on the basis of different assumptions and still arrive at estimates that fall within the range of the answer. The reason is that, in any string of calculations, errors tend to cancel one another. The law of probabilities dictates that deviations from the correct assumptions will tend to compensate for one another, so that the final results will converge to the correct solution (Christian and Baeyer, 1993).

The Pugh matrix is another tool to evaluate and select best design concepts from a set of alternatives. It is a simple but very effective method for stimulating discussion about the pros and cons of concepts under consideration. It requires a thorough understanding of the concepts and their expected performance level on various desired functions. In a Pugh matrix, alternative concepts are written in the columns, and criteria in the rows. One of the concepts is selected as a baseline and all of the other concepts are compared against this baseline.

**Process flow and logical relationship.** All the existing concepts, re-engineered concepts, and new concepts need to be integrated to provide a seamless service process and delivery system and checked against both the axioms of service. Some of the tools, that can help in accomplishing these tasks are process flow charting, the SIPOC model, and workflow analysis.

*Flow charting.* The reason for creating a flowchart is to provide a picture of the steps in a process using standard symbols. There will be a hierarchy of processes for any service design. One needs to understand the hierarchy and process relationship at the same as well as at lower and higher levels. Three basic approaches for developing overall flow are top down, bottom up, or in some cases, a combination of both. In the top down approach, you will start with the macro level flowchart and then develop the detail flowchart for each of the elements. You will continue this process until you have reached the lowest level in the hierarchy. In the bottom up approach, you will start at the lowest level of activities and then move upward by considering the whole flowchart at the lower level as only one element for the next level.

*SIPOC.* Suppliers, inputs, outputs, and customers (SIPOC) is a business interaction model that provides a simple and quick way to develop a macro view of a process with respect to its suppliers, inputs, outputs, and customers. It provides a good summary of the overall process, with input linked with sources or suppliers and output linked with customers.

*Workflow analysis.* Workflow analysis helps in understanding how various departments in the organization will work together to accomplish tasks required by the defined process. A workflow model focuses on the flow of work through the business for a single service process output or input. In most cases, businesses are not organized around processes, and that is why it is important to understand how process activities will cross departmental boundaries and how the flow of communication and information will be managed. In summary, workflow is a map of a service process starting with its initiation, tracing work as it moves from department to department, until service is delivered to the customer.

## DESIGN

Once the concepts are selected and overall process flow is developed, the next step in the process is to develop enough details so that the design can be evaluated for performance and feasibility and risks associated with the design are better understood. Developing a detailed design involves decision making at every step of the way. As described earlier, there is always a hierarchy of processes; consistent with that, there is a hierarchy for making decisions. The sequence of decision making is very important as the feasible solution space for every next decision is smaller or constraints are greater. Every decision choice has its own associated benefits, costs, and risks. The challenge for the team is to make the choices that simultaneously balance the benefit, cost, and risk elements, and are compatible with previous high-level decisions.

It is critical to understand that design is not a linear process, where the preceding step has to be completed before the following step is reached. In most situations, design is an iterative and systematic, but cluttered and complex process. It is supported with tools such as systems engineering design reviews and failure mode and effect analysis (FMEA). For process elements at every level, requirements need to be developed. The systems engineering concepts of *requirement flow down* and *roll up* are very helpful at this stage of the process. Also, a third house of the QFD may be used to establish the relationship between functional and design requirements. This can focus where the design effort should be spent to achieve maximum value for the customer. Understanding the relationship between CTQs, functions, and design elements helps in establishing clear requirements for the critical process elements.

**Design reviews.** It is critical for any good design process to have a proper screening system. Design reviews are a common and very effective approach for ensuring attention to details and avoiding any obvious errors. In most organizations, there is a well defined phase gate review process for the development of new products or services. The idea is: As you move from opportunity to concept to detail and final design, at every stage, minimum corporate requirements, industrial and governmental standards, and societal expectations must be met before proceeding further. Typically, the design team must answer a checklist of questions, and approval must be obtained from a cross-functional team of experts as well as from the management of the company.

**FMEA.** FEMA is a methodology for assessing and minimizing risk in new product or service designs. It is a systematic, team-based approach. FMEAs should establish the following:

· Relative priority among all potential risk factors
· Risk mitigation action plans, for all the high risk factors
· Responsibilities for each action

The FMEA approach requires a critical review by a cross-functional team of all facets of service including process, customer requirements, cost targets, capital investments, market assumptions, legal issues, and resource requirements. It requires the assessment of the chances of occurrence, severity, and detectability for each potential risk. A combined scale for severity, detection, and occurrence is referred to as a risk priority number (RPN). RPN is used for establishing relative priorities. Refer to Chapter 8 for a detailed discussion on FEMA.

## OPTIMIZE

Once the details of a design are available, the process of optimization can be started. In general, any optimization requires a definition of objective, decision parameters, and constraints. The objective of the axiomatic approach to service design is to maximize value to the customer. Constraints are the limits imposed on design options. Decision parameters are those process variables that impact objective function or the constraints and are not fixed. Once we understand the objective, design parameters, and constraints for any given problem, the next step is to formulate a model. Depending on the model, we can select the solution methodology. In most situations, development of the model and selection of the solution methodology will be an iterative procedure. Finally, we need to solve/optimize the model using selected methods. We must understand that the solution of the model need not be the one correct solution for the problem at hand.

In DFSS methodology, the optimize step involves value optimization, validation, and final launch readiness. Optimization may be done by developing a mathematical model and solving it using operations research techniques such as simulation. In some cases, optimization may involve only logical analysis and review of the service design to ensure its inherent robustness and to avoid common pitfalls. Value stream mapping is a good tool for the logical analysis of a process and eliminating wastes. Taguchi Robust Design methodology is a good way of increasing robustness in the service process.

**Simulation.**    Simulation is a technique in which an actual or proposed system is substituted by an abstract model. The simulation model should have the same cause and effect relationship as the system under study. Once an abstract (physical or mathematical) model is developed, it can be used to conduct the experiment and predict the behavior of the system under all or most of the expected situations.

**Scorecard.**    The scorecard is a tool for collecting and analyzing the facts about the service process design performance. The scorecard helps in predicting future performance and optimizing the process. In a scorecard, we are comparing the voice of the customer and the voice of the process to see how we are doing.

**Validation.**    Validation is the final stage of any design process and, if we have been doing a good job in the previous stages, the design changes required at this point should be minimal. While validating, we must focus on the customer usage environment and other noises. A pilot run of the process should validate and verify that the process will meet customer requirements, as defined in the identify phase, and functional requirements, as established in measure phase. The objective of the validation is to ensure that the design provides maximum value to the customer and satisfies all constraints (e.g., cost and wait time), as well as safety and other standards. Finally, it is also important to verify that the final package confirm to the strategic intent of the organization. Validation of the service process may be done either using a simulation model or by pilot runs.

## KEY TAKEAWAYS

Albert Einstein once said, "The whole science is nothing more than a refinement of everyday thinking." The approach introduced in this chapter is also a refinement of everyday or common tools, concepts, and methods used in industry for designing products and services. The following are the key takeaways from this chapter:

· Understanding of DFSS for services and why is it important to use DFSS methodology to design service.
· The axiomatic approach to design as proposed by Suh and following two axioms of service design:
  1. Service interactions should be designed to generate or exchange quality information.
  2. Transactions should process customer interactions efficiently and securely.
· The second axiom implies that if a transaction in business is directly or indirectly not processing output from the customer interaction, it should be eliminated
· Figure 10.4 provides the summary of the DFSS methodology—define, measure, analyze, design, and optimize—and its supporting tools.

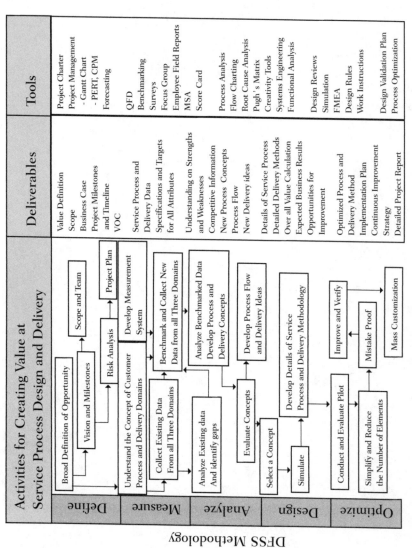

**FIGURE 10.4.** Activities, Deliverable, and Tools for DMADO

# CUSTOMER DRIVEN TRANSACTIONAL PROCESSES

Design for Six Sigma (DFSS) can be seen as a structured process of applying various quality and reliability improvement tools and techniques. DFSS offers a new approach to total quality management with a more integrated approach to ensure customer focus and financial accountability for investment in improvement. A core principle of DFSS is to link customer requirements to product/service characteristics by developing an appropriate transfer function $(y = f(x))$ and then cascade down to a lower level to identify suitable service functions or characteristics. DFSS captures customer needs and organization vision upstream to recognize that decisions made during the design phase profoundly affect the quality and cost of all subsequent activities used to build and deliver the service.

Traditional approaches to assuring quality and reliability often focus on work standards, automation to eliminate human error-prone processes, frequent inspections, and quality improvement teams to empower employees to resolve problems. However, in the present scenario of global competition and changing customer needs and requirements, the traditional approaches are not enough even to survive in the market. With traditional systems, the best you can get is *nothing wrong*—which is no longer good enough. In addition, to eliminate negative quality, we must also maximize positive quality, which creates value and leads to customer satisfaction, and in some cases, customer delight.

To effectively deliver value to customers, it is necessary to understand the customer needs, uses, behavior, and essentially listen to the voice of the customer (VOC) throughout the service development process. Because customers are not apt to voice their requirements properly, it is the responsibility of the organization to explore customer problems and opportunities to build positive quality into their product or service. Quality function deployment (QFD) is the only comprehensive quality system tool aimed specifically at satisfying the customer. It helps organizations to maximize customer satisfaction by capturing both spoken and unspoken needs, translating these into actions and designs, and communicating these end-to-end throughout the organization. Further, QFD provides a powerful mechanism to prioritize customer needs, benchmark them against competitors, and then give direction to optimize those aspects to service and process that will differentiate the organization from competitors and have the greatest edge in the market.

The aim of this chapter is to introduce the application of a few VOC analysis tools and techniques in transactional processes or the service industry to support DFSS efforts. This may well be the most important chapter in this book from the standpoint of the successful implementation of the DFSS approach. This is so because we discuss the following systematic process of capturing VOC, which is essential to building positive quality in the transactional process or service:

· Understanding customer needs and trends (consumer insight)
· Identifying the new opportunities
· QFD in transactional business

One's ability to discard the traditional philosophy of quality improvement and embrace the new approach to building positive quality into the service and transactional process depends entirely on how well these three concepts are understood.

# UNDERSTANDING CUSTOMER NEEDS AND TRENDS

The "voice of customers" is the term to describe stated and unstated customer needs or requirements. The VOC is a strong bias to keep consumers' interest the first priority in designing services that exceed consumer expectations and deliver maximum value to the customer. To deliver maximum value to the customer, it is very important to go much deeper into the consumer's motivations and psychology and get a complete understanding of consumer insight. One way of finding consumer insight is to identify those things that elicit extreme emotional reactions.

Consumer data are foundational consumer information that can help you to build and describe VOC. The VOC is captured in a variety of ways: direct discussion, surveys, customer specifications, observation, warranty data, field reports, etc. This understanding of customer requirements is then summarized in a product-planning matrix or "house of quality."

## CAPTURING THE VOICE OF CUSTOMER

A company doesn't blindly respond to customer needs and opportunities. A business strategy that defines customers and markets to be served, competitors, and competitive strengths provides a framework from which to evaluate potential opportunities. As customers and market segments are identified, appropriate techniques are used to capture the VOC. The techniques used will depend on the nature of the customer relationship, as illustrated in Figure 11.1.

In transactional processes or the service industry, there is no one monolithic voice of the customer. Customer voices are diverse, which leads to a variety of different needs and requirements. Even within one market segment, there are multiple

← ────────────────────────────────────────────────── →

| Direct Relationship with Customers | Indirect Relationship with Customers |
|---|---|
| Relatively few customers | Relatively large number of customers |
| Direct business link with customers | Indirect link (through distributor and retailers) with customers |
| Approaches to collect information<br>— Requirements document, specification<br>— Contract order<br>— Customer meetings<br>— Warranty and repair data<br>— Customer representative | Approaches to collect information<br>— Surveys<br>— Focus groups<br>— Market research<br>— Interviews<br>— Customer service feedback |

**FIGURE 11.1.** Customer Relationships and Requirements Definition

customer voices (e.g., children versus parents). This applies to industrial and government markets as well. These diverse voices must be considered, reconciled, and balanced to design a truly successful transaction process or service.

The interface with the customer should not be the exclusive responsibility of the sales, marketing, or customer service organization. Other functional disciplines involved in service development should also be exposed to customers and establish relationships to facilitate communication. It is important for the individuals involved in capturing VOC to know where the service or product actually gets used, who the customers are, what their culture is, and what kind of service they are looking for, etc. This ensures that service design and development personnel have taken care of a wide variety of perspectives on customer needs, have developed greater empathy about market, and have minimized hidden knowledge and technical arrogance. In chapter 6 we descried SERVQUAL model, that helps in understanding various dimensions of service, in this chapter we will discuss a concept called GEMBA for capturing VOC.

## GEMBA: THE SOURCE OF THE VOICE OF CUSTOMER DATA

The previous section highlighted the importance of getting close to the customer and developing greater empathy with the market and customers. In reality, the selected market segments and customers are the true source of information. The Japanese have coined a word to describe the true source of information—they call it the "gemba." The gemba is where the service becomes of value to the customer. Unlike other customer information-gathering techniques, such as focus groups, here individuals who are involved in capturing VOC employ all their senses to work by using contextual inquiry, video taping, audio taping, direct observation, direct interviewing with the customer, etc. The essence of this exercise is to get complete understanding and knowledge of consumer insight and to understand how to design better service to provide maximum value to the customer.

Going to the gemba requires planning and tools to collect relevant information. An analysis of what is observed at gemba can clarify unspoken opportunities for new service design. Several data collection tools and techniques are available and have been discussed with great detail in the literature. The transactional process table, developed by Mazur (1995) to assist those going to gemba for information collection and organization, is shown in Table 11.1.

TABLE 11.1    Transaction Process Table

| TRANSACTIONAL PROCESS | SERVICE SCENARIO | PROBLEMS AND OPPORTUNITIES | FAILURE MODES |
|---|---|---|---|
| Draw the transactional process flow diagram | Describe service scenario under consideration | Identify potential process failures (what might go wrong) and opportunities for improvement | Describe the failure modes. How problems occur |

Once the customers' needs and requirements are gathered, they then have to be organized. The collection of interview notes, observations, requirement documents, market research, and customer data needs to be distilled into a handful of statements that express key customer needs. The following two tools are most frequently used to organize the gathered information:

- Affinity diagram
- A data dictionary

In addition to "stated" or "spoken" customer needs, "unstated" or "unspoken" needs or opportunities should be identified. Needs that are assumed by customers, and therefore not verbalized, can be identified through preparation of a *function tree*.

The next section describes how to classify the customer needs and requirements to identify the potential opportunities for improvement.

## IDENTIFY THE NEW OPPORTUNITIES

The outcome of the VOC capturing exercise could be a long list of identified customer needs and requirements, and it is very difficult to satisfy all customer requirements. Therefore, to be competitive in the market and achieve maximum customer satisfaction, we must understand how meeting these requirements affects customer satisfaction. Fortunately, you don't have to work on all the customers' requirements to satisfy them. You simply need to understand the relative effect of improving certain types of requirements for customer satisfaction and classify them into different categories. This classification helps in identifying potential requirements to work on to increase value to the customer and customer satisfaction. Many methods are available for investigating the characteristics of customer requirements; for example, one can ask customers to rank-order them. The particular method we will discuss here is based on the work of Professor Noriaki Kano of Tokyo Rika University.

## THE KANO MODEL

Many previous definitions of quality were linear and one-dimensional in nature (i.e., good or bad, small versus large loss to society). Dr. Kano defined quality along two dimensions, which he partially derived from his study of Herzberg's "Motivator-Hygiene Theory." These two dimensions are: the degree to which a customer requirement is met or service performs, and the degree to which the customer is satisfied, as shown in Figure 11.2. This two-axis plot provided an opportunity to classify customer requirements in a more sophisticated and holistic manner. The correlation of quality on two axes led Dr. Kano to three unique definitions of quality, as well as customer requirements, namely: must-be requirements, one-dimensional or performance requirements, and excitement or attractive requirements.

**The three types of requirements.** In his model, Kano (1984) distinguishes amongh three types of product requirements that influence customer satisfaction in three different ways when met. If the level of customer satisfaction is plotted on a vertical axis and the degree that service has achieved a given performance attribute is on the horizontal axis, different types of customer wants and needs can be shown to cause different responses, as shown in Figure 11.3.

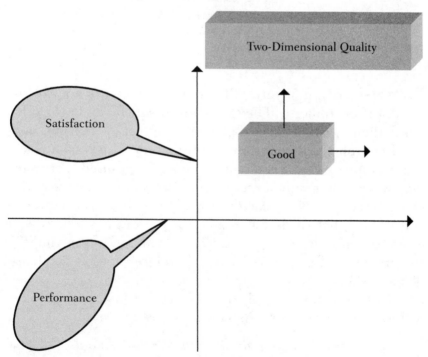

**FIGURE 11.2.** Two-dimensional Quality Definition

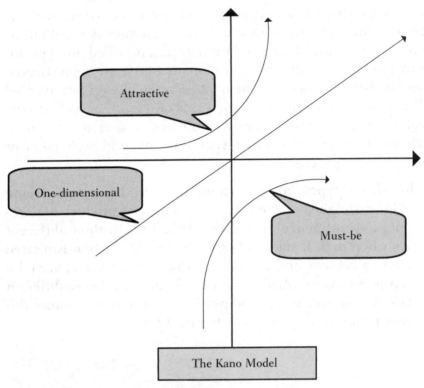

**FIGURE 11.3.**    Three Types of Customer Requirements

*Must-be requirements.*    If these requirements are not fulfilled, the customer will be extremely dissatisfied. On the other
hand, their fulfillment has only limited effect (or in some cases
no effect) on causing customer satisfaction. The reason is that
the customer takes these requirements for granted. For example, when going into a restaurant for dinner, the customer
expects there to be a place to sit. If there isn't, the customer
will be dissatisfied, but if there is a place, no credit will be
given because it is expected. The customer takes must-be
requirements for granted and therefore doesn't explicitly
demand them. Must-be requirements are in any case decisive
competitive factors, and if they are not fulfilled, the customer
will not be interested in the service or product at all.

*Performance or one-dimensional requirements.*    These
requirements generate customer satisfaction proportional to

the level of performance of the service—the higher the level of fulfillment, the higher the customer's satisfaction. The customer explicitly demands one-dimensional requirements. The customers in a restaurant expect their order to be taken promptly and accurately, and the food to be delivered in a reasonable period of time. The better the restaurant meets these needs, the more satisfied is the customer.

*Attractive requirements.* These requirements generate positive customer satisfaction at any level of execution. Attractive requirements are neither explicitly expressed nor expected by the customer. Fulfilling these requirements creates excitement, but if they are not met, however, there is no feeling of dissatisfaction. For example, if the restaurant provides a glass of champagne "on the house," the customer will be pleasantly delighted.

**The advantages of requirements classification.** The following are the clear advantages of classifying customer requirements by means of the Kano model.

- Service requirements are better understood. The service criteria that have the greatest influence on the customer satisfaction, can be identified.
- Priorities for service development. It is not very useful to improve upon must-be requirements that are already at a satisfactory level.
- Potential opportunities for improvement are better identified. This helps in selecting potential requirements that can lead to maximum value creation and customer satisfaction.
- The Kano model for customer satisfaction can be optimally combined with quality function deployment (QFD).
- The Kano method provides valuable help in trade-off situations in the transactional process design.
- Discovering and fulfilling attractive requirements creates a wide range of possibilities for differentiation.

The next section discusses the classification of customer requirements by means of developing a Kano model and questionnaire.

## STEPS TO DEVELOP KANO MODEL FOR REQUIREMENT CLASSIFICATION

**Step 1: Identification of customer requirements: "Going to Gemba."**   The initial input to construct a Kano model and questionnaire is the list of all service requirements customers are looking for. The section on Understanding Customer Needs and Trends gives a detailed discussion on different approaches to determine customer requirements. The questionnaire needs to be developed to ascertain the category of customer requirements determined during this step.

**Step 2: Develop the questionnaire.**   To construct the questionnaire, formulate a pair of questions for each potential customer requirement. The first question concerns the reaction of the customer if the service/product has that feature (functional form of the question); the second concerns the customer's reaction if the service/product does not have the feature (dysfunctional form of the question). The following is an example of the pairs of functional and dysfunctional questions for a particular customer requirement:

By combining two questions in the following evaluation table (Figure 11.5), the customer requirements can be classified.

Based on the responses of the two parts of the question in Figure 11.4, the customer requirements can be classified into one of six categories:

· A = Attractive
· M = Must-be
· O = One-dimensional
· I = Indifferent
· R = Reversal
· Q = Questionable

| If the low carbohydrate food is available during lunch, how do you feel? | 1. I like it that way.<br>2. It must be that way.<br>3. I am neutral.<br>4. I can live with it that way.<br>5. I dislike it that way. |
|---|---|
| If the low carbohydrate food is not available during lunch, how do you feel? | 1. I like it that way.<br>2. It must be that way.<br>3. I am neutral.<br>4. I can live with it that way.<br>5. I dislike it that way. |

**FIGURE 11.4.** A Pair of Questions in Kano Questionnaire

| Customer requirements | | Dysfunctional (negative) question | | | | |
|---|---|---|---|---|---|---|
| | | 1. like | 2. must-be | 3. neutral | 4. live with | 5. dislike |
| Functional (positive) question | 1. like | Q | A | | A | O |
| | 2. must-be | R | I | I | I | M |
| | 3. neutral | R | I | I | I | M |
| | 4. live with | R | I | I | I | M |
| | 5. dislike | R | R | R | R | Q |

Customer requirement is ...

A: Attractive
M: Must-be
R: Reverse

O: One-dimensinoal
Q: Questionable
I: Indifferent

**FIGURE 11.5.** Kano Evaluation Table

Depending on each customer's response to a particular question, one determines into which category a given customer requirement falls, as shown in Figure 11.5. For example, if the customer answers "1. I like it that way" about "low carbohydrate food available during lunch," the functional form of the question, and "5. I dislike it that way" about "low carbohydrate food not available during lunch," the dysfunctional form of the question, and we look at the intersection of the first row and the fifth column and find an O, this indicates that availability of low carbohydrate food for lunch is a one-dimensional customer requirement from the customer's point of view.

It is always advisable to test the questionnaire before sending it to the customer to make sure it is understandable. A test run will help identify ambiguity in wording, typographical errors, or unclear instructions. Refining the questionnaire may require a couple of iterations, but may reduce overall time of data collection.

**Step 3: Collect the data by administering customer interviews.** There are different methods available for collecting data by carrying out the customer interviews. The most favorable method is sending the questionnaire by mail because of the relatively low cost advantage, but frequently the low return rate is a major disadvantage with this method. We recommend personal interviews for Kano surveys. The standardized questionnaire reduces the chances of influence through the interviewer, and the return rate is pretty high.

**Step 4: Evaluation process.** Once the Kano survey has been completed, collect all the questionnaire forms, tabulate them by looking up the classification of each customer requirement on each questionnaire in the Kano evaluation table, and tally it in the appropriate place in the row for that particular requirement on a Kano questionnaire tabulation form, as shown in Figure 11.6.

The results of the individual customer requirement are recorded in the table of results (see Figure 11.7), which shows the overall distribution of the requirement categories. For each row of the table of results—that is, for each customer requirement—the dominant category is selected by the highest tally or frequency of responses. There are various possible analysis approaches but a simple way to classify the requirements is to score each according to the most frequently occurring catego-

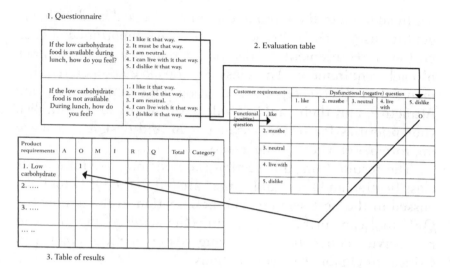

**FIGURE 11.6.** Evaluation Process

| Product requirements | A | O | M | I | R | Q | Total | Category |
|---|---|---|---|---|---|---|---|---|
| 1. Low carbohydrate | 4 | 12 | 35 | 1 | 0 | 1 | 53 | M |
| 2. Low price | 8 | 32 | 13 | 0 | 0 | 0 | 53 | O |
| 3. Fresh packed | 12 | 15 | 21 | 2 | 1 | 2 | 53 | M |
| ..... | | | | | | | | |
| | | | | | | | | |

**FIGURE 11.7.** Table of Results

ry in each row of the table of results. For example, for a requirement for low carbohydrate food, 65% respondents voted as Must-be and 15% respondents voted as Attractive. Based on this result the requirement can be classified as a Must-be requirement.

In addition to the Kano questionnaire, it is also advisable to get the customer's feedback in terms of the importance of each customer requirement to determine the relative importance of each requirement. The classification of customer requirements into three categories: Must-be, One-dimensional, and Attractive, and their ranking order, will help you to establish your priorities for service development and design and make improvements wherever necessary.

The results from this section (i.e., customer requirements classification and their rank order) are fed to the QFD tool discussed in the next section. The next section explores how the QFD tool can further help identify the service functions and new service concepts by cascading customer requirements to Critical to Quality service functions.

## QFD IN TRANSACTIONAL BUSINESS

To efficiently deliver value to customers, it is necessary to listen to the "voice" of the customer throughout the service development process. Quality function deployment (QFD) is the comprehensive quality system aimed specifically at satisfying the customer. It focuses on delivering maximum value by translating both spoken and unspoken needs into actionable service functions and communicating this throughout the organization. Furthermore, QFD takes prioritized customer requirements from the previous analysis and tells us how we are doing compared to our competitors. It then directs us to optimize those aspects of our service that will bring the greatest competitive advantage. QFD helps assure that expected requirements don't fall through the cracks and points out opportunities to build in excitement.

While the QFD matrices (house of quality) are a good communication tool at each step in the process, the matrices are the means and not the end. The real value is in the process of communicating and decision-making with QFD. The house of quality is a kind of conceptual mapping that provides the means for interfunctional planning and communication. It is oriented toward involving a team of people representing the various functional departments that have involvement in service development. The active involvement of all the concerned departments can lead to a balanced consideration of the requirements or "what's" at each stage of this translation process and provide a mechanism to communicate hidden knowledge—knowledge that is known by one individual or department, but may not otherwise be communicated through the organization.

## HOUSE OF QUALITY

Once potential customer needs are identified and prioritized, preparation of the service planning matrix or "house of quality" can begin. The sequence of preparing the house of quality matrix is as follows:

1. Customer requirements are listed on the left side of the matrix, as shown in Figure 11.8. The house of quality matrix begins with customer requirements (voice of customers) obtained by applying consumer insights and is classified using the Kano model to identify potential improvement opportunities. These requirements answer the basic question: What do customers want? If the number of needs or requirements exceeds 20 to 30 items, decompose the matrix into smaller modules or subsystems to reduce the number of requirements in a matrix.

2. Enter the rate of importance in the column "rate of importance" next to the customer requirements column. The importance rating is derived based on the feedback from the customer, as discussed in the previous section. The importance rating for each requirement can also be derived internally based on team members' direct experience with customers or on surveys.

3. Perform performance benchmarking on competitive services for each requirement. Comparison with competitors can help identify opportunities for improvement. Assign the performance ratings on a scale of 1 to 5, where 1 corresponds to a very bad performance and 5 to a very good per-

formance. List these assessments on the right-hand side of the house of quality in the column under "performance benchmarking." This assessment answers the very pertinent question: Will delivering perceived needs yield a competitive advantage?

4. Establish service functions or characteristics essential to fulfill customer requirements. The service function/characteristic should be meaningful and measurable and should directly affect the customer perception. Along the top of the house of quality, the process design team should list those service characteristics that are likely to affect one or more customer requirements. The sign (negative or positive) of each service characteristic shows the direction of improvement. The negative sign for any service characteristic indicates improvement on reduction of that particular measurable value.

5. Complete the relationship matrix, the body of the house, indicating how much each service characteristic influences each customer requirement. The evaluation should be based on the consensus among the team members based on their experience, customer responses, and available data. The teams can use numbers or symbols to indicate the strength of influence on each customer requirement in the relationship matrix. The final relationship matrix would show multiple relationships and the relative strengths of those relationships as well.

6. Carry out technical benchmarking, assessment of service functions, or characteristics on the competitors' service functions. Assign the assessment ratings on the scale of 1 to 5, where 1 corresponds to a very bad performance and 5 to a very good performance. List these assessments at the bottom of the house of quality in the row under "technical benchmarking."

7. Establish preliminary target values for each service function or characteristic. It will provide the direction and base in assessing the performance of any concept against its ability to meet the target values.

8. Complete the correlation matrix at the top of the house of quality by determining potential positive and negative interactions between service functions or characteristics using symbols for strong or medium, positive or negative relationships. Too many positive interactions suggest potential redundancy in "the critical few" service requirements or technical characteristics. Focus on negative interactions—consider product concepts or technology to overcome these potential trade-offs or consider the trade-offs in establishing target values.

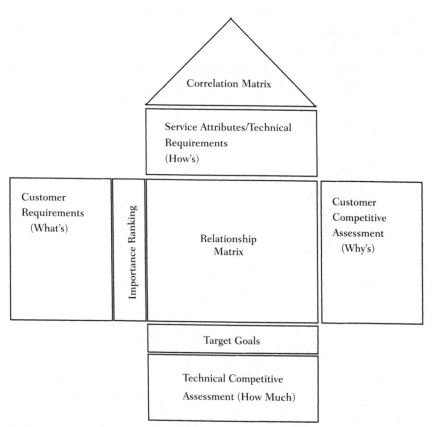

**FIGURE 11.8.** The House of Quality

## QFD PROCESS

How does the QFD help in integrating or linking the VOC into service characteristics and developing new service design concepts? As stated earlier, QFD provides a very powerful mechanism to the DFSS approach to link customer requirements to service characteristics and hence develop an appropriate transfer function. The different phases of QFD and their connectivity build a process to cascade and link business goals and customer requirements to lower most levels of service function and characteristics. The principles underlying the house of quality apply to any effort to establish a clear relation between those service functions and customer requirements that are not easy to visualize. Figure 11.9 demonstrates the mechanism to link and cascade upper level requirements (Y's) to lower level characteristics (X's), and this process continues until the lowest level is reached. These different phases of QFD implicitly convey the VOC through to the service process.

The major advantage of the QFD process is that it relieves no one of the responsibility of making tough decisions. It pro-

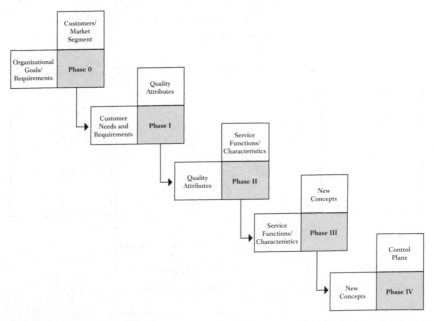

**FIGURE 11.9.** QFD Process

vides the mechanism and means for participants to debate priorities. The principal benefit of the QFD process is quality inhouse. It gets people thinking in the right directions and thinking together (Hauser and Clausing, 1988). For DFSS implementation, the utility of the QFD process has proved to be a quiet revolution.

# A SUGGESTED METHODOLOGY OF INTEGRATING VOICE OF CUSTOMER IN SERVICE

In the earlier sections, we briefly described some of the important tools that are being extensively used to identify the customer requirements, categorize them, and finally identify the service functions or characteristics needed to satisfy those requirements while meeting organizational goals. These tools are well established and very powerful if used effectively and appropriately in combination to capture and integrate VOC in service design and development. We propose a systematic application of these tools to build the VOC in the transactional process or service. To demonstrate the methodology, we will use a case study, described in the next section.

## CASE STUDY

A company, Wholesome Food and More (WFM), has identified an opportunity in the area of freshly packed nutritional meals delivered conveniently to customers. Market studies have revealed a group of potential customers not satisfied with their current choice of meals at work or school.

WFM is a small company with three brick and mortar health food stores and an Internet-based food store. The company was started by two individuals (John and Smith) five years ago with an Internet-based health food store. They later added two stores in strip malls, and they opened a third store six months ago in an upscale mall. All three stores are in one metro area. Wholesome Food has total sales of $2 million per year, of which $600,000 is from the Internet-based business.

Both earlier stores have annual sales of approximately $500,000 each, and in last six months, the store from the mall has generated approximately $200,000 in revenue. In the last four years, business has grown by approximately 20% each year, and the current profit margin is at 8%. WFM has spent 2% of the sales every year in marketing and promotion, and now has an established brand name in the metro area, with a total population of 10 million.

WFM has only two salaried employees (Mark and Julie) in addition to the owners (John and Smith). Each of the three stores has a sales clerk as an hourly paid employee and the new store also has an additional assistant. Julie has the responsibility of managing Internet store orders, as well as updating the website. Mark has the responsibility of purchasing and inventory for all four businesses. John takes care of finances, and Smith has the responsibility of marketing. Both John and Smith spend at least five hours in stores every week to supervise the operations. WFM also has hired a professional accountant, lawyer, nutritionist, and health food expert as consultants on the basis of fixed cost for services rendered. Last year, WFM spent $10,000 on professional services.

WFM has the vision to double the current revenue and increase the profit margin to 10% in the next two years. WFM would like to be recognized as the brand for nutritional, tasty, and affordable food. In the next two years, WFM wants to stay in the metro area; however, its long-term vision is to be a public company and an internationally recognized brand.

To realize its vision for growth and profitability, it is planning the following two activities:

- Start a new service of freshly packed nutritional meals delivered conveniently to the customers
- Growth by opening franchised stores

WFM decided to use DFSS to design these services and hired a Six Sigma consultant to help first design the new service of freshly packed nutritional meals, which it plans to launch in the next six months. WFM plans to design and launch its franchised operations next. WFM has a budget constraint of $1 million to launch this new service, which it expects to borrow from a bank at an 8% annual interest rate.

The following steps demonstrate the application of the suggested methodology of Integrating VOC in service:

1. Define and Prioritize Business Goals

   It is always essential to clearly define the business/organizational vision and goals. The well-defined vision and goals provide clear direction to ensure complete alignment of cross-functional activities and allocation of resources. All other "down the road" activities should support the business goals and vision. That means it is important to keep in mind these goals while designing the rest of these activities. The following are major goals of WFM:

   · Double the current revenue

   · Increase the profit margin to 10% in the next two years

   · Create a brand name for nutritional and affordable food

   · Stay in the metro area for two years

   · Get internationally recognized

2. Define and Prioritize Customer Segments (Markets) Based on Critical Business Goals

   Select the market segment(s) that provide potential opportunities to realize the organizational goals while serving market needs. Decide which market segment will help the company to achieve its goals and will provide opportunities to continuously build on the current business. It is a long-term decision and, therefore, important to warn that while selecting potential market segments, current and future business opportunities should be major decision criteria rather than looking for current opportunities only. For the WFM scenario, different market segments were considered, and finally it was decided to start from the metro area to initially establish the business (see Figure 11.10).

|  |  | Market Segments | | | |
|  |  | Metro area | Rural area | Launch internationally | Mixed market |
| Organization Goals |  |  |  |  |  |
| Double the current revenue |  | ◎ | △ | ○ |  |
| Increase the profit margin |  | ◎ |  | ○ |  |
| Brand name for nutritional and affordable food |  | ◎ | ◎ |  |  |
| Stay in metro area for 2 years |  | ◎ |  |  | ○ |
| Get internationally recognized |  | ○ |  | ◎ |  |

FIGURE 11.10. Mapping of Business Goals and Market Segments

### 3. Collect Customer Requirements Data: Visit Gemba

After selecting the market segment, the next step is to find customer needs and requirements in that particular segment of the market. Analyze customer behavior and unspoken needs and opportunities for new service development. The preceding section provided details on gathering information regarding customer requirements and expectations. This information (customer needs and requirements) will be a major input in the QFD model to select the appropriate service concept and finally design it. A sample of selected customer requirements is shown in Figure 11.11, which also identifies appropriate service attributes (critical to quality service parameters) for customer satisfaction.

### 4. Classify the Customer Needs and Requirements: Kano Model

It is important to understand how meeting these customer requirements will affect overall customer satisfaction. This understanding will help identify opportunities to incorpo-

| Customer Requirements | Service Attributes | | | |
|---|---|---|---|---|
| | Ingredient and spice/variety of foods | Price less than $5 | Service within 15 minutes/door-to-door delivery | Within the reach of large community |
| Nutritional and tasty food | ◎ | △ | ○ | |
| Affordable price | | ◎ | ○ | |
| Quick service response and convenient | | | ◎ | ◎ |
| Variety of foods | ◎ | | | ○ |
| Door-to-door service | ○ | | ◎ | |

FIGURE 11.11. Relationship Matrix for Customer Requirements and Service Attributes

rate in new service design or improve upon the existing one. The preceding section discusses in great detail the classification of customer needs in three different categories (i.e., must-be, one-dimensional, attractive requirements) using the Kano model approach and gives a systematic approach to classifying these requirements.

## 5. Prioritize the Customer Needs

Not all customer requirements, identified earlier, are equally important to customers to satisfy their needs. Therefore, to maximize customer satisfaction, it is essential to identify the one customer requirement that is more important to customers than other requirements. This comparison will help to prioritize all customer requirements in the order of importance and rank them accordingly. This ranking order will help to establish your priorities for service development and design and make improvements wherever necessary.

6. Translate Customer Needs into Service Functions and Solutions: QFD

Once you have identified the customer requirements and prioritized them, it is time to start looking for solutions to satisfy these requirements. As mentioned earlier, in the Six Sigma approach, it is important to develop a linkage between requirements (Ys) and service attributes (Xs) to ensure that we are focusing on the right cause. QFD provides a very simple and effective mechanism to carry out this linkage and cascading process until the end. Other tools are available and can be used for the same purpose. Here we will use QFD to explain this process, but it is not limited to QFD, and individuals are encouraged to explore the application of other tools as well. Rather than going into a very detailed and complex process, we would like to demonstrate this process in a very simple way. We will take the same WFM case study and show the complete process of linking the requirements from a higher level to a lower level.

I. Selecting the market segment to achieve given business goals: The market segment "metro area" seems to have more potential to achieve organizational goals and hence is selected for further analysis. Figure 11.10 gives the mapping of organizational goals and potential market segments.

II. Identify and prioritize the customer requirements for the selected market segment: Figure 11.11 gives the list of customer requirements (left column) for the selected market (metro area) and is prioritized in the order of importance to customers. This list is prepared for a given market and prioritized for the given market segment during steps 3 through 5. The columns represent the service attribute essential to fulfill these requirements. The impact of each service attribute on customer requirements is indicated in the matrix with the help of symbols as very strong (◎), strong (○), and mild impact (△). This helps in developing a relationship between the customer requirement and the different service attributes, and facilitates the cascading process.

III. In the next step, we pick one of the service attributes (service within 15 minutes and door-to-door delivery) to identify different service elements (functions or characteristics) required to achieve it. Figure 11.12 shows the various service elements required to build the given service attribute into the transactional process.

IV. We further take these service elements to develop different concepts, which can integrate various service elements and deliver value to customers. Here value means fulfilling those customer requirements identified earlier while supporting organizational goals. The best concept is selected for detailed design and optimization (see Figure 11.13).

V. Develop detailed work instructions and a control plan for the selected design concept. This will help to continuously monitor the performance of the transactional process and will provide a mechanism to control and improve processes as and when required (see Figure 11.14).

| Service attributes | Service Elements | | | | |
|---|---|---|---|---|---|
| | Scheduling | Transport service | Flexible food processing | Training and consulting unit | Order receiving and processing section |
| Service within 15 minutes | | △ | ◎ | ○ | ◎ |
| Door-to-door delivery | ◎ | ◎ | ○ | △ | ◎ |
| ... | | ◎ | | | |

FIGURE 11.12. Mapping of Service Attributes and Service Elements

| | | Service Design Concepts | | | | |
|---|---|---|---|---|---|---|
| | | Concept A | Concept B | Concept C | ⋮ | ⋮ |
| Service elements | | | | | | |
| Order receiving and processing unit | | △ | ◎ | ◎ | ◯ | ◎ |
| ... | | ◯ | ◎ | ◯ | △ | ◎ |
| ... | | | ◎ | | | |

**FIGURE 11.13.** Mapping Matrix for Service Elements and New Concepts

| | | Control plans | | | | |
|---|---|---|---|---|---|---|
| | | Detailed work instructions | Transition plan | Training plan | Communication plan | Order receiving and processing plan |
| Design concept | | | | | | |
| Concept B: Offer service of complete diet planning and provide home delivery with guaranteed food quality, as in Concept A. | | ◎ | △ | ◎ | ◯ | ◎ |
| ... | | | | | | |

**FIGURE 11.14.** Mapping Matrix for Design Concepts and Control Plans

The preceding substeps describe the process of capturing and integrating VOC into the transactional process or service development. Each step (phase) takes input information from the preceding phase, analyzes it, and generates output to feed to the next step (phase). This process continues until the last phase. Figure 11.15 shows the complete process of VOC, capturing and integrating it into the service/transactional process.

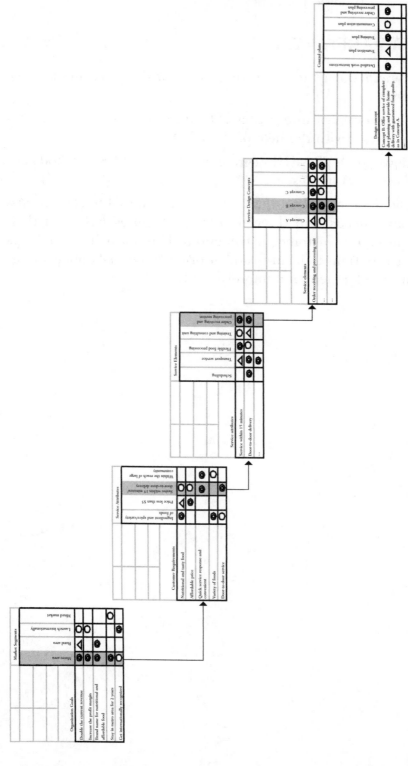

**FIGURE 11.15.** QFD Process to Capture VOC in Transactional Process

327

## KEY TAKEAWAYS

- A methodology for capturing VOC for the given market segment.
- The process of integrating VOC into the transaction process or service with the help of the QFD.
- How QFD phases integrate with the DFSS methodology (DMADO) phases.
- The Kano model provides a very powerful tool to classify customer needs into three different categories based on their impact on customer satisfaction. This classification helps organizations to identify and prioritize potential opportunities for development and improvement.

# DESIGNING TRANSACTIONAL SERVICES FOR SIX SIGMA

Transactional service design is an iterative team-oriented activity to accomplish the business objectives of growth and profitability. This business objective can only be met by creating service interactions and transactions such that the customer is consistently satisfied despite individual preferences and differences. Design for Six Sigma (DFSS) for transactional services relies on the efficient and effective design of interactions and transactions, which together takes care of information generation and processing aspects of the service. Interactions need to be designed so that it allows creativity and balances details with flexibility and accountability. Transactions need to be designed to reduce complexity.

The development of a new service is iterative in nature, but it should not be trial and error. Designing services should be based on axioms and by following the systematic methodology of define, measure, analyze, design, and optimize (DMADO). Everyone understands the need for listening to the voice of customer (VOC), but the difficult task is to ensure the continuous deployment of VOC through the design process. A systematic methodology will ensure the end result is a service that is complete and fulfills the business objectives. Customer sat-

isfaction and keeping every customer the company can profitably serve are the critical factors for success in the service
industry. Reichheld and Sasser (1990) claim in their article
that companies can boost profits by almost 100% by retaining
5% more of their customers. This chapter will go into the
details of the first three steps of the DMADO methodology.
Chapter 13 will discuss the last two steps, design and optimization, including validation aspects.

# DEFINE

The proper start of any project is important for its success.
However, it is critical in the case of a design project. The first
phase of our recommended methodology for transactional service design is "define." The objective of this phase is to provide a
good start by creating a focused definition of the opportunity, a
winning team, and a feasible project plan. At the very beginning,
the project team needs to answer a few basic questions:

· Why do we want to design this new service?
· What is our vision?
· Who is the customer?
· How does it differ from existing alternatives in the market?
· Where will the resources come from?
· When do we want to start and launch (i.e., start and finish dates)?

We recommend using two documents to capture the
answers for these questions: the project charter and the project
plan. The project charter is created to define the opportunity,
vision, risk, team, and stakeholders. The project plan is the
document that will define milestones, tasks, and schedule and
assign resources to achieve objectives.

## PROJECT CHARTER

Figures 12.1 to 12.4 show various sections of a blank project
charter. The project team needs to use the charter as a guiding

document that will help them in answering a few basic questions, which must be answered before the start of a project. The first section of the charter is for identification purposes, The second section defines the opportunity and vision. The opportunity description should be as specific as possible. It should give enough objective information to justify why we are doing this project. Vision provides the link from this project with the strategic priorities of the company. The third section describes the business case, scope, and constraints of the project. The business case provides the justification for spending the company's resources on this project. The scope establishes a reasonable boundary for the work. The fourth and final section describes the timeline and metric/goals.

**First Section of project charter.** The first section has two subsections; the first has identifiers such as project title number, sponsoring department, etc. The second subsection is about the team and stakeholders. It gives important contact information and makes it clear who has what responsibility. Each role has some critical functions that must be performed to complete the project successfully. It is not essential that every role be performed by a different individual. For a simple and small project, many of the roles may be combined and performed by one person.

A champion is the individual who has authorized the start of project after reviewing the business case, and he is the project's link with the company's vision and strategic priorities. The team leader has the overall responsibility of the project and ensures that objectives are being met in a timely fashion within the constraints. The Black Belt (BB) has the technical responsibility from the DFSS point of view and provides the support on tools and methodology application. In many cases, the BB may also perform the role of a team leader. The Master Black Belt (MBB) is the coach for the team; the MBB has a considerable amount of experience in similar projects and is also expert in DFSS methodology and tools. Subject matter experts (SMEs) are the individuals who are experts in specific functions (e.g., marketing, information technology). Their role is to support the team related to their expertise when request-

| DFSS Project Charter Section 1 | | | | |
|---|---|---|---|---|
| **Identifiers** | Project Title | | Project Number | |
| | Department | | Start Date | |
| | Location | | Expected End Date | |
| **Team and Stakeholders** | Role | Name | Email address | Phone # |
| | Team Leader | | | |
| | Champion | | | |
| | Process Owner | | | |
| | Black Belt | | | |
| | Master Black Belt | | | |
| | SME-1 | | | |
| | SME-2 | | | |
| | . | | | |
| | Core Team Member-1 | | | |
| | Core Team Member-2 | | | |
| | . | | | |
| | Support Team Member-1 | | | |
| | Support Team Member-2 | | | |
| | . | | | |

**FIGURE 12.1.** First Section of a DFSS Project Charter

ed by the team. Process owners are those individuals who will own it after completion. Core team members are the people who will do the bulk of the tasks for this project; these people will be part of the project from start to finish. Support team members are required to participate only as and when needed, similar to SMEs. However, these individuals are doers, while SMEs in most cases are like consultants. The following are some questions that need to be answered at the time of deciding the team and stakeholders:

· From which business functions or areas are SMEs needed? At what stage will they be needed?

· What is the approximate time commitment required from each team member and how will their regular work be handled while they are on this project?

· Who is accountable to whom and for what?

· How is the project linked to line management?

· What review structures are in place?

Some of these questions will be answered in the next section when we review champion and team leader roles in more detail. The BB's and MBB's roles are covered in the discussion on DMADO methodology and supporting tools of DFSS, for which these two individuals have the implementation responsibility.

**Champion and team leader's role.**   The champion supports the team by identifying process owner(s) and ensuring their appropriate involvement in the project. The champion has the organizational authority to remove roadblocks. The champion also ensures that the management systems can support the solution across the business and recognizes efforts of the team members. The DFSS project champion is also the representative of the company's executive leadership and shows commitment toward the project. Now we will define the champion's role in the beginning, during, and at the end of the project.

In the beginning of the project, the champion:

· Reviews the business case, team, and stakeholders
· Selects the team leader and process owners
· Allocates or authorize required resources
· Motivates the team and stakeholders

During the project:

· Reviews the team progress at appropriate intervals/milestones
· Makes a decision on changing the course or even stopping the project if it has drifted too far from the original objectives or the company's priorities have changed
· When necessary, intervenes to remove roadblocks
· Ensures cross-functional and cross-business linkages
· Informs the company's executive leadership of the team's progress and learnings

At the end of the project:

· Ensures the recommended design is implemented and the responsibility is transferred from the project team to the process owners

· Congratulates and celebrate the team's success
· Ensures project results are quantified, documented, and captured for corporate memory
· Ensures continued monitoring of key processes and measures

The DFSS project team leader's role is very important and extensive, as these projects are strategically very significant for the company. Typically, team leaders are senior, experienced employees of the company with multidisciplinary skills. They have authority over people working on the project and direct responsibility for the success of the project. Now we will define the team leader's role in the beginning, during, and at the end of the project.

In the beginning of the project, the team leader:

· Drafts the project charter
· Reviews the charter with the champion and process owners
· Works with the champion to identify resources required for the project
· Establishes a good working relationship within the team

During the project:

· Schedules and participates in project reviews with the champion and process owners
· Ensures the timely progress of the project
· Informs the champion of any roadblocks
· Coordinates team activities and communication
· Motivates team members
· Resolves conflicts and manage team dynamics
· Serves as a liaison between team and other stakeholders

At the end of the project:

· Works with process owners to ensure proper implementations

- Ensures that the project is properly documented and results are communicated
- Ensures that lessons learned are captured and communicated
- Works with the champion and process owner to ensure proper recognition and the next assignment for the team members

**Second section of the project charter.** This section answers the questions Why are we doing this project, and how does it fit with the overall vision and priorities of the organization? The opportunity statement should define the need for designing this new service and how it will be different from existing services. It should also provide indications of competitors and potential customers. The opportunity statement describes the market opportunity and addresses questions like the following:

- What is the impact on our business?
- What is the impact on our workforce?
- What will the impact be in the market?
- Will the company's existing customers be pleased?
- Who are the potential target customers?
- What value will this service add to customers?
- Will the company increase its market base for other services?

Vision is the ideal state definition of where the company would like to be in the long-term. Strategic priorities are the mid-term objectives that tell where the organization is focusing today. To make a project successful, the opportunity should be in line with the vision and strategic priorities set by the executive leaders of the company. It is important to establish a strong link between the opportunity, vision, and current priorities. The link subsection should address the following questions:

- Why should this project be done now?
- How does the project connect to the short-term and long-term strategies?

| Why are We Doing It | Vision Statement | | | |
| | Opportunity Description | | Strategic Priorties | |
| | Opportunity's Link with Vision and Priorties | | | |

**FIGURE 12.2.** Second Section of a DFSS Project Charter

· How does this project fit with the overall road map for achieving the vision?
· How does this project contribute to the mission of the organization?

**Third section of the project charter.**   The third section focuses on the business case, the scope, and constraints of the project. The business case should provide a compelling financial story for doing the project at this time. The business case describes potential risks associated with the project and provides financial and strategic justification for the project to management. It needs to provide details of several what-if scenarios. It should provide details on the potential for profit and growth in the short- and long-term and what investment will be required. It also has to consider the impact of the new service on existing business and the overall image of the company. A good business case will help the management team make informed decisions on prioritizing projects, allocating resources, and evaluating the success of the project. It should convince management that project success is likely and help in developing management interest in the project. From a financial viability point of view, the following key factors need to be analyzed:

· Total investment need, including:
  – Fixed assets
  – Working capital
· Cost of the service, including:
  – Fixed cost
  – Variable cost

- Expected revenue generation will depend on:
  - Sales price
  - Sales volume
- Profit: per unit
  - Annualized
  - Total profit for the expected life of service.
- Average return on:
  - Sales
  - Assets employed

From a strategic point of view, the following questions need to be answered and analyzed in the business case:

- What is the impact on the existing customers and portfolio?
- Is there an overlap with any existing service?
- How will it impact the image and reputation of the company?
- How does it fit with the company's long-term vision and short-term priorities?
- Will it help in penetrating new market segments?
- How will it impact the relationship with the community and government?

The scope defines the boundaries of the project within which the team will work. It tells the team about the size of the project and what it covers. For example, if a team is working on designing a service for a bank to process mortgage applications, the team needs to know if this service will be used for home mortgages, business, or both. Does the scope include risk assessment, or will that be treated as an independent subprocess?

The scope helps the project team focus their efforts on the central aspect of the opportunity or the issue and helps avoid diversions. The scope tells the team what elements of service are included and excluded, and what aspect of business is within and outside the project. The scope should be linked with available resources and time. The broader the project scope, the higher the requirement for resources and time. The scope becomes all the more important if there is a bigger project that is divided into subprojects or teams. In such cases, each team needs to understand the boundaries of their work so there is no overlap and nothing falls between the cracks. It is important to define both what is in and what is out of scope on the project charter.

It's important to grasp the fact that scope, though set at the beginning of project, may have to be revised as more information on the project becomes available. The champion and process owners should define the project scope in the beginning. However, if progress made by the team is not as expected, the scope should be reconsidered during project reviews.

Constraints are related to resources and time. Constraints provide the boundary of the solution zone by restricting the availability of alternatives. An example of constraint is the launch date for the service. This may have been fixed by the management or, in some cases, if the service is part of a bigger project, each element needs to meet a deadline to meet the overall target date. Figure 12.3 shows a blank project charter's third section.

**Fourth section of the project charter.**    The last section of the project charter provides information on deliverables, goals, and a project timeline. The deliverables describe the goal of the

| Business Case and Scope | Business Case | | Scope | |
|---|---|---|---|---|
| | Expected Investment | | IN | |
| | Potential Profit | | | |
| | Payoff and Rate of Return | | OUT | |
| | Impact on Current Customer | | | |
| | Growth Potential | | Constraints | |
| | Impact on Company Reputation | | | |
| | Various Scenarios e.g., Worst & Best case | | | |
| | Potential Risks | | | |

**FIGURE 12.3.** Third Section of a DFSS Project Charter

| | Project Goals | | Milestones | Completion Date |
|---|---|---|---|---|
| **Goals and Timeline** | Business Expectation | | First Phase | |
| | | | Second Phase | |
| | | | | |
| | | | | |
| | Customer Expectation | | Review 1 | |
| | | | Review 2 | |
| | | | | |
| | | | | |

**FIGURE 12.4.** Fourth Section of a DFSS Project Charter

project without going into a specific solution. It addresses the following questions:

· What are the business expectations from this project?
· What are the customer expectations from this service?
· How will we know when the design is complete?
· What will constitute success?

The timeline in the project charter includes key milestones and the completion dates of major steps in the project. It is important to break down a complex and lengthy project into milestones that can be accomplished in a shorter time, and that have their own goals linked with the overall project goal. It addresses the following questions:

· What are the critical milestones for the project?
· What deadlines must be met and why?
· What reviews should be scheduled and when?

## PROJECT MANAGEMENT

Most DFSS projects are complex and require proper project management, which will include deciding what tasks need to be completed before each milestone and the relationships between tasks. The first step in project planning and control is deciding a target date of completion and agreeing on deliverables at the project completion. These decisions need to be made by a champion, in consultation with the team leader and process owners.

**Project planning.**    Detailed project planning starts with key milestones and an end date decision. These need to be developed by the team based on the deliverables and end date of the project. DFSS methodology phases, DMADO, or the dates of an internal new service introduction process can be used as milestones. The whole cross-functional team should provide input in scheduling milestones. It is recommended that at every milestone, the team reviews the project plan with the process owners and the champion. Within each milestone, detailed tasks need to be decided and scheduled. The methodology of breaking down a complex project into smaller tasks and subtasks is known as *work flow diagram* or *work breakdown structure* (WBS). Developing a detailed and complete WBS ensures that the team doesn't overlook any significant task or subtask and that the estimate of the overall resource and time requirement is accurate. It is much easier to estimate requirements for a smaller task compared to a whole phase or a complete activity.

Once the WBS is defined and resource requirements are estimated, the next task is to decide the logical sequences. For each task, the team needs to decide the start and end point as well as dependency. Tools such as the Gantt chart, PERT/CPM method, and activity network diagrams (ANDs) can be used to show the relationship between tasks and building the project schedule. The Gantt chart is the most commonly used technique in project management. It is a simple graphical illustration of task sequences. The following steps describe how to draw a simple Gantt chart as a simple bar chart:

1. List all five phases of the DFSS methodology, DMADO, from first to last, down the left of the page along the Y axis.

2. Within each phase, list all the tasks, starting from the highest level of tasks and going to the lowest level of subtasks.

3. At the bottom of the page along the X axis, add the time scale in the appropriate units from the start of the project until the deadline or agreed completion date. The appropriate time unit will depend on the duration of projects. Use quarters, months, weeks, or days.

4. Draw the bars from the estimated start date to finish date for each subtask starting from the define phase tasks.

5. Ensure the dependency between tasks. If a task depends on the completion of other task(s), make sure it is reflected on the Gantt chart.

6. Tasks that are independent can be done simultaneously and adjusted based on resource and other considerations.

7. Based on the start and completion times of the tasks, estimate the start and finish time for the entire phase and then the whole project. Again, draw the bar along the X axis in front of the phase name written on the Y axis.

8. Adjust phase and task times so that the project gets completed within the deadline.

9. Add milestones as diamond points on the bar, such as a regular team meeting, review with the champion and process owners, and phase exit review, as required for the project.

The following is a list of typical tasks that need to be scheduled in addition to DMADO-related tasks:

- Cross functional coordination activities
- Risk management activities
- Regular team meetings
- Communication with the team, process owners, and champion
- Report writing and documentation
- Schedule management
- Project reviews with management
- Change management activities

**Project controls.**    Project control identifies the controls needed to protect the most vulnerable aspects of the project. Good project controls should be based on established organizational practices, but should be flexible enough to accommodate the needs of each project. It is important that all the team members have a good understanding of the organization standard for the controlled elements of the project (e.g., practice for calculating financial benefits and document controls). The project team needs to discuss and plan how it will comply company with standards and keep control on required aspects, as well as how the exceptions will be handled. Document control is often critical for design projects in order to control and manage design changes when subteams are working concurrently. Each document should be marked with who created it, who modified it, and date of creation, modifications, and a version number. It is also important to have an understanding of where and how documents will be stored and who will have access rights.

## VALUE-BASED MEASUREMENT SYSTEM

The objective is to develop an understanding and measure how customers define value so that service interactions and transactions can be developed around delivering that value. All the activities of the measurement phase are geared toward understanding the customer and competition. The output of the measurement phase is information and data that gets analyzed in the analysis phase. The measure and analyze phases are iterative in nature— as data get gathered or generated, analysis starts. Understanding VOC is of utmost importance in a DFSS project, and it is accomplished during the measure phase. At the end of a successful measure phase, the team will know the critical to qualities (CTQs) and their relative importance for this service.

Some of the tools and methodologies used during the measure phase include data collection plan; consumer research using interviews, surveys, and focus groups; Kano model; QFD; and benchmarking. Key questions that will be answered during the measure phase are:

· Who are the potential customers of the service to be designed?
· How much variation is there in the requirements of various groups of customers?

- What are the most important requirements for this service?
- Who are the competitors?
- What will differentiate us in the market?
- What performance targets need to be accomplished to get the customer interested in and satisfied by the service?

## CAPTURING THE VOICE OF CUSTOMER

A concept that is very simple to understand but very difficult to apply in practice is that the service needs to be designed and delivered to satisfy customers' needs and delight them. You will find if we ask those from different functions in the company what are the key requirements for a service to be designed, typically each group will have their preference. Finance is more interested in profit and cost, marketing is focused on the differentiating features, and the research group wants innovative and novel concepts. However, for the success of a service design, we need to understand the VOC. If all the functions are doing their job properly, all the internal voices will be connected with the VOC. Service needs to be designed for customers based on a prioritized list of customer needs. Needs have to be prioritized; otherwise, the design team may not be able to create a feasible solution that meets all the needs within the project constraints. Constraints like available capital, human resources, time, government law, and societal obligations compel the design team to focus on a significant few.

As stated earlier, this is simple concept but difficult to implement. Difficulty lies in understanding the following:

- Who are the true customers?
- How to capture customer voice?
- How to keep up with the dynamic nature of VOC?
- How to translate VOC into actionable specific requirements?
- How to prioritize?
- How to choose between conflicting or opposite requirements?

VOC is important. However, internal functional priorities still need to be managed. Success will depend on the ability of the design team in differentiating between internal requirements that are protectin "turf" from those that are coming from years of understanding and knowledge about a specialized aspect. The most difficult task for the design team is to ensure that a team moves deeper and deeper into the details of the design project, and not farther and farther away from the VOC.

The customer voice needs to be analyzed and deployed in designing every interaction and transaction of the service. It is very common to start a service design project with lots of good information. At the beginning of the project, usually the team spends a significant amount of time on understanding customer needs, translating them, and prioritizing. However, when it comes to designing detailed work instructions and procedures for delivering the service, the team falls into the trap of not upsetting the applecart and doing it the way we have always done it. That is where the DFSS methodology and tools provide the greatest advantage. If a team is following DMADO methodology and utilizing its supporting tools, such as QFD, it ensures that VOC is deployed to every minute detail of the service.

The first axiom of transactional service design states that every interaction needs to be designed to generate or exchange quality information. The quality of information generated or exchanged is directly related to how good the provider is at understanding the VOC. Understanding the VOC must result in customer delight after the interaction, as well as in useful information for the business. The second axiom states that every transaction must be linked with the processing of the output of interactions; this will ensure that VOC is deployed with every transaction at every level.

Let us go back to the basic questions we asked earlier, about what makes VOC activities difficult. In the following section, we will provide answers to these questions. To identify potential customers, we need to probe all types of customers—ours as well as competitor's, current and lost, internal and external, local and global. In addition to probing customers, the team also needs to discuss this with experts and stakeholders. The most likely outcome of this activity will be a realization that there is no single VOC, and different customers or types of customers have different needs and priorities. If there are distinct groups of customers, these groups are referred to as *customer segments.*

If you determine the real differences in the needs of different groups that have potential for influencing the design decision, segmentation is necessary. The key decision here will be to decide whether to address the need of only one or multiple segments. If the team decides to go after multiple market segments, the challenge of the design will be to balance multiple requirements across selected segments. The team will have to go back to the opportunity definition, business case, scope, and the constraints, as described in the project charter, and discuss them with stakeholders. Size, efforts, profit and growth potential, and impact on existing customers are some of the factors that will influence the decision of whether to pursue a specific market segment. It is also important to decide what criteria are best for segmentation. Sometimes there may be a tradeoff between simplicity and ease with relevancy. The simpler ways to segment are based on demography, geography, and the purchasing medium. The more difficult ways are based on situational factors and personal characteristics. There is a strong need to maintain balance between simplicity and relevancy.

Let us take a look at the sources for VOC. There are two broad categories of sources for collecting VOC: reactive and proactive.

**Reactive systems.**    VOC is available without taking specific actions to gather it Reactive systems generally gather data on current and former customer issues or problems. The following are examples of reactive systems:

· Customer complaints
· Web page visits and feedback
· Contested payments
· Analysis reports of loss of sales and market share

- Past sales and internal employee reports
- Commercially available industry surveys

**Proactive systems.** Effort and deliberate actions are required to gather VOC. Proactive systems generally gather data on selected segments, which may include noncustomers and competitors' customers, as well as current and former customers. The following are examples of proactive systems:

- Market surveys
- Interviews
- Focus groups studies
- Competitive analysis and benchmarking
- Feedback forms and data collected during sales visits

## PLAN AND COLLECT THE DATA

First, the team needs to define what data should be collected and then identify what data are available in the company. Collect and summarize existing data, analyze these data, make preliminary conclusions, and decide what else we need to learn. While deciding on the need for additional data collection, consider the following:

- Review the objectives of the project, analyze the existing data, and determine the need for additional data
- Strike a balance between available resources and the need for new information
- Extend thinking beyond existing customers and methods used in the past to understand customer needs

In addition to what data need to be collected, some other key questions that must be asked before starting the data collection activity are:

- What is the right method to collect each type of data?
- What is the correct sample size?

· Who should collect the data?
· When do the data need to be collected?
· How much preliminary analysis is needed?

For collecting data, proceed from higher requirements to detailed requirements, to a level where you can measure the true needs of the customers. Customers may not always state true needs, and it is important to ensure that the VOC data collection focuses on true needs. Define and analysis phase activities becomes iterative at this stage. Collected data need to be analyzed to understand the true needs that underlie the sometimes vague statements of customers. Preliminary analysis will reveal if we need to collect more data. In general, there are two types of customer needs:

**Stated needs.**    Stated needs are simple to collect and analyze, as the customer is willing to and capable of articulating them very specifically. For example, from a car garage, customers expect their car's problem identified and repaired for a low price and within the expected time. These may be stated by the customer in response to a proactive data collection process such as a survey, or by a reactive means such as a complaint.

**Concealed needs.**    Concealed needs are not directly stated by the customer, but the designer needs to extract them either by contextual inquiry or by analysis of VOC data, along with other available information. These are more difficult to understand, but very important for new service design. If you are designing a new service, it is very likely that customers themselves are unaware of their true needs. For example, a stated need in a car garage is quick service, but the latent need may be immediate transportation to go home or to the office, or not getting bored while waiting.

## PRIORITIZATION OF CUSTOMER NEEDS AND BUSINESS GOALS

Once VOC is collected and customers' needs are understood and classified, the next step is to establish priority. Conflict is inherent in prioritization rather than allowing gut feelings solve priority conflicts. They should be resolved on the basis of value to the customer and the company. To prioritize customer needs derived from the collected and analyzed VOC, it is critical to find the relationship between various needs. The team needs to bundle or cluster these needs into groups and then establish hierarchy and priority. Use of graphical techniques like the fishbone diagram, tree diagram, and interrelationship diagram can help cluster various needs into groups. The first step is to eliminate the exact duplicates. The next step is to link cause and effects to pick the root cause and then identify one basic need representing the complete group. The real challenge is how to decide the final list of priority customer needs that will reflect the VOC and balance business interests.

A simple prioritization matrix similar to the first house of quality function deployment (QFD), with weighed business goals and importance ratings on a subjective scale can be used to develop a final list. Figure 12.5 shows a prioritization matrix, which may be used independent of a complete and comprehensive QFD exercise. QFD is explained in detail in the previous chapter. The following are the simple steps for developing a prioritization matrix:

1. Define the relevant criteria based on business goals for prioritization.
2. List all business goals on the top row.
3. Write the weight for each business goal.
4. List all the customer needs in the leftmost column.
5. Evaluate the positive or negative strength of the relationship between the need and each goal on a −5 to +5 scale and write the score in the corresponding cell of the matrix.
6. Multiply weight and strength score, add the total for each need, and write it in the rightmost column.
7. Highlight the significant few needs with high overall score in the rightmost column.

| | Business Goals | Increased Market Share | Penitrate New Market | Less than 'N' Months payback | Improved Image | Innovative Service | Less than 'Y' M Capital Investment | Overall Score |
|---|---|---|---|---|---|---|---|---|
| | Weight | 20 | 25 | 10 | 15 | 15 | 15 | |
| Customer Needs | Lees than 'X' $ price | | | | | | | |
| | Accurate estimates | | | | | | | |
| | Clean Environment | | | | | | | |
| | Knowledgable Staff | | | | | | | |
| | Useful advice | | | | | | | |

**FIGURE 12.5.** Prioritizations Matrix from QFD

Subjective prioritization techniques, such as paired comparison, can be used to establish weights for each business goal.

The outcome of a prioritization activity should be identification of primary, secondary, and tertiary needs. A line graph with scores on the Y axis and needs on the X axis may help to identify natural break points in the primary, secondary, and tertiary needs. Finally, this list needs to be reviewed with the champion and stakeholders and validated with customers and SMEs. This prioritized list of customer needs is a list of CTQs in the DFSS approach.

## BENCHMARKING

This activity focuses on understanding what other organizations are doing to meet the value goals for their customers and company. The benchmarking activity can be performed with direct competitors who are providing similar services or satisfying similar customer needs, or it may be done with a company in a different industry that has one or more similar CTQs. For conducting benchmarking with direct customers, the team can only analyze the information available on services already offered in the market. Assuming competition is also engaged in continuous improvement and designing new and better service, sole reliance on this type of information will put your company in a perpetual catch-up game. A better approach is to identify CTQs for the service to be designed and then identify for each CTQ which industry and company is the best in class and how they do it.

Before benchmarking, consider the legal and ethical aspects to understand the boundaries for collecting and sharing information. The team must identify its own company's policies regarding contacting external organizations, as well as industry standards for ethical benchmarking. Clearly, communication and sharing of information within legal and ethical limits is essential for proper benchmarking. The team needs to reveal upfront how it intends to present and use other companies' information and who will have access to this information and for what purpose.

## ANALYZE

The output from the previous phase is a prioritized list of CTQs that the service to be designed should fulfill in order to provide value to the customer and business. In the analyze step, these CTQs will be converted into functional requirements, and the team will generate the alternative solutions for designing the service. The alternative solutions will be analyzed to evaluate and select the concept that will best meet the functional requirements within time and resource constraints of the project. The key difference between CTQ and functional requirement is that the former is generic and the latter is more specific and usable as a requirement for the design concept.

It is important to understand that concept generation and selection are crucial milestones on the road map of service design with Six Sigma quality. The objective of concept generation is to generate a large number of novel ideas and innovative solutions to meet all the prioritized requirements within constraints. Once concepts are generated, these are analyzed in detail to make sure that expensive and risky decisions are not made. Also, there is a need to have a degree of confidence that the selected concept will perform as expected. Concept generation and evaluation are sequential activities, and it is important that, while generating new ideas, the team suspend judgment. Concept design is primarily a design on the paper. However, few selected alternatives may be "tried out" in greater detail to create a deeper understanding for evaluation. Once an appropriate concept is selected, design details will be developed further and optimized in the design phase. Generating new concepts and using them to develop a design requires more resources and has a lot more risk compared with using available designs. The team needs to identify those functional requirements of the design that have an existing solution. The following two questions need to be answered:

1. Which functions need new concepts, and which existing designs are sufficient?
2. Which functions can be reverse-engineered using the competition's design?

The functions that need new concepts will require controlled creativity and innovation from the project team.

## CONCEPT DEVELOPMENT

Customers expect novel and superior services all the time. To meet this expectation, organizations must be creative and innovative. It is commonly believed that innovations and creative designs are the result of lucky and unpredictable events or serendipity. In reality, most innovative designs are the outcome of a managed process. Certain circumstances may foster innovation, creativity, and original ideas, just as certain conditions may suppress creative activity. This section focuses on those procedures and techniques that will enable a design team to introduce creativity and innovations into concept design. First let us examine the human aspect of creativity in brief.

Most concept development activities are done in groups; very rarely are they done by single individuals, and this is also true for DFSS projects. Team members need to have the ability for independent thinking and intuitiveness, and ideally should come from different functions and backgrounds so they can interpret information in a pattern-breaking manner. The creativity of the team is greater than the individual creative potential of its members. Cross-functional teams are very effective for developing novel concepts. Team membership is very important for the success of concept generation and development activity. Team members should not only match in consistency, but also should reinforce each other in terms of their functional expertise. Mutual respect and shared vision among the cross-functional team members is a must for achievement of the common goal.

The team leader and champion need to understand that the creative abilities of most people are fragile and could be seriously suppressed by disparaging criticism or rigid, restrictive environments. It is important that, to realize the full creative potential of team, a positive environment is created. A communication strategy that enhances rather than suppresses useful creative activity is very crucial for the success of a DFSS project.

## STEPS FOR DEVELOPING THE CONCEPT

Creativity and innovation involve the translation of a design team's unique talents and the vision of the leadership into an external reality that is novel and superior to the existing alternatives. Creativity and innovation should lead to the development of novel service concepts, which will provide value to the customer and business.

Concept development is a crucial activity for designing new services; it requires creativity, cross-functional participation, and a systematic methodology. The team needs to decide for what functional requirement it needs to develop concepts and then either treat them one by one or develop concepts for the whole service. Two approaches for generating concepts are possible:

1. Generate concepts function-by-function
2. Generate concepts across functions

In most cases, the better approach is to generate concepts function by function. Once promising ideas are short listed, the team needs to combine these to form a complete service. In many cases, the idea may be exceptional for an individual function; however, when you combine it with other ideas to offer the whole service, it may not be the one to provide the most value. It is not necessary that all the interactions and transactions of a service are new; some parts of the concept may include existing designs.

Next, we will examine two major steps for concept development, idea generation, and idea evaluation. Finally, we will discuss how to develop an overall process flow of interactions and transactions that will satisfy both the design axioms.

**Idea generation.**    The first step in concept development is to generate a number of alternative solutions without pre-judging them. Alternatives may be generated by using simple brainstorming or more systematic and advanced techniques such as synectic thinking, mind mapping, and TRIZ. At the stage of concept generation, quantity is very important; for example, the literature indicates that drug companies typically require 6,000 to 8,000 or more ideas for every successful commercial new product, while the manufacturer of typical industrial products requires approximately 3,000 raw ideas for one commercial success. The ideas can then be pieced together into alternative concepts.

Capture the ideas by writing complete descriptions as soon as they are generated, and encourage team members to build on existing ideas. It is key that team members build on each others' ideas to create more and better concepts. Combine a few infeasible or not so good ideas together and you get an excellent idea. Ensure that no critical functions have been missed.

**Evaluate.**    Once ideas have been generated, these need to be analyzed for the evaluation and selection of a final few or just one idea. Determine which alternatives will blend with alternatives of other functions, and which will not. Assemble concepts from compatible alternatives. Apply judgment and select the most appropriate concepts. Some other basic questions that can be asked while screening concepts are:

· Does the concept provide intended functions?
· Is this new concept in harmony with the overall service?
· Is this concept elegant?
· Does this concept provide required controllability, modularity, multiplicity, etc?

## PROCESS FLOW AND LOGICAL RELATIONSHIP

The service concept needs to be represented in a flow of inter-actions and transactions. Developing a flow needs the knowledge of many team members, and it requires proper facilitation; otherwise it may become a very confused and unproductive session for team members. The most important part of a service process flow is the logical relationship between various elements. The following steps describe ways to facilitate a cross-functional team meeting organized for the purpose of developing a flowchart.

1. The team leader briefly explains the service concepts and background information.
2. Start the discussion with very high-level activities and develop a flowchart, without details, of how each of the elements will be accomplished.
   a. In some cases, even developing high-level flow may not be obvious. In such cases, brainstorm on what type of activities will need to be performed.
   b. List each activity on a card or adhesive note.
   c. Group the cards that either say the same activities in different words or represent details of the same activity.
   d. Agree on a title for each group of cards.
   e. Arrange the cards in a logical flow.
   f. Document the flowchart using standard flowcharting symbols.
3. For each high-level activity, decide all the inputs and outputs.
4. Develop a detailed flow for each high-level activity, either by brainstorming as a whole team, one activity at a time, or divide the team into subgroups and have each group work on one or a few activities.

# TOOLS FOR TRANSACTIONAL PROCESS DESIGN

This section provides an introduction to a few selected tools useful for the define, measure, and analysis phase. These tools, along with the described activities for each phase of the DMADO methodology, will help in designing a service at Six Sigma quality level.

## PAIRED COMPARISON

Paired comparison is a subjective quantification technique. It is used to establish relative importance among a set of alternatives. It is a common technique in the field of value management, and is described and used by Mudge (1971) in his book on *Value Engineering: A Systematic Approach*. The process is simple; it requires one to choose from a pair of items, of which one possesses a higher level of some specific characteristic (e.g., delivery time, cost, performance). Comparison is done by assigning a rating based on a 0- to 3-point scale. The activity of comparison is iterated over all possible pairs and then a normalized total is calculated for each item. The final normalized percentage value for each item is the score used for relative importance. Paired comparison is very discriminating and easy to apply for those items that are difficult to compare as a whole list. However, it is very time consuming and is not recommended for comparing a large number of items. The following example will provide a good understanding of this technique.

A DFSS project team would like to establish relative importance of the following list of criteria, which they want to use for concept selection:

· Cost of the service concept development
· Time to market
· Innovativeness
· Chances of success
· Positive impact on company image

· Return on investment
· New market penetration

The team has agreed to use the following rating criteria:

0= No difference
1= Slightly more important
2= More important
3= Significantly more important

Figure 12.6 shows the matrix of paired comparison with each of the seven items listed earlier. The activity started by comparing cost versus time (A versus B), and decision on cost is slightly more important and therefore the entry in the first cell is A-1. This process is continued, and the cost was compared with all the other six criteria. It was decided that the chances of success are more important than cost, and there is no difference between cost and new market penetration. The right-hand side of the matrix shows the first raw total for each criteria and then as a percentage of the total rating of all items. The analysis shows that chances of success and return on

| | B | C | D | E | F | G | | |
|---|---|---|---|---|---|---|---|---|
| A | A-1 | A-1 | D-2 | E-1 | F-1 | 0 | 2 | 10% |
| B | | B-1 | D-1 | E-1 | F-2 | B-1 | 2 | 10% |
| C | | | D-1 | E-1 | F-3 | G-1 | 0 | 0% |
| D | | | | D-1 | 0 | D-2 | 7 | 33% |
| E | | | | | F-1 | E-1 | 1 | 5% |
| F | | | | | | F-1 | 8 | 38% |
| G | | | | | | | 1 | 5% |

FIGURE 12.6. Paired Comparison Matrix for Service Concept Selection Criteria

investment are the two most important criteria; cost and time to market are in the middle; and innovativeness, impact on image, and market penetration are the least important criteria for this situation.

## RASIC MATRIX

RASIC stands for responsibility, authority, support, informed, and consulted. It is an important tool for defining roles in a project setting. The RASIC matrix helps in establishing a clear understanding for everyone involved with the project about his role. Table 12.1 shows a RASIC matrix where rows represent various subtasks that may be required for a project and columns represent the role of either a functional representative or specific individual associated with this project. It is the team leader's responsibility to ensure that roles written on the RASIC matrix are followed.

TABLE 12.1    Example of a RASIC Matrix

|  | RESPONSIBILITY | AUTHORITY | SUPPORT | INFORMED | CONSULTED |
|---|---|---|---|---|---|
| Task 1 | Subteam 1 | Champion | IS, HR | Process Owner | Legal |
| Task 2 | Subteam 2 | Champion | IS, Sales | Process Owner | Supplier |
| Task 3 | Subteam 3 | Process Owner | Sales | Champion | Customer |
| Task 4 | Subteam 4 | Process Owner | Finance | Champion | Quality |
| Task 5 | Subteam 5 | Team Leader | IS, Finance | Champion, Process Owner | MBB |
| Task 6 | Subteam 6 | Team Leader | IS | Champion, Process Owner | MBB |

## REVIEW OF CREATIVITY TOOLS AND TECHNIQUES

First, we will discuss the work of some of the most famous authors on the subject of creativity. Alex Osborn's (1953) pioneering book, *Applied Imagination*, talks about "questions as spurs to ideation," and outlines about 75 idea-spurring questions. J. L. Adams' (1986) book, *Conceptual Blockbusting: A Guide to Better Ideas*, contains a good introduction to the topic, as well as games and exercises to limber up mental muscles. This book explores the mental walls that keep one from correctly perceiving a problem or conceiving a solution. Tony Buzan's (1994) mind mapping concept provides a revolutionary new approach to harness the mind's untapped resources. Edward de Bono has written several books and articles on the topic of creativity, the most famous of which is *Lateral Thinking: A Textbook of Creativity* (1970). Robert McKim's (1972) *Experiences in Visual Thinking* provides an approach for being creative with the help of pictures and drawings. Howard Gardner's (1993) model of creativity and mind is explained in his book, *Creating Minds: An Anatomy of Creativity Seen Through the Lives of Freud, Einstein, Picasso, Stravinsky, Eliot, Graham and Gandhi*. Gardner's book provides a synthesis of his ideas within the context of his "seven intelligences model of the mind." This particular model is uniquely "interdisciplinary'" in that each of the following seven intelligences represent a different aspect of human creativity.

1. Logical-mathematical (Einstein)
2. Spatial (Picasso)
3. Musical (Stravinsky)
4. Linguistic-verbal (Eliot)
5. Bodily-kinesthetic (Graham)
6. Interpersonal (Gandhi)
7. Intrapersonal (Freud)

The following information about creativity techniques is based on these books and information collected from several creativity-related Web sites. It provides a brief introduction to some very powerful creativity tools and techniques:

**Synectic thinking.**    This is the process of discovering the links that unite seemingly disconnected elements. It is a way of mentally taking things apart and putting them together to furnish new insight for all types of problems. William Gordon (1961) set forth three fundamental precepts of synectic theory:

1. Creative output increases when people become aware of the psychological processes that control their behavior.
2. The emotional component of creative behavior is more important than the intellectual component; the irrational is more important than the intellectual component.
3. The emotional and irrational components must be understood and used as precision tools to increase creative output.

**Lateral thinking.**    This is about moving sideways when working on a problem to try different perceptions, different concepts, and different points of entry. The term covers a variety of methods including provocation to get us out of the usual line of thought. Lateral thinking is cutting across patterns in a self-organizing system and is strongly related to perception.

**Mind maps.**    Association plays a dominant role in nearly every mental function, and words themselves are no exception. Every single word and idea has numerous links attaching it to other ideas and concepts. Mind maps, developed by Tony Buzan, are an effective means of note taking and are useful for the generation of ideas by associations. To make a mind map, one starts in the center of the page with the main idea and works outward in all directions, producing a growing and organized structure composed of key words and key images.

All of the preceding material is related to creativity and innovation and can be very helpful for idea generation and for the concept development and selection phase of service design.

**Spider charts.**    This is a powerful tool for comparing two or more alternatives based on multiple criteria. Each criterion may have a different unit of measurement. However, on a spider chart, relative performance is plotted, so actual units of measurement are not important; the farther from the center of the chart, the higher the performance. Spider charts are also a very useful graphical representation tool for benchmarking studies. Sometimes these charts are also referred to as *radar charts*. The following steps describe how to construct a spider/radar chart:

1. Collect information about the performance of each alternative against all the criteria.
2. Assign one criterion to each axis in the chart.
3. Decide how many divisions of each axis you want.
4. Plot performance data for each alternative along the axis for each criterion.
5. Join the data points between the axis using a unique color or symbol for each alternative, to generate a performance profile.
6. Analyze the chart and draw conclusions.

Figure 12.7 shows a spider/radar chart for two alternatives. The same seven criteria are used as discussed in the paired

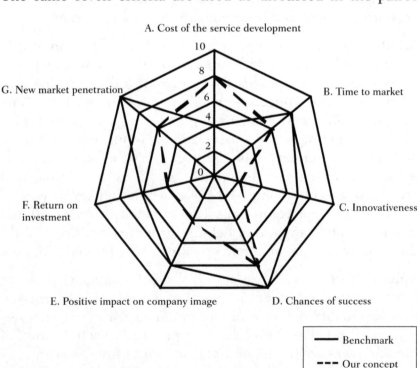

FIGURE 12.7.    Spider Chart Example Comparing Two Alternatives

comparison technique example. The chart shows that in comparison to benchmark, our concept is better in cost; however, we lag behind on return on investment and other criteria.

## FOUR QUADRANT ANALYSIS

This is a simple way of grouping ideas based on two important criteria. It is a macro level analysis, and it provides a starting point in the evaluation, or in some cases, first level of screening. Only two levels for each of the two criteria are allowed. For example, the team may be interested evaluating of service ideas based on time to market and potential for success. The team can choose the X axis, representing time to market, and two levels, short-term and long-term. Similarly, the Y axis representing the potential to success, will have two levels, low and high. Now, review each idea and judge it based on time and success potential and place it in the appropriate quadrant. Figure 12.8 shows a four quadrant for the example described here.

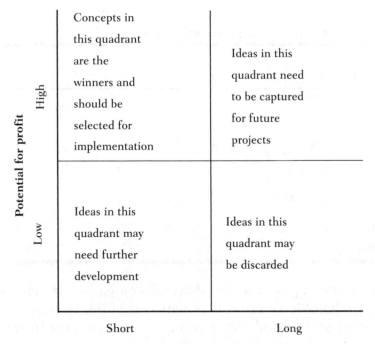

FIGURE 12.8. Four Quadrant Analysis

## PUGH'S CONCEPT SELECTION

Pugh's concept selection matrix is a very useful and simple way of selecting and developing multiple concepts based on multiple ideas. In most cases, the approach requires the selection of a base idea or concept against which all other ideas are evaluated. Typically, ideas are listed along the X axis of the matrix and all the criteria are listed along the Y axis of the matrix. The score in a column of the matrix represents how an idea compares to the base or benchmark idea. Rows represent how various ideas fare against the base idea for a specific criterion. The most commonly used scale is a three-point scale, where "0" represents same, "+" represents better, and "–" represents worse. The last row shows the summation of scores for each idea. While summing, zeros don't count and pluses cancel against minuses. For example, if there are seven criteria and a concept gets one zero, two minuses, and four pluses, the final score for this idea is 2. Table 12.2 shows a Pugh's concept selection matrix.

TABLE 12.2    Pugh's Concept Selection Matrix

| CONCEPTS/CRITERIA | | 2 | 3 | 4 | 5 | 6 |
|:---:|:---:|:---:|:---:|:---:|:---:|:---:|
| A | | + | + | + | – | + |
| B | | + | – | + | – | – |
| C | | 0 | – | 0 | – | 0 |
| D | Base Line Concept | – | – | 0 | – | – |
| E | | + | – | + | 0 | – |
| F | | – | + | + | – | + |
| G | | + | 0 | – | – | + |
| Total | | 2 | –2 | 3 | –5 | 0 |

Pugh's matrix provides a good overall comparison for all the concepts. The objective of this activity is twofold. First, screen the concept and second, develop the selected concept further by combining the strong points of other concepts. The team needs to look at the negatives of the short-listed, high-score concepts, and discuss what is needed in the design to reverse negatives (relative to the base line/benchmark). Make sure that improvement doesn't reverse one or more of the existing positives.

## KEY TAKEAWAYS

In the end, a quote from Blaise Pascal, "What is man in nature? Nothing in relation to infinite, everything in relation to nothing, a means between nothing and everything." This quote also applies and summarizes the content discussed in this chapter. The following are the key takeaways:

- The development of a new service is iterative in nature, but it should not be trial and error. Designing services should be based on axioms and by following systematic methodology of DMADO.
- A good beginning is always essential for the success of any project; it is proven by the success of many DFSS projects that a good project charter along with a feasible project plan is essential.
- Develop an understanding, and measure how the customers define value so service interactions and transactions can be developed around delivering that value.
- The analyze phase needs creativity and there are several tools to help teams to become more creative. The output of the analyze phase is a few good concepts, which will be developed further in the design phase and finally optimized, validated, and launched.

# DESIGN AND OPTIMIZE SERVICE TO ENSURE A ROBUST SERVICE PACKAGE

The aim of transactional service design with design for Six Sigma (DFSS) is to take an idea from concept to reality by creating details of a robust service package that will satisfy both the axioms of service design. The outcome of the DMADO methodology is a service that is developed by listening to the voice of customer (VOC) and delivered to meet the business objectives. As discussed in the previous chapter, this methodology starts with defining the opportunity in the define phase. It focuses on capturing the VOC in the measure phase and generating innovative service concepts in the analyze phase. The outcome of the analyze phase is a handful of viable concepts that have a high potential for satisfying both customers and business objectives simultaneously and can be set in motion for successful completion. At the design stage, companies need to commit significant resources in transforming ideas and concepts into marketable commercial service packages. Before the final introduction into the market, the service pack needs to be optimized and validated. This chapter will provide a detailed discussion on the design and optimize phase of the DMADO methodology.

The design stage of a DFSS project is critical. Its activities will support the project team in creating the detailed design for each transaction and interaction of the service concept. The optimize phase is the final step, and it involves testing and validation of the developed design and finally adjusting it to make a robust service package. The optimize phase includes the activities for preparing the pilot run, executing the pilot, fine tuning the service package, and launching for commercial success.

## DESIGN

The activity of service design involves making decisions or choosing the best possible alternatives for service transactions and interactions. The selection of each alternative has its own associated performance benefits, costs, and risks. The challenge of the service design is to make decisions at various levels of detail. The design team makes the choices that simultaneously balance performance, cost, and risk elements. Decisions at the lower or detailed level must be compatible with previous, high-level decisions. The design phase is supported with tools such as systems engineering simulation and failure modes and effects analysis (FMEA). Developing a detailed service design from concept is a complex process and requires a rigorous approach. In most cases, it will be an iterative process. However, if define, measure, and analyze phase activities were done right, the number of iterations required will be minimal.

The DFSS approach for service design has the following two key differences from the traditional service design approach.

1. Follow the systematic steps of DMADO methodology and don't jump into designing right away.
2. Whenever in doubt or need to check if the decision is correct, go back to the two basic axioms.

The design team needs to take the output generated from the analyze phase in the form of concepts, from the measure phase in the form of the VOC, and from the define phase as to why we are doing it. In many cases, the team may have to go back and repeat or modify activities from previous phases. For example, if the team feels that concepts are not good enough, VOC is not properly understood, opportunity is not focused, or project goals and timing need modifications, they will need to go back and do more work on analyze, measure, or define phase activities.

A service can be designed either bottom-up or top-down. It is recommended to determine the requirements for top-down, but create the design bottom-up. The top-down partitioning is done to get the flow of service functions and requirements to interaction and transaction performance targets. The bottom-up approach is essential to ensure that all the key interactions and related transactions are considered in detail. This approach is based on systems engineering principles. Figure 13.1 shows the concept of combining top-down and bottom-up approaches.

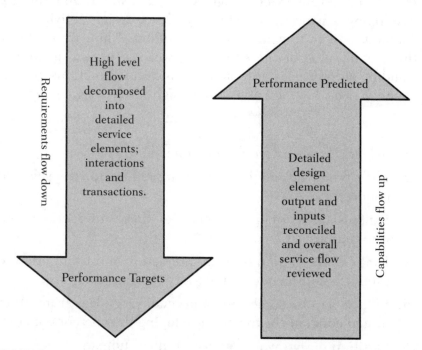

**FIGURE 13.1.** Top-down and Bottom-up Approach Combination for Service Design

The design phase activities can be divided into three sub-phases; high level, detailed, and integration. At the high level, the design team needs to identify where pieces of the design fit or don't fit. These can be corrected before the detailed design stage is reached. At the detailed design level, different parts of the design may be the responsibility of different design teams. The challenge of keeping the design together begins here and continues all the way through the integration. It is important to avoid any duplication of work by various subteams at the detailed design level and entrust the design to the appropriate subject matter experts.

## HIGH-LEVEL DESIGN

In high-level design, you want to develop enough details so that the design can be evaluated for performance and feasibility. Several high-level design alternatives may be analyzed and simulated until a suitable design is selected. The ultimate objective of high-level service design is to establish a foundation for developing a winning service. High-level design activity must enable the team to gain an understanding of how to maximize the overall value of the service, as well as to gain an understanding of interfaces, interactions, and ideal functions of each element. The key activities for a high-level design subphase are:

1. Select a final service design concept.
2. Identify key interactions and transactions for the selected concept.
3. Establish performance targets for these interactions and transactions.
4. Evaluate the design of each interaction and transaction to establish confidence and identify the gap from target.
5. Fill the gap by combining and modifying ideas from other design concepts or further developing the current concept.
6. Partition design work and assign it to subteams.
7. Establish a clear objective and links for each subteam.

8. Conduct a high-level design review to establish confidence in ensuring the following aspects:

   a. Overall service flow is smooth

   b. Service will be perceived by target customers as value-added

   c. Service will meet the business objectives

   d. Service design and implementation is feasible within the given constraints

   e. Service fits well with the long-term vision of the company

High-level design decisions are taken based on the information available from the define, measure, and analyze phase. Key decisions taken at this stage include one final overall concept, approach for decomposition, and commitment of resources. Decisions taken at this stage will limit the choices at detailed design and integration, as well as for the next phase of optimization. The tools and methods that will help in making decisions and completing tasks at this stage are systems engineering, design reviews, QFD II, simulation, flow and block diagramming, and functional trees. Subject matter knowledge of the cross-functional team is critical to make good design decisions.

## DETAILED DESIGN

In a detailed design, the elements of selected high-level design are developed further to enable integration as a complete service package. Detailed design tasks are performed by subteams created during the high-level design subphase. The key activities for a detailed-level design are:

1. Develop details for each interaction and transaction. The following list provides an example of details:

   a. How much or how long?

   b. What must proceed and succeed this step?

   c. Where will it be executed?

    d. Who will perform what task?

    e. What is the risk of not completing it properly?

    f. What are potential failure modes?

    g. What is the effect of failure mode?

    h. How will the failure be detected?

    i. What are recovery plans for failures?

    j. What is the metric for success?

2. Develop a functional boundary diagram for each key interaction and transaction. Figure 13.2 shows a functional boundary diagram.

3. Analyze and simulate the design performance against the requirements for selected complex elements.

4. Ensure that subteams communicate effectively with each other during the detailed design activities.

5. Create detailed documentation of procedures, work instructions, and special instructions.

6. Conduct a detailed design review to establish confidence on the following aspects:

    a. Each service element is adding value

    b. Service elements have enough details for implementation

    c. All the identified risks have been minimized

    d. Ideal function inputs, outputs, controls, disturbances, and links of each element are properly understood and considered in detailed design

Detailed design tasks are performed simultaneously by several teams. Therefore, coordination and communication between subteams is very important at this stage of design. The decisions taken at this stage are related to how each of the interactions and transactions will be executed during the service delivery. Also, it is important to consider potential failure modes and plan to tackle them. These decisions are taken within the constraints of the high-level design decisions taken earlier, but in some cases, teams may find the need to go back and

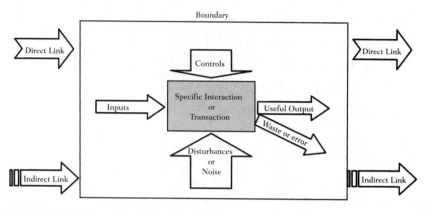

**FIGURE 13.2.** Functional Boundary Diagram

change high-level design decisions. The key tools and methods used at this stage are FMEA, simulation, functional diagram, and design reviews.

## DESIGN INTEGRATION

Once the details of each transaction and interaction are worked out, the next step is to integrate them. During detailed design activity, some of the input and output requirements may have changed. These need to be reconciled by the team before final optimization and validation. The team also needs to take a look at how the service as a whole will be perceived by the customer and whether it meets the objectives established in the earlier phases. A high-level service design flow may be simulated to ensure that it meets customer needs; this is why it is necessary to integrate the bottom-up design. The key activities for integration of design are:

1. Review the flow of service to ensure all inputs and outputs are properly linked and interfaces are considered.

2. Potential issues in delivering the service are addressed, especially the interface between various transactions and interactions.

3. Flow up the capabilities of each service element to estimate the overall capability.

4. Predict service performance.

5. Develop implementation details.

    a. Training needs

    b. Required performance range

    c. Reliability under stress

    d. Safety

    e. Conformance to published standards

6. Conduct a design integration review to establish confidence on ensuring the following aspects:

    a. Interfaces are properly designed

    b. The complete service package is expected to exceed the targets

    c. Will meet all the requirements as established by the team in earlier phases

The integration subphase of the design is very critical for the reliability of the service; as in most cases, failures occur at the interfaces and in the gray area of responsibility. At this stage of design, special emphasis is given to understand how the whole system will work together to create a desired outcome under all possible inputs and disturbances. The detailed robust design of each element is necessary for success, but not sufficient. It is very common that all the elements are designed properly, but still the system as a whole is not functioning properly. That is why in a DFSS service design, we not only pay attention to details, but also ensure that elements are in perfect harmony. Tools and methods used at this stage include systems engineering, simulation, functional block diagram, and design reviews.

The major benefit of this three-phased approach is that the higher level design decisions about the key service elements and how they fit together are made before detailed design work is started. It helps in avoiding rework and costly mistakes. Design tasks are partitioned for manageability and necessary competencies. Finally, the third subphase integrates all the ele-

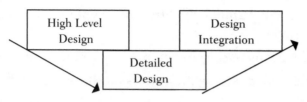

**FIGURE 13.3.** Three-stage Design Process

ments for continuity. This approach results in a robust design. Evaluating high-level design for performance and feasibility before more resources are spent in detailed design is more cost-effective and efficient. Figure 13.3 depicts the three-step design approach recommended in this section.

## DESIGN PRINCIPLES

Throughout the design phase, team(s) need to follow good design practices, which can be derived from two design axioms, as well as lessons learned from the previous design project. These design principles are guidelines that help to produce a higher quality, simpler design. Some common principles are:

- Always consider the VOC at every stage of decision making.
- Eliminate any transaction that is not directly or indirectly linked with interactions.
- Don't take design decisions without reference to the selected concept.
- Consider the interactions between the various elements of service.
- Think if there is a simpler way to perform the transaction.
- Think if any transaction can process input from multiple interactions or provide input to multiple interactions.
- Think about mass customization at the interaction level.
- Think about standardization at the transaction level.
- Automate or outsource transactions that are not core competencies.

- Minimize the number of different people who interact with the customer.
- Make decisions earlier in the process to improve efficiency.
- Make decisions late in the process to improve flexibility.

## SERVICE DESIGN CASE STUDY

The Wholesome Food and More (WFM) case study described in Chapter 11 is discussed here to further explain the service design phase of the DMADO methodology. Let us assume after completing activities of define, measure, and analyze phases, that the WFM team has selected the following two concepts:

1. Offer delivery of the following types of meals for breakfast, lunch, or dinner in offices, schools, and homes:

   a. Low carbohydrate

   b. Low fat and cholesterol

   c. Vegetarian

   d. High energy

   e. Organic

   Each meal will contain complete nutritional facts, ingredients, and impact on health on an easy to read card, with additional healthy lifestyle suggestions. Each type of meal offers at least three styles (e.g., Chinese, Italian, American, Mexican, and Indian).

   Guarantee that the meal was prepared in less than 12 hours before the delivery and, where possible, fresh ingredients were used instead of frozen food.

   Utilize WFM's existing website, stores, and a toll-free phone number to receive orders. Stores will also serve as an outlet to pick up the food as an alternative of home delivery.

2. Offer a service of complete diet planning with the help of a qualified dietitian, nutritionist, and healthy food and lifestyle expert. Recommend a diet based on individual needs, tastes, and preferences; work with a few selected

good restaurants in the area that can provide the recommended diet and are also willing to provide information, home delivery, and the same guarantee as in the first concept.

In the first concept, WFM will need to launch meals under its own brand name. In the second concept, it will utilize the reputation of existing area restaurants. Both concepts meet the opportunity definition of providing a freshly packed nutritional meal delivered conveniently to the customers. These two concepts were developed after screening, combining, and modifying a large number of ideas generated during the analyze phase. Figures 13.4 and 13.5 show the flow for both the concepts.

Flow charts for service concepts provide an overview of the high-level interactions and transactions necessary for providing the service to the customer. The first few interactions and transactions are there to make the customer excited and interested in the service. Next is to interact for collecting information necessary to provide the service. This information is processed to generate more information and goods or services that are delivered to the cus-

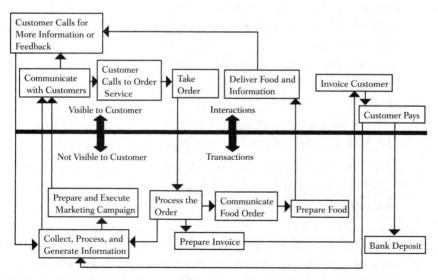

**FIGURE 13.4.** Flow for the First Concept

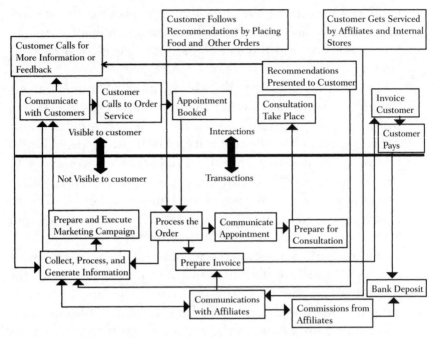

**FIGURE 13.5.** Flow for the Second Concept

tomer. Finally, money and feedback are received from the customer. Flowcharts are the starting point for high-level design, and the first activity is to finalize one single concept that will be further developed, optimized, validated, and launched.

Looking at the two concepts for the service to be designed, the following pros and cons need to be considered:

· The first concept depicted in Figure 13.4 has a simpler high-level flow compared to the second concept.

· However, the first concept will require higher investment, and the company will need to enter into a new area of business.

It is also possible to develop a concept that combines the features of alternative concepts. In our case, it is possible to have a service that is the mixture of the preceding two concepts. The idea of providing consultation for healthy nutritional food and lifestyle can be combined with the first concept and added as a premium feature. Similarly, ideas for recommending restaurants and having their own operation for food can be combined. A new combined idea will be launching their own brand but subcontracting the operation of food preparation. At this stage, techniques like Pugh's matrix are very useful. To make a final decision on which concept to develop further, all the aspects of business need to be considered. At this stage, the role of the champion and process owner are very important. The champion needs to help the team in making the final decision based on the vision, strategy, constraints, and priorities for the overall business. The team needs to develop a business case for both the alternatives. It is important to consider the strengths and weaknesses of an organization and availability of resources.

For WFM, the goal is to grow revenue to $4 million in the next two years and increase the overall profit margin to 10%. If the current businesses continue to grow at 20%, WFM needs to generate $1.6 million from this new service and possibly with franchised operations. This new service needs to be launched in the next six months, with a total budget of $1 million.

The team needs to look at the feasibility of each concept as well as its probability of the success based on strengths and weaknesses. WFM is a recognized name in the area of health food and has an established good relationship with existing customers. WFM has a good brand and success record in the area of service. However, it is new to physical products; currently it has no WFM brand physical product. The team needs to evaluate the feasibility of launching a new service as well as physical product (packed nutritional healthy meal) in the next six months, with a budget of $1 million.

After careful consideration, it was clear that it will not be feasible to launch a service bundled with a product in such a short time. The team needs to focus on launching the service first, as described in the second concept. Once service is established as a second-generation launch, WFM needs to consider launching its own brand of packed nutritional and healthy meals. It has the benefit of lower investment; however, it has the drawback of creating its own competitors.

**High-level design.** Now let us take a look at the activities that the team needs to perform at the high level of design:

*Select the final service design concept:* The second concept is selected based on the logic described earlier: Identify key interactions and transactions for the selected concept. Figure 13.5 shows the key interactions and transactions for the selected concept. Establish performance targets for these interactions and transactions: For each interaction and transaction a critical to quality (CTQ) tree needs to be developed, and for each CTQ parameter, a performance target in terms of sigma level needs to be established. Let us take an example of one interaction and develop a CTQ tree as well as a target.

The first interaction described in the flow is "Communicate with Customer"; Figure 13.6 shows a CTQ tree for this interaction. The main drivers for customer communication are:

· Need to introduce service to target group of customers
· Need for exchanging information with the customer once they establish the first contact
· Need to collect and act on feedback generated

The CTQ attribute of introducing the service is the percent of target group of customers introduced to the service. After introduction, what percent of these potential customers become interested in the service and out of those, who become interested, what percent will establish contact or call. All three CTQ attributes of introducing the service are specific and measurable. The team needs to establish a target for each of the attributes.

The first CTQ attribute for the "information exchange" driver is "Customer received desired information in time." For this type of attribute, the target will be based on VOC. Let us assume that after studying the VOC, the target was determined to be five minutes. As the customer wait increases longer than three minutes, customer satisfaction decreases on an exponential basis. Beyond five minutes, you begin to risk customers and lose future revenue. Another CTQ is: "Information from cus-

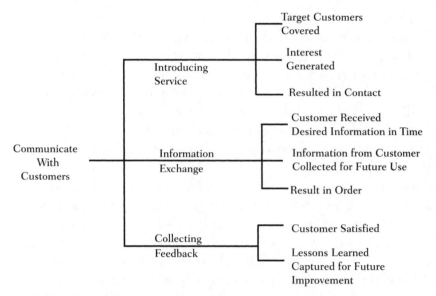

**FIGURE 13.6.** CTQ Tree for One of the Interactions

tomer collected for future use." It is a subjective attribute, and the target may be based on a scale with low and high points defined with the help of an example. The final attribute for information exchange is: "customer calls results." The target for this attribute will be the expected percentage of customer inquiry calls that should result in either first time or repeat orders. For collecting a feedback driver, both CTQ attributes are subjective and need to be further defined with the help of a subjective scale (e.g., 1 to 5 scale where 1 means customer is dissatisfied with the response and 5 means customer is very happy). The target will be the average value of all the feedback above a set limit.

A CTQ tree, similar to Figure 13.6, and targets for each CTQ attribute will be established by the team for all the interactions and transactions for service flow depicted in Figure 13.5.

Evaluate the design of each interaction and transaction to establish confidence and identify the gap from the target. At the high-level design stage, the team needs to establish the concept design for each major interaction and transaction. For example, the interaction of "communicate with customer" has several elements. The first part is related to advertisement and publicity. The team needs to discuss the current strategy of advertisement and decide if it will need modification for the new service. The team will also need to decide: Do we have proper resources for designing this element, or do we need outside expertise? Decisions on advertisement message, target audience, budget, and media need to be taken at this stage. For example, the team may decide that daily train commuters are one of the target groups of customers; to capture their attention, billboards at or near the train station will be used. The decision on the exact form of advertisement, locations, and numbers are not part of high-level design. These need to be covered during the detailed design stage.

Fill the gap by combining and modifying ideas from other design concepts or by further developing the current concept. All the service ideas for conveniently providing a healthy and nutritional fresh meal need to be evaluated by the team. High-level design should be created by adapting ideas from all the available sources. For example, the team needs to look at how various types of companies in the service industry will handle their feedback, and the best way of doing it for this new service. The company's way of handling feedback for the health food retail business may not be the best solution for this service. The WFM needs to study how best-in-class restaurants collect and handle feedback. Sometimes you may have to go out of the industry and look at an unrelated industry that is good in doing that specific aspect.

Partition the design work and assign it to subteams. It is important to decide the criteria for partitioning the work and establishing a subteam for each divided task. For example, subject matter expertise may be the criteria for partitioning the work. Looking at Figure 13.5, you may partition service design into the following six subteams:

1. Marketing
2. Information processing
3. Contract negotiations
4. Consultation format

5. Finance
6. Communication (responsible for designing for this service as well as between subteams throughout the design phase)

Establish a clear objective and links for each subteam. The objective for the marketing team will be to design a marketing campaign to launch and grow this service. The marketing team will need to work with the contract negotiation team, which will be negotiating with meal-providing partners as well as consultants. Similarly, the contract negotiation team needs to work closely with the consultation format team.

The team needs to conduct overall service design review at this stage to establish confidence in high-level design. For review, the project team needs to invite some outsiders to provide a fresh look at the concept of service as well as high-level design. The WFM team may want to select some early restaurant partners and suppliers of advertising service and use them to review the overall concept. Experts from financial institutions who will be providing the required investment may also be part of the review exercise. Finally, owners and current employees of WFM can take another look at the available details and how they feel about the chances of success for this new service.

**Detailed design.**   Once high-level design is approved, subteams need to start working on detailed designs. In this section, detailed design activities are explained with the help of the WFM case study.

*Develop details for each interaction and transaction:* Let us take the example of a consultation format subteam. This team will work on design details for two interactions, consultation and recommendation presentation, and one transaction, preparation for consultation. Let us discuss what the team will need to do for a detailed design on the format of recommendation presentation. The team needs to decide the following:

· How will the recommendation be presented, and will it be given as hard copy only or also a soft copy?

- Will the customer be required)to sign a disclaimer?
- Who should present it to the customer?
- What will be the sequence of recommendations?
- What needs to be included in the various sections (e.g., introduction and conclusions, etc.)?
- Will the recommendation report follow a standard format, or will it be customized for every customer, with certain standard features?

At this stage, the team will also need to look at the function of recommendation presentation, and the potential failure modes, their severity, chances of detection, and recovery plan. For example, one of the functions of presentation may be to protect the business against any liability. This will be accomplished by the signature of the customer on a disclaimer and a presenter properly explaining it to the customer. The team will need to discuss the severity of failing to get a customer signature, chances of its happening, and if it happens, how it will be detected and what actions will be taken after that. It is also important to discuss and decide the metric for success.

*Develop functional boundary diagram:* Figure 13.7 shows the functional boundary diagram for the interaction, "Customer calls to order." The input for this interaction is the call from the customer, and the ideal output is the successful order taken by booking an appointment for consultation. There are several possibilities of error and mostly these errors are caused due to variations in uncontrollable factors or noises, such as suddenly a very high volume of calls. However, the idea of proper design is to set control factors at a level that minimizes chances of error at the lowest possible cost. The boundary diagram also shows the direct and indirect link on both the input and output side of the interaction. The boundary diagram provides a tool for understanding interactions and transactions. It can also be used for establishing simulation experiments.

*Analyze and simulate the design performance against the requirements for selected complex elements:* In some cases, an

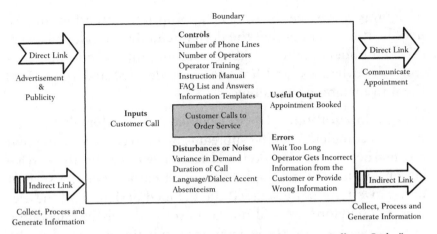

**FIGURE 13.7.** Functional Boundary Diagram for "Customer Calls to Order"

interaction or transaction may be so complex that simple logical analysis may not be sufficient. A proper simulation model needs to be developed and analyzed for making detailed design decisions. In our case, all the interactions and transactions are simple. Therefore, no computer simulation is required at the detailed element design level. Simulation may be required during the optimization phase.

Ensure that subteams communicate effectively with each other during the detailed design activities. As discussed during high-level design activities, there is a strong link among various subteams tasks, so it is essential that all the subteams are communicating. To emphasize the importance of communication, we recommend a subteam responsible for designing communication for the service, as well as ensuring communication between subteams during the detailed design phase.

Create detailed documentation of procedures, work instructions, and special instructions. For example, for handling customer calls and their feedback, it is important to provide standard work instructions to all operators. Also, as a guide, providing answers to frequently asked questions will be very useful.

Conduct a detailed design review to establish confidence on every element of the service. At this stage, detailed design reviews for each element design is a must. The DFSS way of designing a service is to pay appropriate attention to the details. It is essential that the design team has developed a very detailed understanding of ideal function, inputs, outputs, controls, disturbances, and links to each element and considered them in making design choices.

**Design integration.**   Once a detailed design for all the elements is completed, all the subteams need to come together and reconstruct the service flow and review it based on the service design axioms and principles described earlier. For the sake of brevity, we have not discussed the detailed design of all the elements. Therefore, we will not go into the details of integrating aspects for this case study. However, activities described in the previous section for integration need to be performed once detailed designs are ready. The main emphases at the integration is interface design and ensuring the details for implementation. For example, the contract negotiation and consultation format teams need to come together and decide on the unique requirements of experts who will be providing consultations.

## OPTIMIZE

The last phase of DMADO methodology is optimization of the service design. The activities for the optimization phase may be divided into three major subcategories. We will examine each of the subphases in this section.

1. Final value check and robust design
2. Validation
3. Launch preparation

The subphase of final value check and robust design is about examining the service design coming out of the design phase and using lean techniques like value stream mapping to ensure there are no inefficiencies left in the system. The service as a whole package and every element of it must be adding value. In addition, to be lean, every element of the service must be robust. We recommend robust design based on a two-step optimization. Here, the first step is to minimize the variability in the ideal function or output, and the second step is to adjust output as close to target as possible.

In most cases, a validation of service design is done by running a pilot. The objective of the pilot is to observe the outcome of the value optimized service as close to a real-life setting as possible. It is a test of the "whole system" on a small scale. The pilot provides invaluable experience and preview of potential bottlenecks and issues you may face with this service and makes full-scale implementation more effective. Running a pilot should follow the cycle of plan, do, check, and act (PDCA). Plan the pilot, run it, analyze the results, and take actions to improve. Answers to the following questions will describe the PDCA cycle for pilot:

· How do we plan the pilot to ensure that it is realistic?
· What data and information will we collect while running a pilot?
· How do we analyze the data collected from running a pilot?
· What actions do we take if the pilot performance analysis shows areas of concern?

A successful validation subphase will result in an improved implementation plan by anticipating and mitigating risks and complications. It should also help the organization in getting prepared for the changes coming and increased buy-in. The overall benefit of validation is an improved design with fewer surprises.

The final step of DMADO methodology is to prepare for the launch. It is always needed no matter how methodical you have been in your design activity. The difference between a traditional and DFSS launch is that the team is prepared to handle any unexpected events and chances and to minimize the adverse effects of unexpected events. During the define phase, there is an understanding about the boundaries of implementation; however, this needs to be revisited so that the team knows when its work is complete. If implementation involves multiple sites, the team needs to identify how the design will be rolled out to different locations or areas. The pilot plan should be the basis for the implementation plans, using the same PDCA approach. Improvements from the pilot and scale-up issues should be incorporated in the launch plan. It should include the following:

- Detailed work instructions
- Risk mitigation and contingency plans
- Transition plan
- Plan for organizational changes
- Training plan
- Communication plan
- Plan for continuous improvement and be sustenance over time

## TOOLS AND TECHNIQUES FOR OPTIMIZATION

The objective of this subphase is to ensure that the service is optimum and provides maximum value. Lean techniques such as value stream mapping combined with Taguchi robust design methodology is the way to optimize and create maximum value. Lean production is a business philosophy originally made popular by Toyota Motor Company. The objective of lean is to eliminate all types of wastes (*muda* in Japanese). Taiichi Ohno, the "Father of Lean," identified seven types of wastes. These are:

1. Overproduction
2. Waiting
3. Transportation
4. Process inefficiencies
5. Excessive inventory
6. Wasted motions
7. Rework

These wastes are mostly optimized in manufacturing industries. However, all of these are relevant to services. For example, over production for service means the gap between expected demand and actual demand that will result in wasted resources. Similarly, the function of inventory in manufacturing environment is to act as a shock absorber against the fluctuations of demand and other uncertainties. Most of the services designed will have accompanied goods with a poor design in service and will result in excessive inventory of the associated goods.

The technique of lean most useful for service design is value stream mapping. Value stream is the set of all the specific actions required to deliver the service. From the value point of view, three types of actions throughout the process are:

1. Value-added
2. Nonvalue-added; however, unavoidable due to current technology constraints
3. Nonvalue-added and avoidable

All the value-added steps need to be kept and all the nonvalue-added steps need to be eliminated. Those steps that are nonvalue-added but essential need to be properly managed. For all the steps that will be part of the final service, a proper flow should be created.

**Value stream mapping.**   First, develop a flowchart of the service listing what actually happens or, if it is a new service, based on design phase output. Value stream mapping flowcharts are more detailed than the high-level flowcharts shown in Figures 13.4 and 13.5. Flowcharts similar to Figure 13.4 can be a starting point. However, the team needs to develop detailed flowcharts for most of the interactions and transactions described there because most of the inefficiency lies in the detail and not at the macro level. Start at the end point (closest to the customer's pull) and work upstream. The next step is to identify whether process steps add value in terms of customer or stakeholder needs. For each step in the flow, ask the following questions:

- What is the output of this step?
- Who needs the output from this step?
- Does it add value?
- Is it visible and measured?
- Can it be eliminated without harmful effects?
- Can it be combined with another step?
- Is it being performed in the most efficient manner?

Once a service process is analyzed and all types of nonvalue-added steps are eliminated, the next step is to ensure continuous flow. In the service process context, it means all the service steps must be balanced based on the *takt* time. Takt is a German word for pace. In the continuous flow service process, it is the overall service rate required for the complete service to match the customer demand. Takt time is calculated by using the following ratio:

Available service time per day/demand for service per day.

Available time includes all shifts, and excludes all nonproductive time (e.g., lunch, clean-up, etc.)

Continuous flow leads to the concept of *pull scheduling*, in which input is not available for a step until it is ready to receive and act. Another related concept is *just in time inventory*, which is based on pull scheduling. To achieve success with JIT and pull system, these must be supported by some proactive ideas such as total productive maintenance (TPM), proper training, visual controls, 5S, and mistake proofing.

**Robust design.**    Robust design methodology requires selecting specific values of controllable factors such that the desired output is at the target, with minimum variation under all possible noise conditions. Before going any further in the discussion about methodology, certain terms need to be defined. All the factors that affect the output or ideal function can be classified as follows:

*Control factors:* Those factors over which the designer has control and may select any of the feasible levels for these factors depending on their effect on quality characteristics.

*Noise factors:* These factors are either uncontrollable or costly to control or lie in the domain of the customer.

*Signal factors:* These are the inputs from the user or operator, with certain expectations of output or responses for quality characteristics.

Figures 13.2 and 13.7 show all of these factors in the form of a block diagram, also referred to as a 'P' diagram in some texts.

Dr. Taguchi recommends a two-stage strategy for robust design. First, control variability by maximizing signal-to-noise ratio (S/N) and later adjust the response to the target. To calculate the S/N ratio, Taguchi recommends an inner and outer orthogonal array design of experiment. All the control factors are kept in inner array and all the noise factors are kept in outer array. Using this cross array design scheme, experiments are conducted and S/N ratios are calculated for all the runs in the inner array.

Control factors are divided in two groups: The first group is of those factors that significantly effect the S/N ratio. These are fixed at the levels where they provide a maximum S/N ratio. The second group is of those factors that do not significantly affect this ratio. These are treated as scaling factors to bring down response at its target level or to reduce overall cost. Once the level for all the controllable factors is decided, a confirmation experiment may be run to see whether desired results are obtained. Otherwise, scaling factors are used to fine-tune the results.

For more details on Robust Design, refer to the book by Taguchi listed in the reference section.

## VALIDATION

The best way to validate a service design is by running a pilot. There are many ways to run a pilot. It is important to be clear on the purpose of the pilot and to have a good understanding of what we want to know from the pilot and what data we need to collect to evaluate the pilot. The following are some examples relevant for service design:

- Computer simulation
- Test market release
- Limited-time offers
- Limited implementation for one location, one group of customers, or one type of service

As discussed earlier, the basic methodology behind a pilot is the PDCA cycle. For successful planning and execution of a pilot program, support from the project champion is a must. The team needs to review and get the approval from the champion and other stakeholders, as this is the final step before getting ready to launch and the last chance of making any adjustments in the service package. The objective is to be as close to the real-life situation as possible. The team needs to develop a detailed plan and schedule to decide and implement the following:

- What aspects of service need to be piloted?
- What is the best location to run the pilot(s)?
- Who will play what role?
- When will the pilots start, and how long will they run?
- How will the pilots be conducted?
- What are the risks and liabilities associated with pilots?
- What data will be collected?
- How will the collected data be recorded, monitored, and analyzed?
- Who will train the employees selected to provide the pilot services?
- Who are the target customers?
- How will the customers get information on the pilot?

It is essential to recruit a variety of internal and external people for running the pilot. Recruitment efforts should start well before the pilot to ensure a winning team of participants. Internal recruitment will require support from the project champion and communication to other groups or functions of the company. External recruitment will need the support of sales, marketing, purchasing, and others who manage customer and supplier and media relationships.

Training pilot participants to perform tasks for this newly designed service is critical to the success of the pilot. To identify the training and job aids needed, ask the following questions:

- What tasks or process steps are needed in the new service?
- How many of these are completely new or have never been performed?
- What qualifications, knowledge, and skills are required for each of these tasks?
- Who is the best qualified person to perform these activities or tasks?
- What is the gap that needs to be filled with training?
- What is the best way to train them to perform these tasks?
- How do we evaluate the readiness of those who will be performing pilot activities or tasks?

After developing the pilot plan and making all necessary decisions and preparations, it is time to conduct the pilot. Once the pilot starts running, it is important to closely monitor the response and progress, as in many cases, last minute adjustments may be needed. Data need to be collected as planned to assess key indicators and operating/usage conditions. In most cases, time-series data will be collected on all relevant CTQs. Collected data need to be analyzed to provide following answers:

- Are we providing the right service?
- Are we doing it the right way?

· Will it meet our business objectives?
· Will we really delight our target customers?
· What worked, what didn't?
· Did we accurately predict issues?
· Were our estimates on target?

During the pilot run, the team needs to make careful observations of all activities, effects, and interactions to answer the preceding questions. It is important to continue the pilot long enough to collect stabilized baseline performance data. Conduct performance reviews more frequently at the start and reduce frequency later, as performance start to stabilize. During reviews, check both the pilot plan and what actually occurred; analyze the gap to identify the root causes of deviation.

The review team also needs to take a look at how the employees or pilot service providers coped with the unexpected issues. What did they try, and how successful were they? Finally, during pilot results analysis, the review team will make decisions on the following three key questions:

· Are we ready for a full-scale launch?
· What lessons have we learned?
· What improvements are needed?

In most cases, some lessons will be learned and improvement needed. Another round of validation may be required. However, if the team has followed the DMADO methodology and supporting tools, chances are no major improvements will be required. Once design is improved and revalidated, the next step is a full-scale roll out.

## LAUNCH PREPARATION

Launching a new service requires a lot of preparations and last minute checks and balances. A pilot run provides great insights; however, a full-scale roll out will always bring its own challenges. The planning and review activities are similar to those described in the previous section on validation. It is critical that the team spends enough time on deciding the implementation strategy, and plans and updates all the documents and details of design based on the pilot run experience. In the initial launch phase, it is good practice to continue monitoring and making adjustments to the service if needed. After launch activities are over, a formal handover or transition is essential to ensure that the regular organization is prepared to assume the responsibility. The pilot for validation and launch should be a continuum where the PDCA cycle is repeated ideally twice (once for validation and the second time for a full-scale launch), but in reality it may happen more than twice. Figure 13.8 shows this concept of PDCA cycles.

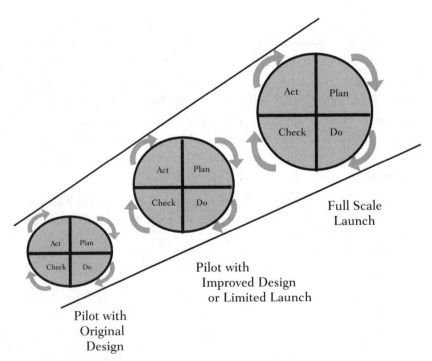

**FIGURE 13.8.** PDCA Cycles for Validation and Launch

## KEY TAKEAWAYS

· The design and optimize phases of the DMADO methodology are further divided into three sub-phases. Design in high level, detailed, and integration and optimize in final value check, validation, and launch preparation.

· The design phase starts with selected concepts from the analysis phase. The first activity is to review these concepts and decide one single concept for design. At the high-level design, sub-phase activities and decisions have broader implications.

· At the detailed design phase, tasks are divided into several teams, each taking responsibility of certain elements and paying attention to details.

· The final steps for the design phase are related to integration, where outcomes from various sub-team activities are put together again, ensuring interaction and interfaces are properly taken care of.

· The overall outcome of the design phase is a complete service package that has all the required details, is seamless in flow, and satisfies both the axioms of service.

· The optimization phase starts with the final value check using lean principles, with the main focus on value stream mapping. Once all the non-value-added interactions and transactions are either eliminated or properly managed, all the key steps are optimized using the robust design approach as recommended by Taguchi.

· The optimized service package is validated with the help of a pilot run using the basic concept of PDCA, and launched with full confidence.

# IMPLEMENTING SIX SIGMA IN SERVICE ORGANIZATIONS

This chapter presents an approach to implementing Six Sigma in the service environment by first understanding the impact of differences between the service and manufacturing organizations. The methodology offers a practical approach to establishing the executive commitment before launching the Six Sigma initiative, a roadmap for successful implementation, and a measurement system to sustain Six Sigma.

Service functions have been an integral part of many corporations. Organizing service functions together into a business entity creates a totally different mindset. A service organization, even with some common business processes, acts differently due to focus on customer requirements and prompt feedback from customers. Service offerings are experienced much faster than products, which sometimes can be stocked in a warehouse or a showroom. Once service is delivered, the customer experiences it and expresses satisfaction or dissatisfaction with the service just received. Therefore, the challenges are different in that problems must be resolved faster.

In a recent analysis of customer satisfaction by the author, it was observed that customer satisfaction is a function of quality, timeliness, and cost. In mathematical terms:

Customer Satisfaction = $F(Q, T, C)$

Reviewing the preceding relationship, if a business can deliver a product or service well within the customer-specified time, the customer will be satisfied. If the product or service is delivered again for a defective item, customers usually do not mind. On the other hand, if one has a great product, and cannot deliver to the customer when needed, the customer does not want it. Therefore, one can deduce that the process cycle time, or the time to deliver service, is more important than the quality and cost of the service. Typically, Q and T are considered prior to the price by a value-conscious customer.

Therefore, the relationship between customer satisfaction and its components can be more appropriately expressed as:

$$\text{Customer Satisfaction} = \text{Price} + \text{Quality} \times \text{Timeliness}^2$$

This relationship explains the significance of timeliness over quality levels. Here the customer expects faster, better, and cheaper services or products. Service organizations, as expounded earlier in the book, have two components, namely the transaction and interaction. The transaction component implies more process dependence for the outcome of high-volume and low-value services. Transaction-heavy industries include fast food restaurants, direct mail, banking, healthcare, insurance, and ticketing. The interaction component implies more personal care and attention dependence for the outcome of low-volume and high-value services. Interaction-heavy industries include sit-in restaurants, specialized healthcare, and personal services. The transaction-focused services are expected to be delivered faster, better, and cheaper, while the interaction-focused services are expected to be delivered better, faster, and cheaper. In other words, the speed of service is more critical in transaction-based services while the quality is paramount in interaction-based services. Comparing the two types of services, one can identify the critical processes in each type of service. Figure 14.1 highlights the uniqueness and commonalities between the interaction- and transaction-based services. One can see that every business consists of both transaction and interaction processes. Therefore, one must scope out the organizational processes before launching the Six Sigma initiative.

| Transaction Based Services | Interaction Based Services |
|---|---|
| High-volume and low-value services | High-value and low-value services |
| Extensive process dependence | Extensive people dependence |
| Fixed customer input | Flexible customer input |
| Invisible to customer | Visible to customer |
| Less responsive to customer | More responsive to customer |
| Faster, better, and cheaper expectations | Better, faster, and cheaper expectations |
| Supported with minimal human interaction | Back up by some transactional processes |

FIGURE 14.1. Transaction and Interaction-based Services

If one looks at the banking operation, the fast food, insurance, or healthcare businesses, they all employ lots of people as well. Normally, the transactional-based businesses are replicated, franchised, or even outsourced to process higher volume, while the interaction-based businesses are smaller or fewer in number, as customer care is a more critical requirement for them. Examples are sit-in restaurants, boutique shops, and specialty healthcare practices.

Service businesses, similar to the nonservice businesses, have processes such as marketing, sales, purchasing, human resources, operations, management, quality, and regulatory compliance. Calibration is an example that must be performed at almost every manufacturing facility with requirements for inspection, test, or measurements. However, in service, the calibration is not a universally applicable process. Similarly, the quality function in manufacturing operations tends to be more tangible and product specific, while in the service function, the quality function tends to be more subjective and process specific. Another function that differs between the two types of organizations is inventory or material management, which is more suitable to the manufacturing and product-specific businesses. However, in the service businesses, material management is typically replaced with information management. Comparing the inputs to the manufacturing and service businesses, one can see the differences in the process inputs, as shown in Figure 14.2. Understanding the process inputs reflects potential areas of problems that can be addressed by the implementation of Six Sigma methodology and helps set the right expectations. It appears that transactional processes have more information- and machine-related problems, while the interaction processes would have more people-related problems. The likely errors are listed in Figure 14.3, which shows that due to high people-related problems or opportunities for improvement, the interaction processes may have higher employee turnover too.

Implementing Six Sigma in the service environment, whether in transaction- or interaction-heavy organizations, one must consider distinct aspects of the business. This begins with understanding business performance in terms of a business's leadership style, corporate culture, systems and processes, values and decision making, organizational structure and politics, and performance and customer satisfaction. An analysis of var-

| Manufacturing Inputs | Service Inputs |
| --- | --- |
| Material | Information |
| Machine | Tools/systems |
| Method | Approach |
| Technical skills | Interpersonal skills |
| Quality measurements | Time measurements |
| Physical environment | Work environment |

**FIGURE 14.2.** Service versus Manufacturing Inputs

| Transaction Errors | Interaction Errors |
| --- | --- |
| Information accuracy | Lack of information integrity and accuracy |
| Information integrity | Misapplication of tools |
| Product quality errors | Inconsistent approaches |
| Product delivery delinquency | Poor customer service |
| | Delinquent service |
| | High employee turnover |

**FIGURE 14.3.** Transaction and Interaction Errors

ious organizational elements is performed to identify opportunities for improvement. For example, in a restaurant, there could be an opportunity for increased sales by improving customer service, creating incentives, or an effective marketing plan in the predefined market segment. Or, the profits could be improved through better inventory management in another restaurant. The business opportunity analysis consists of identifying business profit streams and identifying streams giving away profits or value. At end of the analysis, one must be able to distinctly identify the extent of potential improvement in profitability through the Six Sigma initiative.

# ESTABLISHING COMMON UNDERSTANDING

In some cases, the CEO or a key executive commits to Six Sigma, but not all executives commit to Six Sigma for their own reasons. Because the Six Sigma initiative requires dramatic improvement in performance throughout, a significant change in practices, and lots of creativity and innovation, all executives must develop synergy through a common understanding about the Six Sigma initiative in the organization. Such a cohesive understanding can be developed through executive training of Six Sigma, interviewing executives for understanding their drivers and distracters, and capturing their ideas for personalization. The corporate wide Six Sigma initiative, whether in a small company or a large company, must enlist diverse inputs and offer linkages to departmental and leadership objectives. Harmonization among executives plays an important role at the time of Six Sigma implementation practices that include the intent, methodology, tools, and measurements.

## DMAIE AND DMADO

When Six Sigma was launched the first time at Motorola, there was no Define, Measure, Analyze, Improve, and Control (DMAIC), or the Six Sigma Black Belts. Both of these aspects of Six Sigma have been developed to institutionalize the Six Sigma methodology. They do not address the cultural and strategic alignment of various resources. Based on the initial application of DMAIC in the design function, Define, Measure, Analyze, Design, and Validate/Verify (DMADV) was developed for implementing a design for Six Sigma. Standard DMAIC and DMADV must be tuned in to the needs of the service industry for incorporating service-specific nuances such as people emphasis, customer service emphasis, creativity, and cultural propagation of Six Sigma, instead of just controlling a process after improvement.

The Define, Measure, Analyze, Innovate, and Embed (DMAIE) addresses activity-based creativity in developing solutions, and cultural institutionalization of Six Sigma initiative as well as the methodology. Similarly, while applying design for Six Sigma in service, Define, Measure, Analyze, Design, and Optimize (DMADO) must highlight the need for balancing various conflicting corporate needs and constraints. Optimization in the development phase must pay dividends later because it would take into account potential sources of variability or inconsistency up front. While providing the Six Sigma training, one must emphasize innovation, embedding a solution, and optimizing a service design.

Having understood the objectives of the Six Sigma initiative and the requirements for customization of a Six Sigma toolbox for the service environment, one can look into developing a plan. The plan for Six Sigma must address the following issues.

## SIX SIGMA ROAD MAP IN SERVICE

Leadership orientation and buy-in begins with a common understanding of Six Sigma. To make a conscious decision about implementing Six Sigma, the executive team must be trained. With a good 30,000-foot level view of Six Sigma, the potential for Six Sigma can be determined in order to make the commitment to Sigma. Various steps needed to implement Six Sigma are shown in Figure 14.4. The focus of the roadmap is to implement Six Sigma top-down, based on needs. In other words, first identify the potential opportunity for improving corporate profitability, then define projects for significant improvement in performance and profitability. Benefits of Six

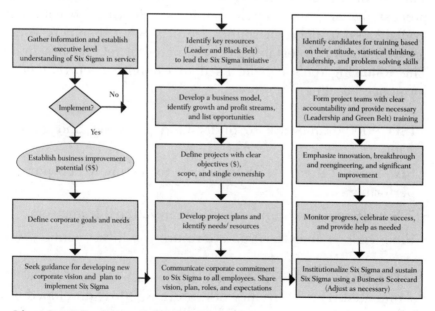

Reference: Gupta, *Six Sigma Performance Handbook*, New York: McGraw Hill, 2004.

**FIGURE 14.4.** Six Sigma Implementation Road Map

Sigma must be clearly understood in quantifiable and measurable terms.

## (RE) DEFINING THE VISION

If an organization has been successful, the service business will benefit from implementing Six Sigma through process efficiency, error reduction, and re-engineering. If an organization has been experiencing significant problems, Six Sigma may impact directly the policy of an organization and waste of resources. In case of transaction processes, the process itself can be improved. In case of interaction processes, the information and approach can be improved. The organization committing to Six Sigma must search its soul and figure out how to integrate Six Sigma into the organization's culture.

To define its new vision, the designated leader must gather information about the company's current implicit or explicit visions, values, and beliefs. If nothing is documented, an audit of the organization must be conducted to establish a baseline. The audit includes interviews of the management team, process owners, and sample employees. In case no data are available—unusual as it sounds, surprisingly it happens—one must find a way to establish an estimated or relative baseline. The following tools can be used to establish a current performance baseline:

- ISO 9000 framework to understand processes and gather data
- Six Sigma Business Scorecard to estimate corporate performance
- 7-S methodology to understand organizational framework
- Financials to analyze corporate returns on investments

The ISO 9000 and Six Sigma Business Scorecard are well documented. The 7-S framework (Stevens, 2001) developed by McKinsey & Co., consists of the following seven elements for assessing the strategic and soft attributes of an organization:

1. Shared values—for mindset
2. Structure—for accountability
3. Strategy—for planning
4. System—operations
5. Staff—culture and teamwork
6. Skills—capability and competency
7. Style—corporate personality and chemistry

Shared values are guiding principles of an organization that are subconsciously used in making decisions. The values could be as simple as mutual respect for people, integrity and accountability, sense of urgency, teamwork, or customer first. At first sight, any of these values appear to be a nicely crafted collection of words. However, when any one of them reflects an organization's mind and heart—its basic daily focus—these words carry a totally different meaning and significance. Collins and Porras in their book *Built to Last* (1996) have concluded that organizations that have survived more than 50 years and done well have one thing in common. They all have defined their core values. Even in hard times, values are preserved and practiced.

The 7-S framework provides useful guidelines for analyzing the strategic attributes of an organization. For an organization to be effective in launching a corporate initiative, the seven elements must be aligned and fit with each other, or each "S" must support the other "S's."

While establishing a baseline of performance, establishing casual relationships, getting feedback from employees, and gathering input from the leadership and analyzing results from operations, one can put the corporate puzzle together. One can start to understand why things happen the way they happen in the organization. The analysis gives insight into the corporation's collective mindset and sources of aberrations in it.

With the understanding of corporate values, strategy, leadership, processes, results, competitive environment, technological advances, future customer expectations, and challenges, one can start asking questions about what the corporation would like to become in coming years, three years, five years, or beyond. Though it is difficult to visualize beyond one year in today's fast-paced environment, one must still think in the future to set direction and commitments. A team of executives is brought together, soul is searched collectively, and direction is set or adjusted from the current course. This results in a set of abiding statements for vision, belief or values, goals, and key strategic initiatives.

## EXAMPLES OF CORPORATE GUIDING STATEMENTS

The following are examples of corporate guiding statements.

### DANNY'S RESTAURANTS

Great Food, Great Service, Great People...Every Time!

### INTERNAL REVENUE SERVICE

Provide America's taxpayers top quality service by helping them understand and meet their tax responsibilities and by applying the tax law with integrity and fairness to all.

### AMERICAN AIRLINES

American Airlines and American Eagle are in business to provide safe, dependable, and friendly air transportation to our customers, along with numerous related services. We are dedicated to making every flight you take with us something special. Your safety, comfort, and convenience are our most important concerns.

### MAYO MEDICAL SCHOOL

Mayo Medical School will be a national leader in medical education that graduates knowledgeable, skillful, and compassionate physicians who will enter postgraduate training prepared to assume leadership roles in medical practice, education, and research.

## THE COCA-COLA COMPANY

The Coca-Cola Company exists to benefit and refresh everyone it touches.

## FEDERAL AVIATION AUTHORITY

To provide the safest, most efficient, and responsive aerospace system in the world, and to be the best federal employer, continuously improving service to customers and employees.

## MCKINSEY & COMPANY

To help our clients make distinctive, substantial, and lasting improvements in their performance and to build a great firm that attracts, develops, excites, and retains exceptional people.

## NORTHWESTERN UNIVERSITY

Northwestern University is committed to excellent teaching, innovative research, and the personal and intellectual growth of its students in a diverse academic community.

## ADVOCATE HEALTH CARE

The mission of Advocate Health Care is to serve the health needs of individuals, families, and communities through a holistic philosophy rooted in our fundamental understanding of human beings as created in the image of God.

# PLANNING TO GET THINGS DONE

Once the vision and mission are clear, the organization's leadership must define its goals and targets to be achieved with the implementation of Six Sigma. To achieve goals and targets, key strategic initiatives must be identified and implemented through effective execution. The executive team develops a strategic plan to achieve goals and targets. These plans are sometimes difficult to explain. Therefore, the planning process must focus on getting things done, instead of identifying activities to be completed. There is a vast difference between a strategic plan and plan for getting things done.

A typical scenario, a company hires a consulting company for performance improvement. After the initial assessment, opportunities for improvement are identified; vision, beliefs, and goals are developed; and Six Sigma projects are identified. However, the support from the management team became scarce. For some reason, the management priorities changed and progress was delayed. It is known that about 10% of strategic plans are well executed. Therefore, one must also focus on how to execute the plan, instead of just develop a plan.

To execute a plan, the Six Sigma leader must understand the informal system of getting things done in an organization. The informal system may include specific individuals, supporting resources, or executive priority, and authority of individuals. To get things done, first there must be a plan consisting of an action item, responsibility, critical resources, and estimated date of completion. Starting the execution of a plan is a first step. To start execution of a plan, the first follow-up meeting must be scheduled promptly to review progress and to understand challenges that would hinder the project success. After a couple of follow-up meetings, or one third to half-way through the project, initial energy, enthusiasm, motivation, and follow-up become weak, and the project starts to slack. To sustain the momentum and activities to make progress, the key individuals must be continually informed and their help sought aggressively. To maintain the commitment of other team members and the management, the value proposition must be continually emphasized and highlighted. A strong project management is a requirement to implement the Six Sigma initiative successfully. The plan to implement Six Sigma initiative successfully must include the following:

· A visible and active participation from the CEO or equivalent leader of the company
· A designated and authorized executive, whose compensation is tied to the progress of Six Sigma, to ensure progress

- Team members committed with scheduled and available time to work on the project
- A strong team leader to maintain focus on the progress of the project
- A weekly and committed follow-up meeting or report

## ESTABLISHING MEASUREMENTS

Service aspects of the business, transaction, or interaction have become more important because of alliances, partnerships, and strong relationships with suppliers. Outsourcing, which is another form of a supplier–customer relationship, has accelerated the role of service even more. Just like nonservice operation, there is a customer expectation to deliver service faster, better, and cheaper (cost-effectively). Six Sigma projects and initiatives must support customer expectations to provide faster customer service with better quality and lower price. Reducing time to deliver a service requires that one maps the operations and measures the performance level in terms of time and quality. To reduce time to deliver the service means identifying processes that add no or little value, or those that are operating suboptimally and take longer than required. One should ask the question, "Why are we doing the unnecessary task, or do we need this step?" Of course, we need to manage resistance to preserve the old and bring in the new.

Establishing effective minimal measurements requires planning and analysis of information, approach or process, tools or systems, and people skills. One can identify critical input, in-process, and service characteristics to monitor customer feedback—parameters that, if not correctly or effectively managed, would adversely affect service performance and customer satisfaction. Once the critical parameters have been identified, measurements to assess effectiveness of a process can be determined by answering the following questions:

1. What is the purpose of a "service" function?

2. What are expected deliverables (people, skills, services, value, or reports)?

3. What are measures of goodness for key deliverables?

4. What are the opportunities for error in performing the "service" function?

5. What improvement activities are carried out in the "service" function?

6. What is the current estimated performance level in terms of errors per deliverable?

7. What is the Sigma level of your key "service" function?

These questions can be answered for each key process in an organization. Answers to these questions will assist in identifying meaning for all performance measures. Once the measurements have been identified, goals for improvement can be established and Six Sigma projects can be launched. For a project to be successful, the results must be visible, and goals must be aggressive. Setting aggressive goals requires creativity and innovation in developing a unique solution.

Another benefit of establishing aggressive goals is that progress can be substantial and employees can share benefits of the improvement; that in turns breeds more improvement.

## SMALL WIN PROJECTS

To launch a Six Sigma initiative, it is critical that corporate-wide implementation is planned in detail. However, before jumping in the full stream, it is safer to practice the methodology in selected pilot areas, or so called "small wins" for Six Sigma. The selection of pilot projects must be based on their probability of success, estimated time to complete, required resources, and resultant savings. Working on small wins helps the learning curve and allows time for people to accept the new approach to doing things. Otherwise, the resistance prevails, the initiative dies, and the wait for the new initiative begins.

Making small wins successful is critical to the success of the corporate-wide initiative. We should deploy most interested, enthusiastic, entrepreneurial, and learning individuals on the projects. The initial projects must also be assigned to strong and successful leaders to ensure they lead these projects to completion. The challenge remains to complete the project in time, before the time runs out and management loses its interest in the Sigma initiative.

Periodic review and publicity of project progress must become a well defined task. Publicity of success is the best way to teach and grow interest and excitement in the Six Sigma initiative. Small wins must generate demand for more projects to be added to for implementation of Six Sigma. Small wins must be greatly celebrated without setting false expectations. Employees must see that Six Sigma is attainable and is important to the organization for its sustained profitability and growth. Depending on the improvement, small wins can be recognized publicly by the CEO for exceptional results.

## INSTITUTIONALIZATION

The success of small wins must accelerate the institutionalization of the Six Sigma initiative. It means identifying more projects, including more divisions of the business, and identifying more people to work on Six Sigma projects.

The most significant preparation for corporate-wide implementation would be to harmonize, integrate, and align Six Sigma with other corporate initiatives. Six Sigma must become an implicit or explicit element in the corporate vision, mission, and values. The leadership must look into the stretch goals, performance-based compensation, consequences for nonperformance, and periodic reporting of progress. The managers must be involved in guiding their employees or providing training. The challenge is who and how much training should be provided to raise awareness and develop internal competency.

The awareness training is usually provided to all employees. They must see how Six Sigma helps to achieve business objectives, and how it affects them in doing their jobs. Most important, employees should know what they are supposed to do to make the Six Sigma initiative successful. Too many times, we have heard that people do not know what they are supposed to do and get punished for not achieving results.

It has been seen that progress has been limited when learning all the tools without learning the purpose of the project and intent of the tool. Therefore, the focus of the training must be to create the Six Sigma thinking first, before dumping all the tools in the employees' toolbox. Six Sigma requires that lots of improvement is achieved very fast, requiring a passionate commitment by leadership and employees who are directly involved in the implementation of Six Sigma. They must be committed to find a solution to achieve the desired level of improvement. Nothing less would be acceptable to them.

Many companies launch thousands of projects under the Six Sigma umbrella. There is no need to initiate too many projects. The important aspect of institutionalization of Six Sigma is to ensure employees understand Six Sigma, what it means to the corporation, what it means to them, what they need to learn, what their goals are, and how they would practice Six Sigma in their areas. Employees must also see that if the results are not achieved, the corporation will suffer, and so will the employees. Too many times, the leadership is unable to communicate consequences of nonperformance and focuses on incentives only.

## PREDICTIVE BUSINESS MANAGEMENT

Service organizations are more process dependent, and the deliverables may be a combination of tangibles and intangibles. Customer satisfaction feedback is more real-time—at the time of delivery, rather than later, as in the case of a product. Sometimes, service deliverables are difficult to inspect and verification is performed by the customer. Therefore, service organizations must implement predictive business management principles, where the performance is predictable and preventive actions can be taken proactively.

To implement a predictive business management method, an integrated performance scorecard is required that establishes effective checks and balances that if one element of the process or business is not performing to the expectations, a flag can be raised and action can be taken before the damage is done. A Six Sigma Business Scorecard (Gupta, 2003) has been developed for managing corporate performance. Accordingly, the leadership is responsible for inspiration, managers for improvement, and employees for innovation.

The purpose of the Six Sigma Business Scorecard is twofold: 1) to identify measurements that relate key process measures to a company's profitability, and making the opportunities so visible that they are difficult to ignore, and 2) to accelerate the improvement in business performance. Optimizing profitability and growth is the primary purpose of the Six Sigma Business Scorecard.

The Six Sigma Business Scorecard provides a new model for defining a corporate Sigma level, aligns with the business' organizational structure, and includes leadership accountability and rate of improvement. One of the main tenets of Six Sigma is to measure what is important. Since Sigma is an important initiative for an organization, its progress toward the goal must be measured. Knowing what is the Sigma level, the leadership can adjust its strategy and tactics to stay the course with lots of improvement very fast.

The Six Sigma Business Scorecard is a great model for establishing a common target in terms of corporate defects per unit (DPU), defects per million opportunities (DPMO), and Sigma. The corporate Sigma level is determined based on the Business Performance Index (BPIn), which is a combined outcome of 10 measurements. The 10 measurements are as follows:

1. Number of employees recognized by CEO
2. Profit
3. Rate of improvement
4. Employee recommendations for improvement
5. Total spent/sales

6. Supplier's defect rate
7. Operational cycle time variance
8. Internal process defect rate in Sigmas
9. New business/total sales
10. Customer satisfaction

The BPIn assigns the significance of each measurement and determines its contribution towards the total performance of the company. BPIn provides a one-number measure of corporate performance, and in terms of Sigma, it can be used as a benchmark for driving future improvements. The benefits of the Six Sigma Business Scorecard include the following (Gupta, 2003):

· Provides a target for performance improvement
· Enables a business to drive dramatic improvement
· Promotes the intellectual participation of all employees
· Forces changes in an organization on a continual basis
· Reduces costs and improves profits

At the department level, the measurements are established in terms of quality (Q) or wellness, responsiveness or timeliness (T), and cost-effectiveness (C). Then, each department plans to improve its performance in terms of Q, T, and C.

The continual and combined efforts of various project teams leads to lots of improvement very fast in terms of response, goodness, and cost. The customer loves to share the benefits, value, and new features that result from dramatic improvement.

## KEY TAKEAWAYS

· Establishing common understanding among executives is a prerequisite for successful implementation of Six Sigma at any firm.

- The focus of the Six Sigma roadmap is to implement Six Sigma from the top down, and based on needs.

- Six Sigma Business Scorecard, ISO 9000, financials, and employee surveys are available to establish a performance baseline.

- Leadership should plan for getting things done after identifying measurable and achievable goals and targets.

- Success of early projects accelerates the institutionalization of the Six Sigma initiative.

# SIX SIGMA IN SERVICES

Speed, flexibility, and operational efficiency are the essential ingredients of success in today's high velocity marketplace. So far, we have studied the role of services, Six Sigma, and DFSS methodology focused on services and transactions. Now we need to see how this methodology leads to specific benefits when applied to the commonly faced issues in the service functions.

In this chapter, we will explore the application of the Six Sigma methodology and tools learned in this book in the following areas:

1. Banking

2. Insurance

3. Healthcare

4. Restaurants

5. Transportation

A quick recap of the power of Six Sigma in these areas is depicted in Figure 15.1. The figure shows the potential of problems if the respective service operates at a particular Sigma level. It may be noted that these are *not* the actual but potential performance figures at a particular Sigma level. It can be seen that the higher the Sigma, the lower the probability of errors that can be severe for an organization.

| Sigma Level | Banking (Per Branch) | Insurance (Per Regional Office) | Healthcare (Per Hospital) | Restaurants (Per Outlet) | Transport (Per Airport) |
|---|---|---|---|---|---|
| 2 | One error in loan processing per day | 966 errors in premium follow-up per day | 106 prescription errors per day | 483 wrong items delivered to customer | 916 misplaced baggage items per day |
| 3 | Three errors in fund transfer per day | 105 errors in claim payment per day | 11 wrong medical tests per day | 105 delays in serving the customer | Three wrong landings per day |
| 4 | One error in account opening in 20 days | 19 errors in responding to the customer queries per day | Two misplaced personal items per day | 10 poor quality items delivered to customer | Two wrong frequent flyer program statements per day |
| 5 | Three errors in statement in 1000 days | 21 wrong statements in per month | One wrong test result per month | Two in five delayed cleaning of area | Two wrong tickets per three days |
| 6 | One customer complaint in 70 years! | One customer complaint in two years! | One customer complaint in 2.3 years! | One customer complaint in 100 days! | One customer complaint in 100 days! |

**FIGURE 15.1.** Sigma Level and Corresponding Error Rate

The following are the assumptions for various sectors, in calculating the figures:

1. Banking
   a. 24 million customers
   b. 5,790 branches
   c. 25% of customers apply for loan
   d. 1% of customers request for fund transfer
   e. 50,000 new accounts are open per day
2. Insurance
   a. 16 million customers
   b. 14 regional offices
   c. 50% of customers apply for claims
3. Healthcare
   a. Two million patients per year
   b. 16 hospitals
   c. 50% of patients undergo medical tests

4. Restaurant
   a. 47 million customers per day
   b. 30,000 outlets
5. Transport
   a. 65 million passengers per year
   b. 60 airports
   c. 2,800 flights per day
   d. 10% of passengers are members of a frequent flyer program

# BANKING

Six Sigma works well in the banking industry because it has a process-orientated environment. The banking industry in general looks to Six Sigma to do the following:

1. Improve customer relationships
2. Increase customer retention
3. Streamline processes
4. Increase employee satisfaction within the organization

Citibank, Chase Manhattan, and Bank of America have success stories of Six Sigma application in the banking industry. There are many more in the pipeline for the application of Six Sigma. This not only implies that the application of Six Sigma is crucial for all other banks to stay competitive, but also makes a way for the industry to operate with increasingly demanding customers.

Let's now look at the each element of the Six Sigma application in the banking industry.

## Corporate Vision

Normally, when a Six Sigma project is initiated, the bank is already running its operations and would like to improve certain areas dramatically fast. The purpose of reviewing the vision is to understand it and then evaluate the strategies to achieve the vision. The end objective is to have a fresh look at all the features of the bank operations to verify the consistency and then initiate the projects that are directly aligned with the vision. This process would involve the top management, which is vital for any Six Sigma project's success. Also, this verification may unearth some of the issues, which were probably relevant earlier, but are no longer applicable.

The beauty of Six Sigma is that there is no end to the improvements you can make. Once an organization gets a taste of the type of process improvements that Six Sigma enables, it becomes addicted to a culture of process improvements. Success breeds success, and Six Sigma provides plenty of success.

A bank may typically define its vision as:

- "To become number one bank in the local community"
- "To become number one bank in the state"
- "To become number one bank in the region"
- "To become number one bank in the country"
- "To become number one bank in the world"

For the purpose of this example, consider a bank that has a vision to become the "number one bank in the local community."

## CORPORATE STRATEGY

Once the vision is defined, the corporate strategy is chosen suitably. The strategy defines the means of achieving the vision. The elements of corporate strategy include:

- What would be the target market? For example:
  - High income group
  - Middle income group
  - Low income group

- How would the market be approached?
  - Advertisement
  - Internet
  - Viral marketing (viral marketing depends on a high pass-along rate from person to person)
- How many locations should be created to serve the customers?
- Where shall the locations serve the best purpose?
- What technology shall be used?
- How would the organization look like?
- What shall be the sources of funds?
- What are the projected financial statements?
- How would the management team be created and developed?

In this example, the bank has targeted middle to low income people, believes in viral marketing, and has a secured website. It has five branches in suburban locations, near the target market. All branches are headed by branch managers and have 5 to 10 staff, depending on the location.

## Operations Strategies

Once the corporate strategy is in place, the operations strategy is developed. The elements of operations strategy include:

- What shall be the processes of operations?
- What shall be complete organization of the resources?
- What would be the capacity?

The bank in this case has 60% of its employees in the front desk and 40% in the back office to support the customer operations. The bank has implemented a software system and has trained its employees.

## Define Phase

The define phase will identify the projects that may be candidates for the application of rigorous analysis. Defining a service defect is one of the most challenging aspects of applying Six Sigma to service-delivery systems. Until we reach agreement on what constitutes a service defect, your Six Sigma effort will not work.

We need to define what's critical to the customers and confirm that the core processes are aligned to those requirements. At the same time, we must understand the key business issues for our company and align the voice of the customer with them.

Banks have applied Six Sigma to decrease the number of reversals they need to make by identifying the factors impacting reversals, then seeking to eliminate or reduce those factors. Banks have also successfully used Six Sigma to reduce fraud.

Now let's come back to our example and apply Six Sigma principles to improve the customer satisfaction in bank branches. But before the Six Sigma project is initiated, it is important to create the right team. It is preferred to have a cross-functional team to maximize the effectiveness. Its members must be familiar with the Six Sigma process and at least the leader should have the complete understanding of the Six Sigma methodology and tools.

We either have in-house data or know the biggest problems faced. However, there is a caution. Just about everyone in financial services has a "gut feeling" that technology makes customers happier and banks more profitable. Some of the most accepted assumptions about customer behavior can turn out to be wrong. As discussed in previous chapters, various tools used for identifying the needs include:

1. Surveys
2. Focus groups
3. Promotional campaigns

The key banking activities that affect the customers include:

· Account opening
· Deposit

- Withdrawal
- Loan processing
- Credit decision
- Credit card approval
- Customer statement preparation and dispatch
- Customer query resolution
- Fund transfer—local, international
- Address changes
- Web-based support
- Customer complaint

We need to identify the factors that result in customer dissatisfaction. The issues are mostly identified by survey/focus groups. It's important to be precise in identifying how we are failing to deliver what the customer wants. Wait times, errors, unfriendly staff, and inaccurate customer information are some of the factors that come to mind. Once we determine the factors that affect wait time, we then work to minimize their impact. Tools like gap analysis may help identify the defects.

The following are the prominent gaps that need to be addressed while trying to define the problem:

- The gap between customer's expectations and management's perception
- The gap between management's perception and service quality specifications
- The gap between service quality specifications and actual delivered service quality
- The gap between actual delivered service quality and communication
- The gap between customer's expectations and customer's perceptions

After the issues have been identified, they need to be measured with the available data and then ranked in the order of their impact to the customer. This helps to decide the number of Six Sigma projects required to be initiated and their schedule using the available resources.

Let's assume that two Six Sigma projects are identified:

· Reduce the waiting time in the bank
· Reduce errors in the customer statement

In this text, we shall elaborate on the first project as an example.

The first activity after the project identification is to perform the stakeholder's analysis. As we recall from the text, the purpose of the stakeholder's analysis is to enlist all project participants whose influence is vital for the project's success and align their interest to the desired level to make it happen.

Figure 15.2 shows the stakeholder's analysis for the project identified. It may be interesting to note here that while for some participants the involvement may be low, their influence and interest is vital for project success.

After the stakeholder's analysis, the define phase is concluded with the preparation of the project charter. The project charter will be required for both of these projects separately.

The elements of the project charter would include:

· The business case for undertaking the project
· The problem definition

| Who | Required Involvement | Present Interest | Required Interest | Influence on Project Success |
|---|---|---|---|---|
| Chief Executive Officer | L | H | H | H |
| Chief Financial Officer | L | M | H | H |
| Vice President Marketing | H | H | H | H |
| Branch Manager | H | H | H | H |
| Front Desk Personnel | H | L | H | H |
| Team Leader | H | H | H | H |

FIGURE 15.2. Stakeholders Analysis—an Example

- The project scope
- Goal statement
- Roles of team members
- Timeline and key deliverables
- Resources for the project

The project charter for the identified project may be as follows:

**Business case.**  The average waiting time for customers during the weekdays is two minutes, and during the weekends, it's six minutes. The standard deviation of the wait time during weekdays is 1.3 minutes, while during the weekends, it's 4.5 minutes.

The high average and standard deviation of the wait time has caused frustration among customers and they are switching to other banks. In 2003, the bank lost 15% of its customers, and the resulting loss was approximately $100,000. If uncontrolled, the bank would lose many more this year, and the resulting financial losses may be much more.

By undertaking this project, the target is to reduce customer loss by more than 90%. This means the resulting savings shall be $90,000 per year.

To check this trend, the project team plans to analyze the whole process and perform certain experiments. This would result in the following additional cost elements:

1. Training of one Black Belt candidate @ $20,000 = $20,000

2. Training of four Green Belt candidates @ $3,000 = $12,000

3. Purchase of statistical software like Minitab for five persons = $5,000

4. Overtime to certain employees during the project period = $30,000

5. Customer participation scheme (motivation for feedback) = $20,000

The overall additional expense would be close to $87,000.

While the cost of 1, 2, and 3 may be allocated to the other projects, the overtime cost shall belong to this particular project. Let's assume that trained personnel completes five projects and uses the same software; the allocated cost for this project would be:

$7,400 ($37,000/5) + $30,000 + $20,000 = $57,400

Hence, we can say that by spending $57,400, we shall achieve a savings of $90,000 per year. This proposal therefore warrants an investment.

**The problem.**   In 2003, the bank lost 15% of its customers and the resulting loss was approximately $100,000.

**The project scope.**   The project shall be conducted at one of the five branches where customer loss was maximum. Learning, however, shall be implemented at the rest of the four branches.

**Goal statement.**   To reduce customer loss due to the problems associated with waiting in the bank by 90%. By reducing customer loss, the bank would move a step closer to its vision to become the top bank in the local community.

**Roles of team members.**   The team member roles are defined in Figure 15.3. These are broad roles, and there may be times when a lot of flexibility may be expected to ensure the project success.

**Project's timeline and key deliverables.**   The broad timeline and key deliverables of the project are listed in Figure 15.4.

**Resources for the project.**   The following resources would be required to complete the project:

· Black Belt training for candidate
· Green Belt training for four candidates
· Statistics software for five candidates
· Additional time of the regular employees is required for training and experimentation

| Who | Role |
|---|---|
| Chief Executive Officer | Communicate the project importance |
| Vice President, Marketing | Provide overall guidance |
| Black Belt | Lead the project team through all project stages |
| Green Belt | Lead different project modules and perform necessary experiments |
| Branch Managers of the Bank | Provide necessary overall support particularly in data availability and collection |
| Front Desk Personnel | Implement the team findings and share customer experiences |

**FIGURE 15.3.** Role of Team Members

| Project Stage | Time Frame (days) | Key Deliverables |
|---|---|---|
| Process mapping—As is | 5 | As is process map |
| SIPOC | 5 | SIPOC |
| Process analysis | 4 | Recommendations |
| Experiments | 10 | Results of experiments |
| Conclusion of experiments | 4 | Action plan |
| Implementation at the branch and monitoring | 30 | New data under revised conditions |
| Analysis of the effectiveness of the solution | 4 | Decision on further implementation |
| Implementation at remaining branches | 45 | Results of implementation |
| Analysis of the results from all other branches | 5 | Gap analysis and action plan (if any), or else conclusion |

**FIGURE 15.4.** Project Timeline and Key Deliverables

After the project charter is ready, the team moves to the next phase—Measure.

## MEASURE PHASE

Now is the time to translate customer expectations into measurable characteristics of the processes. Once we understand customer expectations, we must fulfill them by measuring the processes' effectiveness and efficiency. While we use the terms efficiency and effectiveness frequently, it's a good idea to recap their meaning and difference in context to Six Sigma projects:

"Effectiveness" is providing the solution to the problem of defects that the processes produce.

"Efficiency" is regarding the consumption of resources such as the time and money that the processes consume in meeting customer needs.

For the purpose of our project, we need to obtain the following information by conducting the customer survey:

· What is the most frequently used service by visiting the bank?
· What is the most frequently used service by visiting the website?
· Which other services would you be willing to use by visiting the website?
· What is the most frequently used service by visiting the ATM?
· Which other services would you be willing to use by visiting the ATM?
· Which of the following would be most valuable to you while you need to wait at the bank lobby:
  – Read a business magazine
  – Read an entertainment magazine
  – Read jokes/humor
  – View the latest news

The feedback obtained would help mitigate the problem in both ways: reducing the frequency of arrival to the bank and keeping customers productively engaged when they are into the bank. This has been elaborated as follows:

· This survey would help identify the actual reasons for visiting the bank and then the possibility to reduce the frequency can be explored by providing those services through the website or ATM to the customer's full satisfaction.

· The bank would know what to do when the customer is compelled to visit.

The following methodology may be implemented to administer the survey:

· Provide the survey form to all customers at the bank entrance with a pen and a clipboard.
· Request them to fill it while waiting for the service and inform them of the incentive of submitting the form at the service counter.
· Thank the customers for the survey and provide a gift certificate to the customers who submit the form at the service counter.

The most likely responses to the survey are as follows:

1. Most frequently used services by visiting the bank:
   a. Deposit the checks
   b. Clarify the charges to the bank statement
   c. Issues pertaining to the loan application
2. Most frequently used services by visiting the website:
   a. Balance confirmation
   b. Make transfer from one account to the other
3. Preferred additional services by using the website:
   a. More explanation to the charges in the bank statement
   b. Responsiveness from the website regarding the clarifications sought
   c. Loan management and administration
4. Most frequently used services by visiting the ATM:
   a. Cash withdrawal
   b. Make transfer from one account to the other
5. Preferred additional services by using the ATM:
   a. Loan management and administration
6. Customer preferred to read jokes/humor while waiting

## ANALYZE PHASE

The first task in the analyze phase is to prepare a process map and then prepare SIPOC (supplier, input process, output, customer). Just to recap quickly, the purpose of the process map is to see how the process is run under present circumstances, and the purpose of SIPOC is to link the process to its inputs, the suppliers to the inputs and the outputs, and customers to those outputs.

The "As-Is" process map is shown in Figure 15.5, and the SIPOC is shown in Figure 15.6.

**Cause and effect diagram.**    Now that we know how the process is performed and its inputs and outputs, we now move to the next level—analyzing the reasons for the high average waiting time and its variability. To prepare the cause and effect diagram, it is a good idea to brainstorm with the process performers and process owners. Let's assume that Figure 15.7 shows the cause and effect diagram for the problem under consideration. After the brainstorming session, the following is the rank of causes:

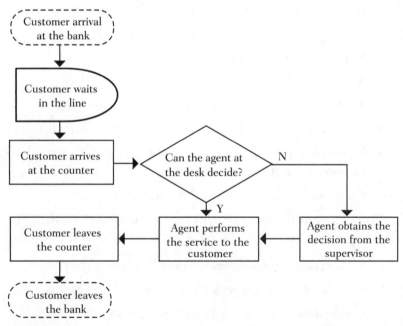

**FIGURE 15.5.** "As-Is" Process Map

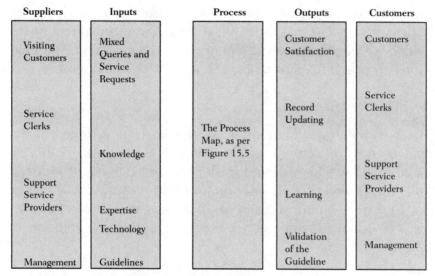

**FIGURE 15.6.** SIPOC for Serving the Customer

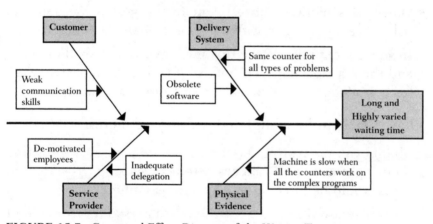

**FIGURE 15.7.** Cause and Effect Diagram of the Waiting Time

· Same counter for all service requirements of the customers
· Inadequate delegation
· Weak communication skills of the customers
· De-motivated employees
· Slow machines
· Obsolete software

After we know the causes, and also after we heard from the customers in the measure phase, we know the direction in which we need to proceed. We need to make various combinations and perform experiments to see which works best.

## INNOVATE PHASE

After completing the analysis, we know what we need to do. But before we actually implement the actions full scale, we need to know which should work best. To establish it, we need to conduct experiments and obtain the best outcome.

In our problem, we need to undertake the actions in phases. The following are the reasons for splitting the solution in phases:

· Quick wins should be achieved first.
· Quick wins motivates investment in the next phase and also help solve the problem with a greater degree of confidence.
· In subsequent phases, the financial investment shall increase and the management decision shall be based on past success for confidence and increased cash from success.

### Phase One—to be completed within one month

· Modify the process of serving the customers by segregating them right at the entrance; customers seeking assistance in loans would be directed to a specific counter.
· Designate a waiting area for the customers and provide light entertaining magazines and joke and humor books.
· Develop the guidelines for the employees and empower them to make all the decisions to resolve the customer issues.

At the end of phase one, the process flow diagram would look like that in Figure 15.8.

### Phase Two—to be completed within three to six months

· Provide training to key employees in the language frequently spoken by the majority of non-English speaking customers.

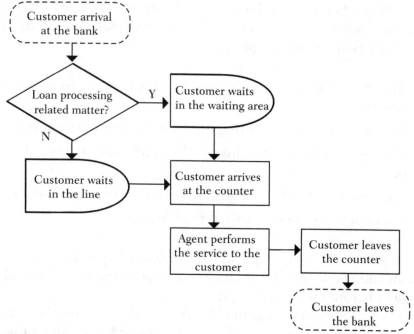

**FIGURE 15.8.** Modified Process Map

- Identify the reasons for employee demotivation by talking to them one-to-one.
- Provide online real-time support to customer queries by out-sourcing the function, and define the quality parameters for it.
- Provide the customer code with password to access the loan information online.

### Phase Three—to be completed within one year

- Explore the new machines and latest software to improve the productivity.

**Hypothesis testing.** The success of each of the phases needs to be evaluated by monitoring the following metrics:

- New waiting time
- Customer feedback on their satisfaction
- Any loss of customer during the period

These data need to be compared with the data prior to the phase implementation. The following statements need to be tested based on the sample data:

· The new waiting time during the weekdays is lower than before.
· The new waiting time during the weekdays has lower variance.
· The new waiting time during the weekends is lower than before.
· The new waiting time during the weekends has lower variance.
· Customers are more satisfied than before.

Once we know that any of the preceding hypotheses is true, it should result in a lower loss of customers and should be validated before implementing it.

Let's assume that Figure 15.9 depicts the data on the waiting time and customer satisfaction before and after phase one. We need to test whether there is an improvement with 95% confidence. We now have five hypotheses to be tested, as follows:

| | |
|---|---|
| Null Hypothesis 1: | The mean waiting time during the weekdays after phase one is more than or equal to that before phase one was implemented. |
| Alternate Hypothesis 1: | The mean waiting time during the weekdays after phase one is less than before phase one was implemented. |
| Null Hypothesis 2: | The variance of the waiting time during the weekdays after phase one is more than or equal to that before phase one was implemented. |
| Alternate Hypothesis 2: | The variance of the waiting time during the weekdays after phase one is less than before phase one was implemented. |
| Null Hypothesis 3: | The mean waiting time during the weekends after the phase one is more |

than or equal to that before phase one was implemented.

Alternate Hypothesis 3:     The mean waiting time during the weekends after phase one is less than before phase one was implemented.

Null Hypothesis 4:     The variance of the waiting time during the weekends after phase one is more than or equal to that before phase one was implemented.

Alternate Hypothesis 4:     The variance of the waiting time during the weekends after phase one is less than before phase one was implemented.

Null Hypothesis 5:     The customer satisfaction after phase one is less than or the same as before phase one was implemented.

Alternate Hypothesis 5:     The customer satisfaction after phase one is more than that of before phase one was implemented.

For the purpose of this illustration, we shall test the last hypothesis. The other hypotheses may be solved based on this illustration.

**Testing of this improvement in customer satisfaction:**

H0    :    $X_m \leq X_p$
H1    :    $X_m > X_p$

Where $X_p$ is mean value of customer satisfaction index before phase one and $X_m$ is customer satisfaction index after phase one.

*F – Test*

We know from statistics that before analyzing the improvement, *F – Test* is conducted to compare the variances of the two samples. *F – Test* confirms that the two variances are statistically the same.

**Before Phase One**

|  | Sample Size | Mean | Standard Deviation |
|---|---|---|---|
| Waiting Time (min.) Weekdays | 100 | 2 | 1.3 |
| Waiting Time (min.) Weekends | 40 | 6 | 4.5 |
| Customer Satisfaction (out of 10) | 70 | 7 | 3.0 |

**After Phase One**

|  | Sample Size | Mean | Standard Deviation |
|---|---|---|---|
| Waiting Time (min.) Weekdays | 20 | 1.5 | 1.0 |
| Waiting Time (min.) Weekends | 8 | 3.5 | 2.0 |
| Customer Satisfaction (out of 10) | 10 | 9 | 3.5 |

**FIGURE 15.9.** Process Output Results: Before and After Phase One

$$\text{Here, F Statistic} \quad = \quad (3.5)^2 / (3.0)^2$$
$$= \quad 1.3611$$

We know from *F – Test* that F Critical for 95% significance is close to 2.0.

Since F Critical is more that the F Statistic, the variances are statistically the same with 95% significance. Now we shall compare the improvement based on the two samples.

*t – Test*

We know from statistics that *t – test* is conducted to verify the improvement. To confirm the improvement, we need to calculate the t – statistic.

Here, the t – statistic shall be calculated as follows:

$$t \quad = \quad \frac{X_m - X_p}{S_{pl}\sqrt{(1/n_m + 1/n^p)}}$$

Where
$X_m$ is the mean of the customer satisfaction after phase one
$X_p$ is the mean of the customer satisfaction before phase one
$S_{pl}$ is the pooled standard deviation calculated as follows:

$$Spl = \sqrt{Spl^2} \text{ and}$$

$$Spl2 = \frac{(n_m - 1) S_m{}^2 + (n_p - 1) S_p{}^2}{n_m + n_p - 2}$$

Where

$S_m$ is the standard deviation of the customer satisfaction after phase one and

$S_p$ is the standard deviation of the customer satisfaction before phase one.

In our example, when we plug in the values in the preceding formulaes, we find:

$S_{pl}$ = 3.0619
t = 1.8431

The critical value of t corresponding to a confidence level of 95% and degree of freedom of 78 (i.e., infinite) is = 1.645.

Because the value of t statistic (1.8431) is more than the critical value of t (1.645), the null hypothesis is not supported by the data. Hence, it may be concluded with 95% confidence that there is an improvement in customer satisfaction after the completion of phase one.

It may be noted here that there is a need to perform these tests at the end of each phase to confirm the improvements.

## EMBED PHASE

Once we are sure of the improvements, we need to continue with the changed process and train the employees in the modified process. The overall effect of all the process improvements becomes self-evident by sustained efforts and keen monitoring by the team.

It is important that the data are captured and reported periodically to monitor the effectiveness of the plan. The suggested format for data reporting is listed in Figure 15.10.

| Metric | Month | Last Month | Last 12 months |
|---|---|---|---|
| Mean wait time—weekdays (min.) | | | |
| Std. Dev. of wait time weekdays (min.) | | | |
| Mean wait time—weekends (min.) | | | |
| Std. Dev. of wait time weekends (min.) | | | |
| Average customer rating | | | |
| Std. Dev. of customer rating | | | |
| Loss of old customers | | | |

**FIGURE 15.10.** Report Format for Each Month to Monitor the Effectiveness of the Change

Once the project has been successful at one location, the phase-wise implementation may be made simultaneously at the remaining four branches. Four Green Belts may now lead the implementation process by adopting one branch each, under the overall guidance of the Black Belt. In the new branches, the same monitoring systems are implemented to capture the data.

# INSURANCE

In the previous section, we saw an example of the application of Six Sigma tools in the banking sector. Now we shall explore the possibilities of the application in the insurance sector.

Instead of going through the complete process in detail, as we went through in the case of banking, our focus now is on the key aspects of the service features that interact with customers and the corresponding metrics that may be used to define the defects and the sources of data for decision making.

In the insurance sector, the key customer processes, with customer interface, include:

- Claim settlement
- Policy premium follow-up
- Customer acquisition
- Online support, e-mail management
- Customer service

For each of these processes, the definition of defect needs to be agreed on from the customer perspective. Figure 15.11 depicts possible measures of defects and their data sources. Eventually, customer expectations supersede/modify the industry standards.

After identifying the top issues, the DMAIE can be applied as discussed in the case of banking.

| Insurance Processes | Measure of Defect | Source of data |
|---|---|---|
| Claim Settlement | Days to complete the process cycle (from the date of claim to final settlement) | 1. Industry standards<br>2. Ask the customer |
| Policy Premium Follow-Up | Number of days in advance the customer is approached for reminding | 1. Industry standards<br>2. Ask the customer<br>3. Past record |
| Online Support | Delay in responding to the queries | 1. Ask the customer<br>2. Past record |
| Customer Service | Amount of time spent to get the desired information on one issue:<br>1. Number of times contact was made<br>2. Total amount of time spent by the customer | 1. In-house call log<br>2. Ask the customer |

**FIGURE 15.11.** Key Insurance Processes, Defects, and Data Sources

# HEALTHCARE

Like banking, healthcare is another area that has successfully applied Six Sigma and tasted the benefits as a result. The possible reason may be the intense customer interaction and physical presence in the service area in both the cases, which impacts the customer's perception strongly. The key here is to manage and balance the expectations among the three Ps:

1. Patients
2. Payers
3. Physicians

In the healthcare sector, the key customer processes that have been successfully refined and/or have the potential of refinement include:

· Turnaround time for various medical tests at the hospital
· Surgical cycle time
· Length of stay
· Patient care
· Cost reduction of various processes to make it competitive in light of insurance pressures and customer expectations
· Resource utilization
· Capacity availability
· Technological upgradation
· Overall quality improvement as a strategy for survival

For each of these processes, the definition of defect needs to be agreed on from the customer perspective. Figure 15.12 depicts possible measures of defects and their data sources. After identifying the top issues, the DMAIE can be applied as discussed in the case of banking.

## RESTAURANTS

Restaurants have the lowest lying fruits to be picked through the application of Six Sigma. The reasons include:

· Presence of big food chains with big names, but none seem to have attempted this tool to uncover the opportunities for improvement.
· Restaurants have inherently customer-intensive processes.
· Restaurant's biggest competitor is the customer herself.

| Healthcare Processes | Measure of Defect | Source of Data |
|---|---|---|
| Turnaround time for medical tests | Days to complete the test cycle (from the date of customer request/need to final report by the medical specialist) | 1. Industry standards<br>2. Ask the customer |
| Patient care | Patient's dissatisfaction with the hospital:<br>1. Excessive wait<br>2. Not getting required attention at the time of need<br>3. Courtesy of staff | 1. Industry standards<br>2. Ask the customer<br>3. Past record |
| Cost reduction | Costs exceeding the expectations and/or more than that of the competition | 1. Industry standards<br>2. Ask the customer |
| Overall quality improvement | 1. Accuracy of report<br>2. Accuracy of billing<br>3. Amount of customer time spent in obtaining the desired information. | 1. In-house log<br>2. Ask the customer<br>3. Industry standards |

**FIGURE 15.12.** Key Healthcare Processes, Defects, and Data Sources

- Customers have wide choices, but high loyalty (long lines and excessive wait doesn't deter the loyal customers).

The key processes, capable of producing customer dissatisfaction, include the following:

- Customer arrival
- Customer serving
  - Time
  - Accuracy
- Food preparation
- Customer billing
- Customer feedback and closing

For each of these processes, the definition of defect needs to be agreed on from the customer perspective. Figure 15.13 depicts possible measures of defects and their data sources. After identifying the top issues, the DMAIE can be applied as discussed in the case of banking.

| Restaurant Processes | Measure of Defect | Source of Data |
|---|---|---|
| Customer Arrival | 1. Inconvenient parking<br>2. Inadequate space for wait<br>3. Ambiguous/excessive wait<br>4. Courtesy of staff | 1. Industry standards<br>2. Ask the customer |
| Customer Serving | 1. Delay in waiter attention<br>2. Courtesy of staff<br>3. Delay in arrival of food<br>4. Quality of placement of food items<br>5. Quality of food items | 1. Industry standards<br>2. Ask the customer<br>3. Past record |
| Customer Billing | 1. Error in billing<br>2. Excess time of error correction | 1. Industry standards<br>2. Ask the customer |
| Customer Feedback and Closing | 1. No action on feedback<br>2. Courtesy of staff | 1. In-house log<br>2. Ask the customer |

**FIGURE 15.13.** Key Restaurant Processes, Defects, and Data Sources

# TRANSPORT

We are all familiar with the transport services and tend to encounter some issue or other. We rarely feel content with the quality of service received. Let's look at the opportunities offered by this area of service. For our purpose, in transport, we cover the following service areas:

· Car rental service
· Railroad
· Airline

Transport, like restaurants, is an area that has a number of low hanging fruits and the following similarities:

· Presence of big players, but none seem to have attempted this tool to uncover the opportunities for improvement.
· All modes have inherently customer-intensive processes.
· Industry's biggest competitor is the customer himself.

The major difference, however, comes in the last point—the customer has wide choices and *no* loyalty.

## CAR RENTAL SERVICE

The key customer processes, capable of producing customer dissatisfaction from the car rental service, include the following:

· Customer arrival/pickup
· Vehicle maintenance
· Vehicle return
· Customer billing
· Discounts and deals
· Customer complaint and closing
· Customer departure/drop off

For each of these processes, the definition of defect needs to be agreed on from the customer perspective. Figure 15.14 depicts possible measures of defects and their data sources.

| Car Rental Processes | Measure of Defect | Source of data |
|---|---|---|
| Customer Arrival/ Pickup | 1. Inadequate signs at the airport arrival<br>2. Long/uncertain wait for pickup<br>3. Improper line formation at the service counters<br>4. Courtesy of employees | 1. Industry standards<br>2. Ask the customer |
| Customer's Perception of Vehicle's Performance | 1. Response time to customer complaint regarding the vehicle condition<br>2. Response time to the incidence reported | 1. Industry standards<br>2. Ask the customer<br>3. Past record |
| Vehicle Return | 1. Inadequate signs for return area<br>2. Excessive verification at the time of return | 1. Industry standards<br>2. Ask the customer |
| Customer Billing | 1. Error in billing<br>2. Long wait to obtain the bill | 1. Industry standards<br>2. Ask the customer |
| Customer Complaint and Closing | 1. Time to close the general customer complaint<br>2. Complaint resolution and customer satisfaction | 1. Industry standards<br>2. Ask the customer<br>3. Past record |
| Customer Departure/Drop | 1. Excessive customer wait for drop | 1. Ask the customer |

**FIGURE 15.14.** Key Car Rental Processes, Defects, and Data Sources

After identifying the top issues, the DMAIE can be applied as discussed in the case of banking.

## RAILROAD SERVICE

The key customer processes, capable of producing customer dissatisfaction from the railroad service, include the following:

· Customer arrival
· Comfort of journey
· Train maintenance
· On time performance
· Deals and discounts
· Customer complaint and closing
· Customer departure

For each of these processes, the definition of defect needs to be agreed on from the customer perspective. Figure 15.15 depicts possible measures of defects and their data sources. After identifying the top issues, the DMAIE can be applied as discussed in the case of banking.

| Rail Road Processes | Measure of Defect | Source of Data |
|---|---|---|
| Customer Arrival | 1. Inconvenient parking<br>2. Ambiguous/excessive wait<br>3. Courtesy of staff | 1. Industry standards<br>2. Ask the customer |
| Comfort of Journey | 1. Late arrival<br>2. No communication of delay<br>3. Air conditioning/heating system not working | 1. Industry standards<br>2. Ask the customer<br>3. Past record |
| Customer Complaint and Closing | 1. Error in ticketing<br>2. Excess time of error correction | 1. Industry standards<br>2. Ask the customer |
| Customer Departure | 1. Courtesy of staff | 1. Ask the customer |

**FIGURE 15.15.** Key Railroad Processes, Defects, and Data Sources

# AIRLINE SERVICE

The key customer processes, capable of producing customer dissatisfaction from the airline service, include the following:

· Customer arrival
· Comfort of flight
· Aircraft maintenance
· On time performance
· Baggage handling
· Deals and discounts
· Frequent flyer program
· Customer complaint and closing
· Customer departure

For each of these processes, the definition of defect needs to be agreed on from the customer perspective. Figure 15.16 lists possible measures of defects and their data sources. After identifying the top issues, the DMAIE can be applied as discussed in the case of banking.

| Airline Processes | Measure of Defect | Source of Data |
|---|---|---|
| Customer Arrival | 1. Inconvenient parking<br>2. Ambiguous/excessive wait<br>3. Courtesy of staff | 1. Industry standards<br>2. Ask the customer |
| Comfort of Flight | 1. Late arrival<br>2. No communication of delay | 1. Industry standards<br>2. Ask the customer<br>3. Past record |
| Baggage Handling | 1. Excessive delay in baggage arrival<br>2. Misplaced baggage<br>3. Inadequate compensation | 1. Industry standards<br>2. Ask the customer<br>3. Past record |
| Customer Complaint and Closing | 1. Error in ticketing<br>2. Excess time of error correction | 1. Industry standards<br>2. Ask the customer |
| Customer Departure | 1. Courtesy of staff | 1. Ask the customer |

**FIGURE 15.16.** Key Airline Processes, Defects, and Data Sources

## KEY TAKEAWAYS

- Six Sigma finds widespread application in various service segments.
- Understanding opportunities and measurements for various service processes for creating value through Six Sigma.
- A methodical approach to implementation of various phases and tools of Six Sigma ensures successful project completion.

# SIX SIGMA IN OUTSOURCING

## Outsourcing—A Core Component of Your Strategy

This chapter rationalizes role of outsourcing in today's global business environment, the need for an effective strategy for outsourcing, and the role of Six Sigma in outsourcing. This chapter covers the various factors that impact outsourcing and measurements for monitoring outsourcing performance.

Today, outsourcing is a major component of many firms' strategies. A lot of political debate on this issue occurs. The question is, "Is this good or bad for our economy?" We are all very familiar with outsourcing as we all outsource various activities in our daily lives. Depending on our skill set, availability of time, potential alternative use of our time, and capacity to earn, we make outsourcing decisions every day in our lives. Some of us change the oil in our cars ourselves, while others believe that for the $15 to $20 most automotive service places charge, it is not worth our time to undertake it ourselves and we outsource it to someone else. Others have their lawn trimmed by someone else or use a cleaning service to clean their home.

Most of us do not wish to generate our own electricity, or produce gas for cooking and heating. We outsource these services to the utility companies. Similarly, we buy telephone and

Internet services from telecommunication companies, and medical services from healthcare providers. In the preceding examples, the outsourcing decision is simple, as we do not have the capability or the option to perform the services ourselves (e.g., services provided by utility companies).

When we have the capability or the option to perform the service ourselves, the decision criteria change. You can cook a meal for yourself or eat out; you can change the oil in your car or go to an automotive service garage; and you can mow and fertilize your lawn or have a landscaping service take care of it for you. These decisions are likely based on how much time you have available, your interests, alternative uses for your time, and last but not least, the cost of these services. If you are debating whether to mow the lawn or to outsource it to an external service provider, the cost factor may have a significant influence on your decision. If all other factors remain the same, the decision may change, depending on whether the service provider can do it for $5 or $100 per mowing. This example illustrates the same considerations that are involved in an outsourcing decision for a firm's internal process.

## WHY CONSIDER AN OUTSOURCING STRATEGY?

Many reasons for outsourcing exist. Organizations outsource to:

· Save money
· Access new skills
· Supplement capacity
· Focus on their core competencies
· Have an alternative source of capabilities as part of risk mitigation strategy
· Manage their processes more effectively

Many times, businesses and outsource providers seem to fit each other like a hand in a glove—one lacks skills to operate efficiently, and the other specializes in delivering those very skills, usually at a lower cost.

As a client, the organization outsourcing its processes is initially very apprehensive. If the outsourcer has multiple clients, how do we ensure that we get the attention when we need it? Each client wants to be among an outsourcer's most important clients, if not *the* most important client. This is usually a highly unrealistic demand from a vendor's perspective. As a result, the vendor tries to provide confidence through the promise of service and service level agreements (SLAs). To be realistic about their expectations from outsourcers, clients should focus on benefits like reducing costs and improving performance.

Does the size of the outsourcing vendor matter when you have an SLA? Experience would suggest—yes. While most vendors create client teams to service each client, larger outsourcing vendors will typically give preference to the larger customers. Midsize businesses that want a strong "partnership" relationship with their vendor may be more successful with relatively smaller outsourcing vendors that are more eager for their business. Unless you represent a significant portion of a vendor's revenue, businesses that seek a high level of attention may be disappointed by larger vendors.

From a vendor's perspective, when a customer represents a large portion of its revenue, it increases the risk. Outsourcing vendors have to make significant upfront investments. It is important for them to have longer term contracts so that they are not left "high and dry" in case the customer decides to move the business elsewhere. Initially, the vendors seek some marquee accounts at the earlier stages in their development—and large accounts provide some stability. However, in time, they need to develop a risk mitigation strategy.

## OUTSOURCING STRATEGY

Is there a single strategy that, if followed, would guarantee success? Probably not. Most outsourcing today is driven by the attraction of cost saving to become competitive. However, as discussed earlier, there can be many other drivers. A sourcing strategy must always be based on factors applicable to each individual case. There are many common elements and a wealth of best practices, but no "one size fits all" sourcing strategy exists.

There are at least two basic outsourcing models:

- *Lift and shift:* Here, the entire process or function is transferred to the vendor. The vendor takes over the people, processes, systems, and all related assets "as is," and manages them on behalf of the customer. The vendor promises to provide the same level or better service with improvement in cost of the operation. This model made outsourcing popular. Ross Perot introduced the concept for the IT function of the various organizations and, as a result, created EDS (Electronic Data Systems) that today generates $22 billion dollar annually in revenues.

- *Transitioning:* Here, the vendor does not take over the assets, but helps transition the processes to its own efficient but customizable system. Here, the corporation still holds on to its investment in the people and systems, and once the process is transferred to the vendor's technology-driven platform, it eliminates these resources if they cannot be absorbed within the organization or redeployed. This model will typically result in more significant cost reduction (compared to the lift and shift model) for the corporation, but takes more time to transition.

One of the most important decisions must be: "Do you plan to outsource to a third party or to yourself (in another location)?" And then, "Which model do you wish to adopt?" These decisions in themselves are driven by multiple considerations, such as:

- Prior experience in operating from remote locations
- Culture of the organization
- Criticality of the processes
- Confidentiality and privacy issues surrounding the data
- Volume of work to be performed

Once the decision to do it yourself (usually in another country), or to utilize another vendor has been taken, other elements will affect your strategy. These considerations are frequently unique to each organization and include:

- Prior experience in outsourcing—that in turn will drive the pace and aggressiveness with which an organization pursues outsourcing.
- Prior relationship with the vendor (if going external).
- Knowledge of the country (today most outsourcing includes some component of offshoring or else you may be missing an opportunity).
- Maturity of the process you plan to outsource.
- Industry, as the business requirements and practices vary in each industry.
- The organization's business culture.
- The legal framework.
- Financial health and structure.
- Taxation status and treaties with other nations.
- Competitive pressures.

A review of these factors could lead to a different solution or a different approach. Hence, there is no predetermined single strategy.

To take this a step further, the strategy is also not static—it will continue to evolve and change with time. Over time, the business changes, the competition increases, the internal knowledge base becomes stronger, and the technology advances. In addition, the economic and political environment changes. This means that an outsourcing strategy is not a fixed event, or a make-or-buy decision. Nor is it an evolution toward a predefined steady state. As the business environment changes, the outsourcing strategy will need to be reevaluated continuously. It is a journey, not a destination.

Outsourcing is not without its inherent risks. On the one hand, you have potential benefits through reduced costs, and higher performance through specialization. Offshore outsourcing reduces costs even more when the labor cost is less in the offshore country when compared to the host country. This strategy can also be very effective if you are struggling with unions or environmental regulations. Go offshore and don't worry about these issues any more. But, is it that simple?

When you outsource a function, you also outsource some control and opportunity to quickly change. Even though you outsource a process, you retain most of the risk—and you are no longer in control of that risk. For example, if your payroll is outsourced, and if the employee salaries did not get deposited in their accounts due to some fault of the vendor, no employee will be satisfied with an explanation that it is the fault of the outsourced vendor. For your safety, in your contract requirements, you can spell out the risks and specify the risk mitigation requirements. However, in case of nonperformance, your only relief is the penalty you may have negotiated. This increases the importance of the relationship with your vendor—and it is critical to your success.

When Forrester Research asked a sample of IT executives why they had not yet moved work overseas, security and the vendor's financial stability were ranked as the top two reasons. Other concerns included:

· Resistance from IT staff or business executives
· Overhead of managing the offshore counteracts the cost saving
· Culture and language differences
· Company's lack of project management skills

Corporations that have been successful have identified some key success factors. These include:

· Clarity of objective
· Facilitation of the offshoring process: Dedicate a point person

- Management buy-in: Strong commitment is required
- Use of project management tools: Use them on large projects
- Grow into it slowly: Pilot with easier projects before taking on major ones
- Ensure that the right processes are offshored: Not every process fits the profile

Considering that this offshoring trend is likely to continue, *if it is worth doing, it is worth doing it well!*

**Vendor landscape.** A significant shift over the last year has occurred in the vendor landscape. Also, corporations are using professional consulting firms like Hewitt Associates to help them select vendors, countries, cities, and site locations. Many studies by firms like A.T. Kearney and McKinsey still place India as the leading location for outsourcing.

A short list of vendors from last year for outsourcing to India consisted primarily of Indian external service providers (ESPs), like Infosys, Tata Consultancy Services, and Wipro. Today, the short list still includes Indian vendors but has become much more heterogeneous, as many large traditional global service providers such as Accenture, Electronic Data Systems, and IBM are now formidable contenders.

What has led to this expansion of the vendor list? Over the last year, the global ESPs have invested millions in expanding their delivery bases to India and other regions, in an effort to create a global delivery model that would incorporate a combination of onsite, onshore, nearshore, and offshore services.

Many large traditional global service providers like Accenture, BearingPoint (formerly KPMG Consulting), Capgemini, Ernst & Young, Computer Sciences Corp., Deloittes, Electronic Data Systems, IBM, and Keane are aggressively enhancing their global delivery strategies. Growth has come largely from acquisitions, joint ventures, other partnerships, and on a smaller scale, organically. For example, IBM recently acquired Daksh, a major service provider based in India servicing a number of U.S. multinationals from India. These global service providers have even developed rate structures in which many of their options for resources in India are competitive with ESPs headquartered in India (e.g., Infosys, Tata Consultancy Services, and Wipro). In the meantime, the Indian ESPs have invested heavily in trying to build their global brand and extend their offerings in the United States, Europe, and Asia Pacific.

Given the trend of vendors toward becoming stronger global players, if you were to choose a vendor, is one group of ESPs preferable to the other? There is no correct answer. A winning ESP can be from the United States, Europe, or Asia Pacific. However, the customer will be the real beneficiary, as the vendor will be able to provide a seamless end-to-end solution. The vendor should be able to provide you with the right skill set, providing the right set of services at the right time in the right quantity on the right "shore" at a competitive global price. Which vendors can provide this capability? Today, no one can do it all, but the leading vendors are moving in this direction at an accelerated pace.

Before going through a vendor selection process, it is important to determine what is your objective and what is the result you desire from offshore outsourcing. Some examples in IT may include:

- Access to better network, systems, and applications management platforms.
- Improved IT availability and performance.
- Reduction in IT management costs.

Currently, a large number (in excess of 300) of vendors in North America provide remote monitoring and management services. As a result, so far, businesses in North America did not have to look far to find a reliable local vendor. The vendors that offer IT remote monitoring and management services include carriers, systems integrators, outsourcers, and niche players. Due to the proximity to the customer, the offshore vendors have made little progress to date. However, the prices quoted by these offshore vendors are compelling and have gained trust of the corporations through their performance in previous engagements. As a result, what has been an exclusive domain of the local businesses so far is starting to face global competition. This trend is likely to be encountered in every sphere of outsourcing opportunity.

## Impact of Outsourcing

Outsourcing has a significant impact on any organization. Objectives such as lower cost and improved efficiency will certainly improve a corporation's bottom line. However, the impact goes way beyond that, touching almost every aspect of the organization. The most significant is the impact on people. What happens to the employees who are currently responsible for the process and also, how will the transition to a remote location be managed? Some may fear the loss of control. In addition, improved efficiency will result from process as well as system changes. Let us take a more detailed look at some of the more important issues.

### People-related impact

*Changing people skills.* When work is outsourced, one of the most immediate impacts is on the skills requirements within the organization. As an example, when an IT department is focused on delivering the service, it requires technically focused staff. When work is delivered by one or more ESPs, however, there is a significant reduction in the technical skills required. Going forward, the skills needed become a blend of technical capability with business acumen and relationship management. You still need to:

- Understand the technical issues faced by the business.
- Align IT with the business strategies and ensure proper support and outcomes.
- Delicately manage the relationship with both your customer and the vendor.

This combination and complexity of skills in one single individual is rarely required otherwise.

It is important that a role-based organization evolves from this change, with greater emphasis on project management, contract management, and relationship management, in addition to the operational service management issues (such as service level management). New recruitment is often required for this skill set. Another change in the skill set is driven by the need to transition the processes to the vendor. This creates demand for strong process definition skills and talent with focus on business process expertise, business intelligence, high-level process design, and acceptance testing. The vendor can also provide some of these skills (if using an outside vendor).

*Cross-cultural communications.* The challenges of managing global virtual teams are enormous. As a result, potential risks are much greater for organizations with limited international experience at the corporate office. Differences in time zones, interaction styles, and even everyday slang can lead to flawed communications and expectations, jeopardizing outsourcing projects and undercutting potential cost savings.

When offshoring, one of the biggest areas of discomfort is the lack of knowledge about the other culture. One way managers can reduce these risks is to ensure that such initiatives leverage the talents of people who have the multicultural "fluency" often found in expatriates. They have the expertise needed to perform the task at hand, and in addition, have an understanding of the customs, language, and workplace practices that are essential for effectively integrating offshore services into the overall enterprise. They make a very special contribution by acting as liaison between the corporation and the offshore service provider, by focusing on cross-cultural communications, collaboration, and improving management control of the outsourcing initiative.

The use of expatriates is no doubt expensive. However, the cost of *not* using them may be even higher. Corporations should develop an effective "cross-border strategy" that clearly links the use of expatriates with the enterprise's overall business and sourcing strategies. Any lessons learned should be recognized as they emerge and be used to influence the ongoing evolution of the strategy. The expatriate's role should focus on empowerment, by facilitating the development of more effective cross-cultural relationships and teamwork.

*Political and social impact.* When jobs are outsourced, particularly to offshore locations, jobs are lost in the United States. But how is that different from the loss of jobs when we lose a large contract to a lower-priced competitor? If the Boeing Corporation and Lockheed Martin are bidding for a particular project, they have to build up a team to deliver that project. When one loses the bid, however, the team is dismantled and let go—and this happens every day. Unfortunately, today, this competition is coming from other economies around the world.

Today, the airline industry is going through significant restructuring. Most major airlines have gone through Chapter 11 reorganization. If you consider the employees who still have jobs with any of the major airlines, wages and benefits have likely been cut significantly. Are they upset with the current economic situation? I suspect they're very mad, but they learned to accept it, as they cannot do much about it. And, there is no political upheaval because this is a result of local competition.

One way to protect against foreign competition is to create import and protectionist barriers. In the long run, however, protective legislation would harm a country's competitiveness. For example, in Germany (where the auto industry labor unions have been very protective), after facing a competitive threat from lower cost producers from around the world, it has finally succumbed and allowed more jobs to move to lower-wage countries to sustain German "industrial competitiveness."

This outsourcing trend will have huge political and social impact for a few years to come. These outsourced jobs are not only low-level data entry jobs, but include some high-paying white collar jobs that feed significant funds into local, regional, and national economies. This outsourcing trend has the potential to create a snowball effect of economic growth without job creation, which ultimately affects the political climate.

From a worker's perspective, if you are one of the unlucky ones to have lost your job, you cannot buy goods and services

and are more prone to vote against current political adminis-
trations. But the bigger impact is a huge shift in the social and
economic structure of this country. Few people in this country
benefit from sending jobs overseas unless those displaced can
be retrained to take on alternative positions. Among potential
gainers will be company executives who get larger bonuses and
shareholders who receive bigger dividends.

Some analysts will suggest that another potential impact of
the loss of these white-collar jobs could be to ignite anger and
resentment between the "rank-and-file," leading to a desire for
a "pro-union" sentiment at a corporation. If one analyzes the
situation more carefully, however, one realizes that while it is
natural to trigger such sentiments, it is a bitter pill that we
must swallow, or else entire corporations will suffer from competitive threats, resulting in much larger job losses.

Over the last few months, there has been significant media
coverage about the backlash in the United States against offshore outsourcing. Much of this is focused on Indian service
providers, due to India's prominent position in the global outsourcing services market. The American corporations are
socially responsible organizations and would not transfer jobs if
it were not essential. Unfortunately, it is critical to their survival. The pressures of being a public company, where high
investment returns are demanded by the stock market and
financial performance is being measured every quarter, it
leaves little choice to the leadership of the corporations.
Secondly, the blame may be somewhat misplaced and political.
The jobs are being lost by Americans of all nationalities, as well
as a large number of immigrant workers who were here on temporary work visas. Finally, the trend today is that the American
service providers are setting up offshore centers. It is not the
corporations from foreign countries that are getting the outsourcing contracts and taking the work away.

### Process-related impact

*Vision of future processes.* With the increasing maturity of
ESPs, in the not too distant future, organizations will use a
combination of standardized outsourcing, customized out-

sourcing, and in-house systems and solutions for their various processes.

To optimize benefits, end-to-end process support in this multisourcing environment, enterprises, and multiple service providers may be required. Such an operating environment will require sophisticated system integrators. In many instances, the service provider will have to take responsibility for acting as a business solution aggregator.

*Increased clarity of processes.* Business process outsourcing forces you to document and be more explicit about your processes, as there is a need to explain it to others. Distance creates the need for clearer communication. Finally, if one is going to enter into contracts and service-level agreements, a more in-depth understanding of the service is required. Whether you decide to outsource or not, there is value in being explicit about your processes. Unfortunately, in many organizations, this discipline is not enforced until one has to outsource the process.

*Approach to the outsourcing process.* Many organizations still fail to manage the deal in a way that delivers a satisfactory outcome. It's not enough to bring in a third party and say, "OK, over to you."

Corporations that have been more successful at outsourcing and/or offshoring have managed with a simple fixed price and risk transfer approach. Outsourcing or offshoring can be managed effectively either from a "managed services" approach or with an SLA. Corporations have been successful with both approaches. Of course, it is incumbent on the organizations intending to outsource and the service provider to ensure the success of outsourcing or offshoring. An effective program governance structure to address each program management process is essential to ensure a successful outcome.

*Relationship management.* Corporate satisfaction with outsourcing deals is currently running at around 50%. The gap in companies' internal capabilities, and failure to invest properly in relationship management skills and processes, is a leading cause for project failures.

Poorly managed relationships cost companies billions of dollars per year. Firms should invest at least 5 to 10% of the value of an outsourcing deal in building and developing the skills of an internal team to manage outsourced relationships effectively. The potential risks are greater for organizations with limited international experience. Differences in time zones, interaction styles, and even everyday slang can lead to flawed communications and expectations, jeopardizing outsourcing projects and undercutting potential cost savings.

*Virtual teaming.* Most conversations around global sourcing, or offshoring, focus on the benefits of cost reduction, the risks of employee backlash, and the potential long-term effects on intellectual capital. However, one of the biggest challenges is that of leading and managing globally sourced virtual teams, and it is typically not addressed in any depth. Offshoring, by necessity, requires work groups of individuals from different companies, from different places, and with different cultures and languages, to work together effectively. Virtual teaming is hard, and virtual teaming with diverse intercompany members is harder still.

What does it take to be successful at virtual teaming? Fundamentally, virtual teaming may be only about 10% technology, and the remaining 90% is human behavior and communication. Globally sourced teams, like any team, must build a culture of trust, accountability, and transparency to work effectively. If managers and their colleagues treat the offshore sourcing members as strictly "contractors," then working together virtually will be problematic.

Virtual team leaders must focus on building trust by creating a strong sense of shared purpose, delineating clear roles and responsibilities, and insisting on information transparency, within the boundaries of confidentiality and other "nondisclosure" policies. Further, virtual team leaders should minimize asynchronous communication (avoid endless email threads), and strive for frequent and regularly scheduled phone or video conferences. The use of collaborative applications that facilitate virtual meetings, maintain and catalogue meeting outcomes, and other products that facilitate teamwork are highly desirable.

## Technology-related impact

*High technology as an enabler.* Who would be willing to outsource if you had to send information to another location to be processed and then wait a few days to see the results? Growth in high technology is probably the biggest enabler and driver of the current outsourcing trend. It is at the core of many benefits. Perhaps the most important benefit has been to make the world a small place and a neighborhood of people we know. Technology removes barriers and extends our reach. Millions of people are communicating with each other—anytime, anywhere—as quickly as they can dial their telephone.

Technology offers the opportunity to serve the customer 24 hours a day and seven days a week. Because work can carry on while we are sleeping in one part of the world, it helps us increase our turnaround and service to the customer.

*Managing performance through SLAs.* The use of SLAs first became popular a couple of decades ago when the concept of shared services became popular. As the business units lost control over their supporting services with an intent to consolidate them with the promise of a more efficient and comprehensive service, SLAs were developed between the shared service organizations and the operating business units. Similarly, in outsourcing and offshoring, SLAs are a common way to set expectations and manage performance. What does an SLA cover? One can find large variation, from a fairly simple one-page SLA, to a highly complex and detailed document that may include roles and responsibilities, performance targets and how they will be computed, as well as any performance bonuses and penalties. The intent is to reduce ambiguity and misunderstanding between the concerned parties.

There has been discontentment with SLAs as well—a gap that may well be exposed by outsourcing. Many poorly constructed SLAs focus more on the technical aspects and less on the business need. For example, system users' perspectives are set by the service they actually receive, not what is written in the SLA. In most cases, this service exceeds the written SLA. If the SLA asks for a 99% uptime, it means that the service provider is providing acceptable service as long as the down-time is less than 1%. If that 1% downtime happens to be when you are running your most critical applications, or when you are operating against very strict deadlines, the customer will not be happy even though the service provider attained its service commitment. Typically, when an outsourcing arrangement is put in place, the vendor will ensure that resources are dedicated to the aspects of service stated in the SLA.

Recently, we had a situation where the IT function was responding to the demands of the users within 30 minutes compared to an SLA that required them to provide services within two hours. When the load increased and the same users were getting service in 50 minutes (instead of the 30 minutes they were used to), the users started complaining of poor service. When the SLA was reviewed, the users realized that they were still getting a great service, instead of the poor service they had been complaining about. However, if the service providers are asked to commit to the faster turnaround (which the users have become accustomed to), then they would increase the price to meet the more aggressive service level demands.

Let us take another real-life example. A critical business system recorded no downtime over an extended period of time. When the SLA was reviewed, management responsible for the users asked for this high availability level to be guaranteed. From the service provider's perspective this guarantee would require a major upgrade in hardware and software—hence, increase in service cost. Experience suggests that it is harder to agree on a pragmatic SLA once you have received a higher level of service.

Service level agreements have not been effective for many organizations to show its value to business units. Frequently, service level agreements are too technically focused. Business managers and users usually don't care about 99% of anything. They want the services they need to be available when they need them. In the preceding example, the IS organization and the service provider may be very pleased, thinking, "We're doing such a great job." But customer satisfaction may be gradually and consistently decreasing as the complaints continue to increase.

*Global versus local network service provider.* When offshoring work, another critical factor that requires serious consideration is the cross-border communication. How do we have an effective information sharing process across thousands of miles? Deciding whether to single-source (use one network service provider) or use several providers (multisource or best-of-breed strategy) depends on several factors. These include:

· Number of locations
· Location of sites
· Required quality of services
· Redundancy
· Number of vendors providing the service
· Most important, their capability to manage the networks

Managing multiple contracts (in foreign jurisdictions) with a variety of service providers in a different country or in several different countries is not fun.

The simplest solution may be to use a single service provider globally even though this may not be the lowest cost alternative. If you are on a fixed budget, you may have to go through some negotiations before you can get a realistic price to compare against other alternatives. Prices can vary greatly from region to region. Domestic service providers are usually cheaper than international providers, who could be charging as much as 50% higher prices.

If you go with a multivendor strategy, then you are likely to face a finger-pointing game when the communication network is down, as it is not easy to determine where in the network the breakdown has occurred. Each vendor will have to test its own network (and you are hostage to their priorities). Hence, it is likely to take longer to fix your nonfunctioning network. Also, in this case, you would still have to manage the global network yourself or through a global integrator, who may then charge you more because it is getting a small share of your networking budget. Therefore, network service provider decisions cannot be made solely based on the prices you can negotiate. There is a need to optimize the solution based on cost, complexity of the network, ease of managing it, and service level required (depending on criticality of the applications/information shared).

*Conclusion.* Outsourcing has a much more profound effect on people, process, and technology than one may realize at first glance. The number of variables is high, and level of complexity increases with distance and cultural differences in the land of outsourcing.

## ROLE OF SIX SIGMA IN OUTSOURCING AND OFFSHORING

Considering the benefits of outsourcing, the trend is here to stay. In most processes, we are likely to find that there is a sharing of responsibility with an outsourced vendor. How can Six Sigma have a role in that scenario? It has a dual role. Traditionally, Six Sigma methodology has been utilized while transitioning processes to an outsourcing service provider. The opportunity here is to develop a more streamlined process and then to gain higher performance commitments from the service provider. The other opportunity is in the outsourcing process itself. As the former varies with the process transferred to the service provider, lets us look at the opportunity to leverage Six Sigma in the outsourcing process itself.

Next, potential service providers should consider utilizing Six Sigma to enhance their service offer and to gain competitive advantage. In a recent press release, Blue Cross and Blue Shield of Florida stated it was reviewing and replacing vendors based on an extensive bidding process. Vendors were being evaluated on multiple criteria including network, service, quality, and price.

Various organizations that have shared their experience with offshoring indicate that the net savings are typically a function of:

- Process improvement through reengineering
- Economies of scale due to consolidation
- Labor cost reductions

The streamlining effort to improve and optimize the process during offshoring is limited and typically contributes a 5 to 10% reduction, as per a study conducted by Booz Allen Hamilton Inc. A more concerted effort using the rigorous Six Sigma methodology could significantly enhance this potential. Why isn't the real potential realized during the transition? The attractiveness of the labor arbitrage dominates the decision process. If we look at the elements of operation that change and need to be reconsidered during offshoring, the value of using a more disciplined and rigorous analytical tool can be more evident. Let us consider the following factors that change significantly during the offshoring process (see also Figure 16.1):

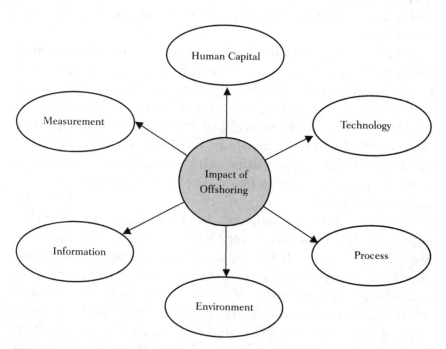

**FIGURE 16.1.** Factors that Are Significantly Impacted During the Offshoring Process

- Human capital
- Technology
- Process
- Environment
- Information
- Measurement

We described the impact of some of these factors on the organization outsourcing process earlier. Let us look at these factors from a different lens.

**Human capital.**    A key driver in moving work offshore is the labor cost savings. But it is rare that we ever employ the same quality of workforce as we employ in the United States—it is usually much more educated. More than 40% have advanced degrees, such as Masters' or Doctoral degrees, or professional qualifications equivalent to a CPA. These highly qualified professionals are attracted to the multinational corporations and feel a sense of pride working for them compared to the local businesses.

Motivation to work for foreign corporations includes opportunity to travel overseas (e.g., for training or on assignment), better benefits and working environment, and specific targets and incentives—resulting in a significantly more productive workforce that proactively seeks to enhance the processes. The experience of some organizations has indicated that the combination of higher qualifications and the right environment can motivate these workers to be even twice as productive as their U.S. counterparts!

General Electric Company, American Express, and other companies have long known the value in supporting this highly qualified talent with Six Sigma initiatives to deliver performance improvement in processes that are four to five times what may be expected otherwise. When we look at the *human capital* considerations of offshoring, let us not forget that our workforce is far more qualified, and hence, can be expected to perform more complex tasks and deliver superior results. No doubt, an investment in training these professionals may be required.

**Technology.** The next major component of change is systems that support the process. The systems available to the offshore entity are at least as good as they are in the United States. Today, most offshore operations use the same systems as their U.S. counterparts to ensure consistency and transparency. The U.S. corporations usually obtain global licenses for their software. Many times, the local license for the same software is less expensive because of lower margins on which companies operate in these countries.

With advancements in telecommunication technologies, global connectivity is not as expensive as it used to be some years ago. Especially with countries that are almost 12 hours out of sync, corporations can get double the work done in the same day—because these countries can process the work while we are sleeping in the United States, and the results are available the next morning. For example, the work that you would have performed the next day and then delivered the day after, can now be ready to be delivered to your customer the very next day, reducing your processing time in half (from a customer's perspective) and improving your turnaround time by 100%!

Many technical support centers are based in locations around the world that allow the customer to call any time of the day or night and they can speak to a live person, which is becoming rare within the United States due to cost pressures. These support centers are linked around the world on the same software—the customer's history is available to the customer service representative around the world. Hence, when you call, they can start from where you left off on the last call. Depending on the application, this 24 *by* 7 service can provide a real competitive advantage. Or, if this was a customer's requirement, then providing this service in the United States would be prohibitively expensive. Through offshoring, the technology today allows you to add at least one, if not two extra shifts to serve the customer in a seamless manner without a proportional increase in cost. That is a real productivity gain!

The Six Sigma methodology can assist in arriving at the right balance between onshoring and offshoring, as well as in developing a strategy that will ensure that you at least maintain the quality and performance, regardless of where the service is delivered from.

**Process.**   Many multinationals have taken their existing processes to the offshore countries, have taken advantage of the talent there to achieve breakthrough improvements, and have then replicated those processes back in the home country and around the world. Just like the labor arbitrage in processing, offshore operations can afford to throw a lot more brains to address a problem. Generally, they are fairly savvy technically.

Smooth and disciplined transition of a process to offshore location is critical to its success. Best practice companies can typically transition their first process in three to four months, and after some experience, perform the same transition in four to six weeks. A Six Sigma initiative would evaluate the change in the human capital, enabling systems to drive a revised process that can deliver significantly greater benefits. The Six Sigma methodology can also help in redefining new targets for enhanced performance.

**Environment.**   It is not unusual for us to consider that the rest of the world should do things similarly to the United States. If we have developed a way to do things, it should be easily replicable in other parts of the world. However, the environment in offshore locations can be significantly different. Even an offshore location as close as Canada is very different from the United States.

Elements that make the environment different include:

· *Culture:* Many Eastern cultures have much stronger hierarchical organizations with significantly more centralized decision making.
· *Language:* It is much easier if the offshore location is an English-speaking nation. If not, communication issues become very visible. Even in an English-speaking environment, a strong accent can hinder communication.

- *Time difference:* This has a significant impact on working hours. Many of these places have staff working on shifts and at all kind of odd hours to support processing when no support is required in the United States, and also at times when support is required.

- *Commercial laws:* In case of a dispute, laws of which land will prevail? In addition, laws relating to intellectual property rights vary significantly across the world.

- *Labor practices:* The local hiring practices and compensation requirements can be very different. Most nations have a much larger component of benefits included as part of the standard employment agreement compared to the United States, where employees have a lot more choice to select what they want.

- *Industry practices:* Certain practices such as shift work are unique to the business process outsourcing industry.

- *Regulatory requirements:* Many nations offer special export zoning that places certain restrictions on how the business is supposed to operate in exchange for providing special benefits, such as tax holidays.

- *Tax regulations:* Transfer pricing regulations have come to significant prominence in most countries, and thus need to be considered in the cost equation.

This is not a comprehensive list, but should provide an understanding that there can be enough variation from one country to another that the offshoring must be considered with a deeper understanding of the environment. Above all, we need to be open to changes to suit the environment and optimize the process based on that environment. Can Six Sigma be useful in addressing these issues? The Six Sigma analysis would consider the different factors such as behaviors, laws, practices, and cultures in its analysis to drive to an enhanced solution under the new environment.

**Information.** At the core of outsourcing is information—whether we are developing software, performing data entry, processing forms, designing new products, or conducting research. The common element across each of these activities is the information that is processed and transmitted across global boundaries. Whenever, information is involved, it raises concerns with respect to privacy and security of data—because it is so easy to share it with others today due to advancements in technology and communications. At the press of a keystroke, one can transfer a customer list or personal information about an individual. In addition, there is a huge market for information. As a result, there are individuals and organizations just waiting to lay their hands on information that they can sell to others. Anyone who uses the Internet is familiar with spam emails. Where do all these individuals obtain your email address? Many times, these spams are originating in countries that do not have very stringent privacy laws, so the concern is genuine. Some countries (e.g., in Europe) have laws prohibiting transfer of personal information across borders. As in the case of environmental factors, the Six Sigma analysis would consider the requirements to comply with the regulatory and privacy needs and arrive at a more optimal solution.

**Measurement.**    Distance creates the perception of loss of control, especially when an offshore location is thousands of miles away. So, what provides comfort against this loss of control? The mutually agreed upon set of targets and incentives that would motivate the staff overseas to deliver on the agreed performance level provide comfort. Most firms create SLAs that specify performance requirements in detail.

What would be the threshold on which these performance expectations should be based? Is it the historical performance of the process before it was outsourced? If you consider the changes that are occurring, particularly in the human capital, technology, and environment, historical performance can at best be a target that should not be exceeded. Considering the number of variables you are dealing with, ideally, you need to reassess and revise the performance target to be more aligned with the new operating realities.

Corporations like General Electric Company, American Express, Hong Kong and Shanghai Banking Corporation, and Citicorp have reset their targets after offshoring their processes. The not-to-exceed target is a good starting point, but does not help drive the phenomenal efficiencies that should be achievable. Most of these corporations have used Six Sigma as a tool to reengineer their processes and achieve breakthrough performance improvements.

## KEY TAKEAWAYS

- Outsourcing involves tradeoffs between risks and benefits.
- Six Sigma methodology should be implemented throughout various components of the value chain.
- Human capital, technology, processes, environment, information and measurement are important aspects of outsourcing.

# MANAGING
# HUMAN CAPITAL

Managing human capital incorporates the management of one of the three critical voices in a business, the importance of employee satisfaction in Six Sigma, various human capital principles, globalization and its impact on human resource (HR) management, key human capital practices, a few human capital measurement tools, and several best-practice examples. This chapter is focused on how to get the best from everyone in the organization while deploying Transactional Six Sigma.

In the current global economy, successful enterprises must listen to the voice of the employee, the customer, and the process. When managed in an integrated fashion, all three voices will help an organization succeed in streamlining their operations, which will lead to enhanced operational and financial performance (see Figure 17.1). The customer voice can be obtained through voice of the customer management as described in previous chapters (VOC, Vora, 2002b), whereas the process voice can be determined through process audits. In this chapter, we will develop the idea of listening to the employee's voice, including the employee satisfaction survey (ESS), in greater detail. According to Walt Disney, "You can dream, create, design, and build a most wonderful place in the world—but it requires people to make the dream a reality" (Nelson, 1997).

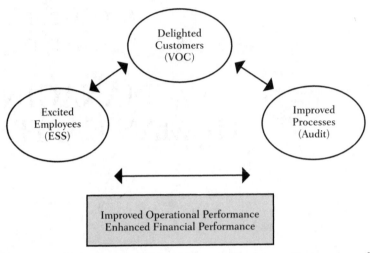

**FIGURE 17.1.** Management of Critical Voices—Employees, Customers, and Processes

## IMPORTANCE OF EMPLOYEE SATISFACTION IN TOTAL QUALITY MANAGEMENT

The Employee-Customer-Profit Chain Model (Rucci et al., 1998) developed and deployed at Sears (a major United States retailer) is similar to a balanced scorecard for the company. With the use of extensive data, they discovered that an employee's ability to see the connection between his or her work and the company's strategic objectives was a driver of positive behavior. Also, two dimensions of employee satisfaction—attitude toward the job and toward the company—had a greater effect on employee loyalty and behavior towards customers than all other dimensions. Its vision became "Sears, a compelling place to work, to shop, and to invest (3 Cs)." The vision was combined with its shared values, "Passion for the customer, Our people add value, and Performance leadership." Their Employee-Customer-Profit Chain model (see Figure 17.2) showed that a 5-point change in employee attitudes would drive a 1.3-point improvement in customer satisfaction, which in turn would drive a 0.5% improvement in revenue growth.

**FIGURE 17.2.** The Employee-Customer-Profit Chain at Sears (Rucci et al., 1998)

A seminal study done by the US General Accounting Office of 22 Baldrige Award winners or site visit companies (from 1988 to 1991) showed major improvements in employee satisfaction parameters, as shown in Table 17.1. In addition, customer satisfaction and operational performance improved, which led to better financial performance. The balanced scorecard approach of Kaplan and Norton (1996) also advocates a similar focus on learning and growth (employees), customers, internal/business processes, and financials.

**TABLE 17.1**    Employee Satisfaction Results of TQM Companies (18 Companies Responded)

| PARAMETER | AVERAGE ANNUAL % IMPROVEMENT | FAVORABLE INDICATOR | UNFAVORABLE INDICATOR | NO CHANGE |
|---|---|---|---|---|
| Employee Satisfaction | 1.4 | 8 | 1 | 0 |
| Attendance | 0.1 | 8 | 0 | 3 |
| Turnover | (6.0) ↓ | 7 | 3 | 1 |
| Safety and Health | 1.8 | 11 | 3 | 0 |
| Suggestions Received | 16.6 | 5 | 2 | 0 |

Source: GAO NSIAD 91-190, May 1991.

Next, we will examine various employee value principles that are applicable in a global economy.

# HUMAN CAPITAL PRINCIPLES

## CUSTOMER-SUPPLIER RELATIONSHIP

One of the simplest principles is the customer-supplier relationship (AT&T, 1988), as shown in Figure 17.3. This relationship should be viewed from two perspectives—external and internal. To delight the external customer, there must be harmony among all internal process steps, so that everyone is treated as internal customer and supplier. Looking at it from the employees' perspective, a systematic understanding of requirements, expectations, and satisfaction is essential to design internal processes to meet employees' needs. With internal cooperation and collaboration, it is easier to focus on delighting the external customer. Though it sounds simple, this basic principle requires discipline and constant work to implement. With the right focus and dedication, a better customer–supplier relationship can be achieved.

## BUSINESS EXCELLENCE MODEL

Next, we will look at the Business Excellence Model (Vora, 2001), as shown in Figure 17.4. The model calls for four basic steps to achieve business excellence. The steps are: ideas, actions, results, and celebrations. To achieve business excellence, we need ideas. These ideas come from employees in the organization. For select promising ideas, we need a team of people who will take actions leading to results. We need to

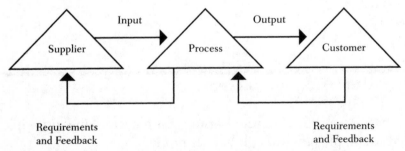

**FIGURE 17.3.** Customer–Supplier Relationship (AT&T, 1988)

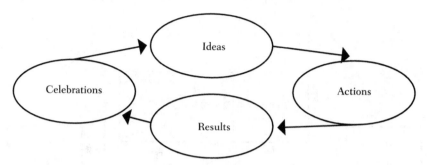

**FIGURE 17.4.** Business Excellence Model (Vora, 2001)

review the results, and if the results are good, then we must celebrate as a team. We can then focus on the next set of ideas, and the cycle continues for achieving business excellence.

## USA Malcolm Baldrige National Quality Award Criteria

The Malcolm Baldrige National Quality Award in the United States (Baldrige, 2004) recognizes organizational performance excellence using seven categories (see Figure 17.5). These Baldrige criteria have been used extensively to assess organizations in their Six Sigma journey.

· Leadership
· Strategic planning
· Customer and market focus
· Measurement, analysis, and knowledge management
· Human resource focus
· Process management
· Business results

One of the important categories is the human resource focus. This category focuses on work systems (organization and management of work, employee performance management system, hiring, and career progression), employee learning and motivation (employee education, training, development, motivation, and career development), and employee well-being and

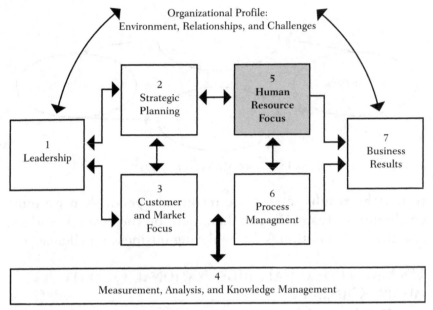

**FIGURE 17.5.** Malcolm Baldrige National Quality Award Framework (Baldrige, 2004)

satisfaction (work environment, employee support, and satisfaction). Positive trends in human resource results are required for this category. This category carries 160 points out of the total 1,000 points for the Malcolm Baldrige National Quality Award.

## EUROPEAN FOUNDATION FOR QUALITY MANAGEMENT EXCELLENCE MODEL

According to the EFQM Excellence Model (EFQM, 2004), shown in Figure 17.6, excellent organizations manage, develop, and release the full potential of people at an individual, team-based, and organizational level. They promote fairness and equity, and involve and empower their people. They care for, communicate, reward, and recognize in a way that motivates staff, which in turn builds commitment to using employee skills and knowledge for the benefit of the organization. These organizations comprehensively measure and achieve outstanding results with respect to their people. Both perception measures and performance indicators are used.

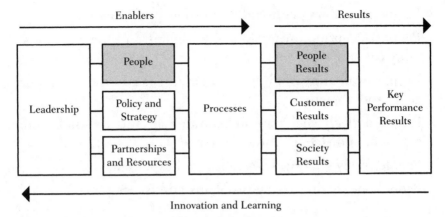

FIGURE 17.6.  EFQM Excellence Model (EFQM, 2004)

## ISO 9000:2000 STANDARDS—INVOLVEMENT OF PEOPLE

The ISO 9000:2000 Standards (ISO, 2004) are based on eight quality management principles. According to Principle 3 (involvement of people), people at all levels are the essence of an organization, and their full involvement enables their abilities to be used for the organization's benefit.

Key benefits of involvement of people include:

· Motivated, committed, and involved people within the organization
· Innovation and creativity in furthering the organization's objectives
· People being accountable for their own performance
· People eager to participate in and contribute to continual improvement

Applying the principle of involvement of people typically leads to:

· People understanding the importance of their contribution and role in the organization

- People identifying constraints to their performance
- People accepting ownership of problems and their responsibility for solving them
- People evaluating their performance against their personal goals and objectives
- People actively seeking opportunities to enhance their competence, knowledge, and experience
- People freely sharing knowledge and experience
- People openly discussing problems and issues

## THE SEVEN HABITS OF HIGHLY EFFECTIVE PEOPLE

Stephen Covey (Covey, 2004; Covey, 1989) has developed the seven habits paradigm as follows:

1. Be proactive.
2. Begin with the end in mind.
3. Put first things first.
4. Think win/win.
5. Seek first to be understand ... then to be understood.
6. Synergize.
7. Sharpen the saw.

Covey describes habits 1, 2, and 3 as personal victory to take us from dependence to independence. Then in the workplace, we practice habits 4, 5, and 6 leading to public victory through interdependence. Habits 4, 5, and 6, are crucial in achieving harmonious relationships among internal customers (employees). Without employees' cooperative and synergistic attitudes and actions, the organization will not succeed in satisfying external customers.

"While reading the book, *Gandhi*, I was profoundly influenced by the account of this principle-centered leader," begins the article by Dr. Stephen R. Covey, author of *Seven Habits of Highly Effective People* (Covey, 2004). "In the book, Gandhi described what he found to be the secret of all success. He said, in effect, 'It's win-win to always seek the interests of all parties.' His leadership role was not to add fuel to the adversarial fighting and feuding, but rather to be a peacemaker and to create solutions for all parties that were better than any of those proposed initially." In the article, Covey relates Ghandi's win-win principals to today's business relationships. "Ultimately, business is all about the relationships between suppliers and customers. Every person is a supplier of his or her talents. Employees are suppliers of their talents and labors. The best companies invest heavily to establish and maintain win-win relationships with internal and external suppliers," says Covey.

**The Win-Win Agreement™.**    The Win-Win Agreement is the heart of employee empowerment. The Win-Win Agreement is a clear, mutual understanding and commitment regarding five things: desired results, guidelines, resources, accountability, and consequences. An important consequence of the Win-Win Performance Agreement is that every person in the organization can answer five questions:

· Why am I here?
· What are my objectives?
· How am I doing?
· Where do I go for help?
· What's in it for me?

It is important to involve people in setting the standards and use discernment more than quantitative measurement when evaluating performance. Allow people to judge themselves after they receive feedback from the stakeholders—all the people they interface with. And make sure the performance agreement is reinforced by structure and systems, with both natural and logical consequences.

Rewards should deal with four basic human needs:

1. Physical and financial (benefits and money)

2. Social and emotional (recognition and relationships)
3. Mental (learning and growth opportunities)
4. Spiritual (expanded stewardship, responsibility, influence, freedom, latitude, contribution, and legacy)

In the short-term, a win-lose compensation system will beat out win-win rhetoric any day. Often in trying to get cooperation, we promote people, systems, structures, and styles focused on competition. We try to establish a value around cooperation, but the paradigm is one of individual success around competition. Consequently, people are not thinking ecologically. They are not thinking win-win. They are thinking win-lose, piecemeal, competitive, and quick fix.

Let us now review globalization and its impact on HR management.

## GLOBALIZATION AND ITS IMPACT ON HUMAN RESOURCE MANAGEMENT

Globalization has set new boundaries for competitive businesses. While geographic distance matters less and less, the possibilities as well as the necessities that result are endless. Some of the new business practices that emerged on a worldwide basis are as follows (Rosen et al., 2000):

· Companies will need *global reach* to serve global customers. Lacking that capacity, companies must build partnerships around the world.
· All markets are *local markets*. Quality, pricing, and service must be globally competitive and locally appropriate.
· *Speed and urgency* will be the norm as companies change strategy and direction continuously. Flexibility and innovation will be their secrets to success.
· *Local distribution* will require a much deeper understanding of local business needs as well as prevailing national culture.

- *Outsourcing noncore services* will force companies to rethink their basic competencies and develop relationships with suppliers to provide other functions.

A major report that examined the changing role of HR function in the context of a globally competitive marketplace identified top HR priorities, the consistency of HR practices, key challenges to consistency, and general challenges for global HR function (Rioux et al., 2000):

**Top HR Priorities:**
- Leadership development
- Recruiting high-quality employees
- Employee retention

**Consistency of HR Practices Amongst Increasing Levels of Decentralization:**
- International organizations use consistent HR practices to develop a common corporate culture and to improve the effectiveness of the HR function.
- Domestic organizations are creating consistent HR practices to improve both the effectiveness and efficiency of the HR function.
- In International organizations, selection practices such as assessment, testing for selection, and Internet advertising vary greatly across locations.

**Challenges to Consistency of HR Practices:**
- Variations in social, political, and economic circumstances
- Different locations/offices have their own way of doing things and are resistant to change
- The perceived value of HR function varies across locations/ offices

**Best Practices to Achieve Consistency Across Locations/Offices:**
- A long-term HR plan to align HR strategies/objectives with corporate strategies/objectives

- A centralized reporting relationship around the globe
- Standardized assessment, development, and compensation practices
- Tied regional accountability to performance management
- Shared HR best practices used in certain locations with all other locations

**General Challenges for Global HR Function:**
- Coordination of activities in many different locations
- Understanding the continual change of the globally competitive environment
- Building a global awareness in all HR departments/divisions
- Creating a multicultural HR team

Next, we will explore key employee value practices.

# HUMAN CAPITAL PRACTICES

## FOCUS ON EMPLOYEES

A clear link has been established, which shows that in an organization, employee satisfaction is a prerequisite to achieving customer satisfaction (Carr, 1990; Freiberg & Freiberg, 1997; Vora, 2004). Without looking after the well-being of your own people through trust and care, do not expect your employees to help your customers. In the 21st century, we have to manage and treat employees as "knowledge workers." If we do not treat them well, they will find other places to share their specific knowledge. We must treat each employee as a human being first, rather than an expendable cost item.

A study by the Gallup organization based on over one million employee surveys and over 80,000 manager interviews from a broad range of companies, industries, and countries boiled down to the following suggestions for an effective focus on employees (Buckingham & Coffman, 1999).

- Select a person ... based on talents (not simply experience, intelligence, or determination).

- Set expectations ... by defining the right outcomes (not the right steps).

- Motivate a person ... by helping him focus on strengths (not on weaknesses).

- Develop the person ... by helping him find the right fit (not simply the next rung on the ladder).

## HUMAN CAPITAL ACQUISITION

**Recruitment.**    This activity requires identifying sources of HR including those both inside and outside an organization (Buckingham & Coffman, 1999; Flamholtz, 1999; Flamholtz, 1990). The activity also refers to attracting possible future members of an organization. The major components of external recruitment are advertising, college recruiting, employee agency services, entertainment, travel, and administrative efforts to track the recruitment effort.

**Selection.**    For a given position, the recruitment advertisement is run through appropriate channels (newspapers, trade journals, internal company postings, website, employee referrals, etc.). This will result in an influx of resumes (paper as well as electronic), which get screened at a high level by an HR specialist to ensure that potential candidates meet the minimum requirements posted on the advertisement for a specific position. The screened resumes are then forwarded to departments and groups where new openings exist. Recruiting managers then shift through these resumes and select a subset of candidates for an interview process.

Depending on the size of an organization, candidate interview teams are put together to thoroughly evaluate candidates who come in for an interview. Here, one should leverage the Gallup study, which stresses selecting a person based on his or her talents rather than skills, experiences, intelligence, and determination alone. By talents, we mean things that come naturally to a person. There are three major types of talents—

striving (sales people), thinking (management people), and relating (customer service and HR people). With the right talents for a particular position, additional skills can be imparted through training and education.

**Hiring and placement.**   This activity consists of bringing an individual into an organization for a specific position and placing him or her on the job. Hiring involves moving and travel for external candidates. Placement includes a variety of administrative efforts for a new person to get started with the organization. Again, the first-line supervisor should work with the new individual by setting the right expectations by defining the right outcomes instead of the right steps. Generally, a first assignment is carefully crafted to allow the new individual to make a contribution, during her learning curve time period. New employee orientation, new employee mentorship, and new employee training are very critical aspects during the placement effort, as described later in this chapter.

Now, we will look at the principles of Microsoft's successful strategic recruiting, as shown below (Workforce, 2004):

- Hire quality candidates: Uphold Microsoft's recruiting religion and be "the guardians of the threshold."
- Ownership: Individuals and teams to assume responsibility and authority.
- Respect: Treat all candidates, hiring managers, and the recruiting team with dignity and respect.
- Open and honest communication: Interactions are open, honest, and direct.
- Quality and innovation: Encourage experimentation and risk taking.
- Commitments are sacred: Always keep the commitment we make.
- Have fun: Humor, team member relationships, self-expression and celebration are critical to an organization's success.

Here is another example from Toyota Motor Manufacturing, Kentucky, USA, emphasizing the human side of manufacturing. The Toyota philosophy consists of the Toyota Production System and the Toyota Way (Bodek, 2003). The Toyota Way is to realize that the human effort is very critical to its success. There are two pillars of the Toyota Way:

1. Continuous improvement: Toyota's success rests on the need of all employees and management to look for and striving for continuous improvement, and never being satisfied.

2. Respect for people and honesty: If you do not have respect for people who work for the company, you are in the wrong business.

A general road map to effectively manage "knowledge workers" in the 21st century consists of human capital management (participation and motivation), and human capital development (learning and growth—Nelson, 1997; Vora, 2002a; Vora, 2003):

## HUMAN CAPITAL MANAGEMENT

**Participation.**   Participation includes new employee orientation, mentoring, effective meetings, communications, and teamwork areas as follows:

### New Employee Orientation
· Overview of the company (provide a big picture)
· Organization structure, company history
· Key products/services
· Key markets/customers
· Strategic direction: vision, mission, and values
· Executive presence is a definite plus

### Mentoring
· Assign two mentors: one social and one technical

- Mentoring cuts down the learning curve of the new hire.
- Win-win for new employees and the company (better assimilation, productivity gain, and positive retention).
- The role of a leader is to mentor others, both within and outside the organization. A successful person is blessed with many mentors throughout his or her life span It is a human obligation to mentor others in your organization, society, and the world.

**Effective meetings.**    Meetings take up a significant amount of time at all levels within an organization, particularly as one moves up in the management chain. A quick poll conducted by the author on effectiveness of organizational meetings always ends up with general dissatisfaction about meetings and its value for the people in attendance. Properly managed meetings lead to significant productivity gain. Also, it is important from a human respect point of view to attend a meeting if you are invited and add value by making contributions.

- Establish the need for a meeting (if a meeting is not needed, use other means of communicating your message such as e-mail, memo, bulletin board, broadcasts, etc.)
- Define purpose (P): Must be a specific purpose to call a meeting
- Define agenda (A): Send an agenda 24 hours in advance (if no agenda was sent, some organizations allow people not to attend a meeting)
- Set a time limit (L): Always start and end the meeting on time
- Invite the appropriate audience (key people needed to make decisions as well as people who need to make presentations must be invited)
- Designate a scribe and a facilitator (rotate a scribe among the attendees for standing meetings; facilitator keeps the meeting on target)
- Conduct meeting evaluation (what worked and what did not work, which leads to better meeting next time)

- Provide a quick meeting summary with key discussion and action items (generally a one page e-mail with a turn around time of 24 hours after the conclusion of your meeting)

**Communications.**   The success of Transactional Six Sigma requires constant communication among all stakeholders in an organization, as follows:

- Executives should gather employee ideas and maintain open communication throughout the organization.
- "An individual without information cannot take responsibility; an individual who is given information cannot help but take responsibility" (Carlzon, 1987).
- Drop in on your employees from time to time.
- Employees deserve to know what is happening and will handle the responsibility better than you can imagine.
- Well-informed employees are good and productive employees, because they feel involved.
- Communicate all information to all employees, all of the time.
- Continuous and supportive communication from managers, supervisors, and associates is too often underemphasized. It is a major motivator.
- Interoffice mail, e-mail, voice mail. Whatever happened to face-mail?

**Teamwork.**   In any organization, to achieve lasting quality improvement, teamwork is essential.

- Two people will produce a better solution than one person working alone.
- All work is teamwork.
- The four stages of teamwork: forming, storming, norming, and performing. The goal is to move quickly from a forming stage to the performing stage.

- People become unstoppable when they are moved by a common vision and have the power and tools to achieve a shared goal.
- Create and maintain an atmosphere of mutual trust and respect among team members.
- Let the team manage risks, control the budget, and get involved in the reward process.
- Focus on recognizing and rewarding teamwork.
- Teamwork between people, departments, and functions is critical to an organizations' success.

**Motivation.**    According to Andy Grove, Intel founder and former CEO, "Motivation comes from within. The most a manager can do is to create an environment in which motivated people can flourish" (Nelson, 1997). In the motivation area, we look at recognition, the suggestion system, the theory of strengths, and empowerment, as follows:

**Recognition.** An informal survey conducted by the author in the last decade (DePree, 1989; Dunn, 1991; Nelson, 1994) consists of the following four questions:

1. In the last six months, how many of you received any recognition?
2. In the last six months, how many of you gave any recognition to others?
3. How many of your organizations have a systematic recognition process?
4. How many of you think you get too much recognition?

Invariably, answers to the preceding questions lead to some positive response for questions 1 and 2. Few hands go up for systematic recognition process question 3, and for the fourth question, the author has yet to find an individual who feels that he or she gets too much recognition, as follows:

- Day-to-day (thank you note, memo, e-mail)
- Informal (team milestone completion—memento)

· Formal (nomination, evaluation, celebration)
· There is never enough praise and recognition
· Nonmonetary but sincere, timely, and meaningful recognition is a great motivator

According to Mary Kay Ash, founder of Mary Kay Cosmetics, "Every person you meet has a sign around their neck that says, 'Make me feel important.' If you can do that, you will be a success not only in business, but in life as well" (Nelson, 1997). If you show people you don't care about them, they will reciprocate. Show them you care about them, and they will return the favor.

**Suggestion system.**    In Japan, a humanistic approach in managing people is encouraged. Japanese companies demonstrate very supportive attitudes toward receiving suggestions and feedback from all levels of employees (Khoo & Tan, 2003). The track record is better for Toyota employees, with an average of 240 suggestions per person per year, compared with about two suggestions per person per year in the United States. The suggestion system stipulates the following:

· Employees have knowledge.
· Create a simple process to gather, acknowledge, and act on suggestions.
· Involve employees who gave suggestions during the implementation of their idea.
· Recognize and reward employees whose suggestions are implemented with positive impact on the organization. In general, return part of the savings back to the person who gave the suggestion.

**Theory of strengths.**    The Theory of Strengths (Buckingham & Clifton, 2001; Clifton & Nelson, 1992) runs counter to human intuition and commonly accepted practices in life. We are always taught to keep working on our weaknesses and not to pay much attention to our strengths. As a result, individuals lose confidence and waste time on things they do not enjoy. Take a case of your child bringing a report card home. If the child gets three As, one B, and one D, the entire discussion will revolve around the one D. Ideally, you want to acknowledge the three As, and spend a little time to talk about a strategy on how to convert the one D into a B.

The Theory of Strengths works well in the arts and sports world. There, the role of a coach is to find individual strengths and encourage the practice of strengths so as to overcome any weaknesses. In real life, we all need to do the following to harness our strengths:

· Focus on strengths and manage weaknesses.
· Find out what you do well, and do more of it.
· Find out what you don't do well, and stop doing it.
· Strengths develop best in the framework of a mission.
· Strengths develop only in relation to another human being.
· Nothing happens until someone expects something of you in ways you can achieve.
· Celebration: The way to recognize good work.

**Empowerment.**    Most of the time, the word "empowerment" has been misused by many. For true empowerment you need the following:

· Responsibility and authority to get things done by employees.
· Include employees in the decision-making process.
· Encourage individual and team initiatives to serve your customers.
· Employees can "stop production" when quality is not up to standard.
· Value and appreciate employee's work, knowledge, and skills.
· Employees must be more self-directed and autonomous on the job.

# HUMAN CAPITAL DEVELOPMENT

The human capital development consists of education and training, performance appraisals, employee satisfaction, coaching, and change management, as follows:

**Education and training (E/T).**   According to Robert Reich, former U.S. Secretary of Labor, "Well trained and dedicated employees are the only sustainable source of competitive strength" (Nelson, 1997).

- Continuous education and training for superior skills.
- Encourage new learning—Leaders must lead by example.
- Develop curriculum at all levels (technical, problem solving, teamwork, continuous improvement, etc.).
- Set objectives to achieve X hours and $Y per year (e.g., Motorola has 50 hours per year requirements).
- Develop a company database for education and training—allow employees to log their hours and cost in the system. This allows a manager to assess a training program of each individual working for him or her.
- Tie education and training to the annual review process.
- Use a variety of methods (in-class, videos, book/journal reading, and cascade training).
- Use education and training to recognize good work.
- Evaluate education and training effectiveness. Four levels of evaluation for education and training (E/T) effectiveness are (Kirkpatrick, 1998):
  - Level 1—Reaction (immediate feedback at the conclusion of E/T).
  - Level 2—Learning (before and after knowledge retention through a survey at start and end of E/T).
  - Level 3—Behavior (six months after completion of E/T).
  - Level 4—Results (one year later, assess value of E/T through a supervisor).

- Levels 1 and 2 are easy to implement. Level 3 and 4 evaluation require significant effort and as a result, do not get measured.
- Learning is a process, not a goal. Each new insight creates another new layer of potential insights.

**Performance appraisals.**    Think about how many of us look forward to a dreadful day of an annual performance appraisal meeting with our boss. In general, a quick poll (Prince, 1994) conducted by the author at various meetings around the world invariably leads to the same conclusion—no one is happy about the current method of performance appraisals. There are several flaws in the current system such as, the process is infrequent, subjective, and in some instances vindictive. Organizations following Six Sigma philosophy should practice performance appraisals on a continuous basis wherein a supervisor should provide feedback to an employee after major events throughout the year. This method allows an employee to recover from any mistakes, learn from them, and make adjustments for subsequent events. The idea of feedback is to improve employee performance and add value to achieve organizational goals. There is a new movement to use a 360-degree feedback system where an employee receives feedback from customers, suppliers, peers, and superiors. All the inputs are compiled to arrive at employee areas of strengths and areas for improvement. Management should be judicious in the use of the 360-degree system with integrity and not to use it as a weapon to punish employees. To be effective, performance appraisals should:

· Be more frequent (more than once a year)
· Be objective (criteria: quality, customer satisfaction, people management, financial performance, etc.)
· Focus on positives
· Be constructive and developmental
· Have a variety of input (360-degree feedback—customers, suppliers, peers, superiors, etc.)

- Encourage and reward teamwork
- Focus on the systems perspective (do not blame individuals)

**Employee satisfaction.**   Because employees are internal customers, the organization must determine whether they are satisfied or dissatisfied with any specific issues. At AT&T Bell Laboratories Switching Systems Business Unit, Vice President Donald Leonard used to arrange "Donuts with Don" meetings on a regular basis. In this meeting, about 25 to 30 employees without the middle management would interact with Don Leonard and share their views on the work environment, issues, and opportunities for improvement. Don would take those informal suggestions and work with his direct staff to improve the work environment. On an annual basis, the organization should conduct an employee satisfaction survey as follows:

- Gather the pulse of employees—annual surveys (ideally by a third party).
- Use data at a department level to make improvements on a few key issues of dissatisfaction.
- Involve employees in the solution of the key issues.
- Communicate issues, actions, and results with employees throughout the organization.

**Coaching.**   An ideal role of a supervisor/manger is to be a coach for their employees. Being a coach means there is a responsibility to find out employee strengths and use those strengths in making proper work assignment to harness their strengths for the organization. This is a common practice in the sports world and it works. To be an effective coach, one must do the following:

- Create leaders at all levels.
- Empower your people, and give general direction.
- Stay out of the way and support risk takers.
- Do not shoot people who make mistakes, shoot people who do not take risks.

**Change management.**    In the current turbulent global environment, change is inevitable. To survive and thrive in the chaotic world, one must understand, appreciate, and participate in managing change. Leaders should:

· Share their excitement about change with employees.
· Involve employees in making decisions that affect them and their work.
· Change only what needs to be changed.
· Be honest and timely with good news and with bad news.
· Be clear and consistent about change expectations.
· Create supportive work environments to foster desired behaviors and outcomes.

Next we will describe a few human capital measurement tools to develop a colleague profile, employee satisfaction survey, and a human resource scorecard.

## HUMAN CAPITAL MEASUREMENT TOOLS

### COLLEAGUE PROFILE

Develop a colleague profile (Bounds, 1996) to reinforce attitudes and behaviors that are important to the mission of your organization. This profile should be used to identify strengths and constructive areas for growth among the management team. A few items to consider are:

· Customer first (internal and external)
· Providing leadership
· Performance planning and feedback
· Coaching
· Working with others

- Encouraging open communications
- Empowering people
- Business knowledge
- Continuous improvement
- Supporting the mission

## Employee Satisfaction Survey

Well-orchestrated employee surveys lead to high return rates (Gilbert et al., 2003). Here is how to conduct them:

- Establish clear goals and objectives
- Develop a communication plan
- Brand the survey process
- Allocate sufficient resources
- Define roles and responsibilities
- Demonstrate management commitment
- Ask the right questions the right way
- Collect data the right way at the right time
- Take clear follow-up actions
- Review and audit the process

## New Hire Survey

To deal with employee turnover, Ceridian's HR staffing department created a quarterly survey in 2002 to gather information from each new hire regarding their satisfaction of the hiring process, orientation, training, and impression of the manager (Workforce Online, 2003). There are 31 questions, which include, "How satisfied are you with how the job was described during the interview process compared to what you are actually doing?" The questions are broken into four major areas as follows:

- The interview process (preemployment)—eight questions

- New hire introduction—nine questions
- New hire training—seven questions
- Job specific satisfaction—seven questions

## EMPLOYEE ENGAGEMENT QUESTIONNAIRE

In 1999, The Gallup Organization created a feedback system for employers (Buckingham & Coffman, 1999; Workforce Management Online, 2003b) that would identify and measure elements of worker engagement most tied to the bottom line, such as sales growth, productivity, and customer loyalty. After hundreds of focus groups and thousands of interviews with employees in a variety of industries, Gallup came up with the Q12, a 12-question survey that identifies strong feelings of employee engagement. Results from the survey show a strong correlation between high scores and superior job performance. The 12 questions are as follows:

1. Do you know what is expected of you at work?
2. Do you have the materials and equipment you need to do your work right?
3. At work, do you have the opportunity to do what you do best every day?
4. In the last seven days, have you received recognition or praise for doing good work?
5. Does your supervisor, or someone at work, seem to care about you as a person?
6. Is there someone at work who encourages your development?
7. At work, do your opinions seem to count?
8. Does the mission/purpose of your company make you feel your job is important?
9. Are your associates (fellow employees) committed to doing quality work?
10. Do you have a best friend at work?

11. In the last six months, has someone at work talked to you about your progress?

12. In the last year, have you had opportunities at work to learn and grow?

## THE COST OF TURNOVER

A comprehensive worksheet was created to address the cost of employee turnover (Workforce Management Online, 2003a). It includes direct costs, such as the cost of background checks, as well as indirect costs, such as lost productivity. The worksheet shows the "green money," or actual cost of turnover, and the "blue money," or softer costs of turnover. The worksheet consists of:

· Notice period
· Vacancy period
· Hiring/orientation period
· Hidden costs

The sum of the preceding costs gives the total replacement cost.

## HUMAN RESOURCE SCORECARD BASED ON COMPANY'S STRATEGIC FOCUS

The HR scorecard (Beatty et al., 2003) is a quantitative measurement tool that should complement an organization's balanced scorecard, particularly in the area of learning and growth. The HR scorecard should address the following:

· HR strategic focus: Operational excellence, product leadership, and customer intimacy
· HR competencies: Administrative expertise, employee advocacy, strategy execution, and change agency
· HR practices: Communication, work design, selection, development, measurement, and rewards

- HR systems: Alignment, integration, and differentiation
- HR deliverables: Workforce mindset, technical knowledge, and workforce behavior

The HR deliverables complement the business balanced scorecard as the learning and growth component.

## KOLBE CONATIVE INDEX

From Plato to Freud, philosophers and psychologists have recognized three distinct parts of the mind: the cognitive (governing thoughts), the affective (governing emotions), and the conative (governing actions). Kathy Kolbe (Kolby, 1990) was the first to develop a practical theory to define the ways in which we instinctively perform simple and complex tasks. She has tested her ideas on people at all levels of major corporations on four continents, on a professional basketball team, in school systems, and on families—and has found the same startling results. The more you understand, trust, and—most important—use the power of your own creative style and that of people around you (coworkers, teammates, family, and friends), the more you will succeed in any endeavor, personal or professional. Kolbe identifies four modes of action, which each individual has in varying amounts. You generally use one or two modes primarily and avoid other modes. The four modes are:

- Fact finder: People who are careful to research a problem before responding.
- Follow thru: People for whom systems are the answers.
- Quick start: People who are spontaneous and innovative in their solutions.
- Implementor: People who relish demonstration and tangible results.

You can take the Kolbe index test online to determine your dominant style of action (http://www.kolbe.com and click on Kolbe A™ index). There is a cost associated for this index, presently it costs $49.95. Having the Kolbe Conative Index (KCI), you can understand that you do not have to change the way you act to improve your performance. Your best performance comes from doing what comes naturally to you. The KCI is also a great facilitator in recruiting and retaining people, and in building and improving team performance by matching a team member's action mode with requirements for the role in a team.

Here are some of the best-practice examples from the Malcolm Baldrige Award winners for the human capital creation in a global economy (Baldrige, 2003; Bounds, 1996).

## BEST-PRACTICE EXAMPLES

- AT&T Universal Card Services—now part of Citigroup (1992 Baldrige Award, Service): Best-in-class in employee recognition and suggestion system (spent 2% of payroll on Recognition and Suggestion System).
- Federal Express Corporation (1990 Baldrige Award, Service): Survey/Feedback/Action management evaluation system. From 1985-1990, 91% of employees "proud to work for Federal Express."
- IBM Rochester (1990 Baldrige Award, Manufacturing): Invested heavily in Education and Training, the equivalent of 5% of its payroll.
- Milliken & Company (1989 Baldrige Award, Manufacturing): "The 'Pursuit of Excellence' process continues to evolve after beginning the journey in 1980. Over 90 'opportunities for improvement' were submitted per associate for 1999, providing a means by which all associates can contribute to improving the process. Milliken associates participate in over 14,000 teams each year. The Sharing Rally is another avenue by which associate involvement is created, and Milliken & Company has now held over 160 Sharing Rallies." *Craig Long, Director of Quality.*
- Pearl River School District (2001 Baldrige Award, Education): PRSDs overall staff and faculty satisfaction rate

increased over the past four years, from 89% to 98% for staff and from 86% to 96% for faculty.

· SSM Health Care (2002 Baldrige Award, Health Care): "Employees Drive Success"—At SSMHC, the turnover rate for all employees was reduced from 21% in 1999 to 13%, as of August 2002.

· Trident Precision Manufacturing, Inc. (1996 Baldrige Award, Small Business): Empowered their employees to make process improvements using the suggestion system. 95% of the improvement at Trident came from its own resources and people.

In addition, the innovative Human Resource Management (HRM) strategy in place at Herman Miller Inc. (one of the world's largest manufacturers of office furniture) has three primary goals: 1) Building employee capabilities, 2) Building employee commitment, and 3) Improving the professional capabilities of the HR function itself. HR believes that attracting and retaining best-in-class talent requires continuing business success, corporate-wide commitment to living the company's core values, superior employee growth, development and participation opportunities, and superior opportunity for employee-owners to share in the financial success of the business (McCowen et al., 1999).

## KEY TAKEAWAYS

· The voice of the employees plays a crucial role in enhancing operational and financial performance of a service firm.

· Human capital principles play central role in various business excellence models and award criteria.

· Globalization is adding increasing pressure to attract, retain, and develop valued workforce.

· Human capital measurement tools are being developed and are finding increasing use in service settings.

# APPENDIX

TABLE 1  Corporations Making up the Nasdaq 100 (As of May 20, 2004)

| COMPANY NAME | SYMBOL | % OF INDEX (ADJUSTED) |
|---|---|---|
| Adobe Systems Incorporated | ADBE | 0.9173 |
| Altera Corporation | ALTR | 0.9915 |
| Amazon.com, Inc. | AMZN | 0.9714 |
| American Power Conversion Corporation | APCC | 0.2987 |
| Amgen Inc. | AMGN | 2.8292 |
| Apollo Group, Inc. | APOL | 1.3880 |
| Apple Computer, Inc. | AAPL | 1.2163 |
| Applied Materials, Inc. | AMAT | 1.4080 |
| ATI Technologies Inc. | ATYT | 0.3185 |
| BEA Systems, Inc. | BEAS | 0.2759 |
| Bed Bath & Beyond Inc. | BBBY | 1.2766 |
| Biogen Idec Inc | BIIB | 2.0281 |
| Biomet, Inc. | BMET | 1.1655 |
| Broadcom Corporation | BRCM | 0.7560 |
| C.H. Robinson Worldwide, Inc. | CHRW | 0.2878 |
| Career Education Corporation | CECO | 0.5584 |
| CDW Corporation | CDWC | 0.4875 |
| Cephalon, Inc. | CEPH | 0.2320 |
| Check Point Software Technologies Ltd. | CHKP | 0.4929 |
| Chiron Corporation | CHIR | 0.9784 |
| Cintas Corporation | CTAS | 0.7882 |

**TABLE 1**    Corporations Making up the Nasdaq 100 (As of May 20, 2004) *(Continued)*

| COMPANY NAME | SYMBOL | % OF INDEX (ADJUSTED) |
|---|---|---|
| Cisco Systems, Inc. | CSCO | 4.6681 |
| Citrix Systems, Inc. | CTXS | 0.3414 |
| Comcast Corporation | CMCSA | 2.3866 |
| Compuware Corporation | CPWR | 0.1515 |
| Comverse Technology, Inc. | CMVT | 0.3045 |
| Costco Wholesale Corporation | COST | 0.7810 |
| Dell Inc. | DELL | 2.8438 |
| DENTSPLY International Inc. | XRAY | 0.3229 |
| Dollar Tree Stores, Inc. | DLTR | 0.2432 |
| eBay Inc. | EBAY | 3.4542 |
| EchoStar Communications Corporation | DISH | 0.6995 |
| Electronic Arts Inc. | ERTS | 1.2851 |
| Expeditors International of Washington, Inc. | EXPD | 0.3967 |
| Express Scripts, Inc. | ESRX | 0.4660 |
| Fastenal Company | FAST | 0.3132 |
| First Health Group Corp. | FHCC | 0.1268 |
| Fiserv, Inc. | FISV | 0.7836 |
| Flextronics International Ltd. | FLEX | 0.8298 |
| Garmin Ltd. | GRMN | 0.2791 |
| Gentex Corporation | GNTX | 0.2460 |
| Genzyme General | GENZ | 1.0165 |
| Gilead Sciences, Inc. | GILD | 1.1739 |
| Henry Schein, Inc. | HSIC | 0.2357 |
| Intel Corporation | INTC | 5.4955 |
| InterActiveCorp | IACI | 1.9450 |
| Intersil Corporation | ISIL | 0.2511 |
| Intuit Inc. | INTU | 0.8362 |
| Invitrogen Corporation | IVGN | 0.2810 |
| JDS Uniphase Corporation | JDSU | 0.4375 |

**TABLE 1**    Corporations Making up the Nasdaq 100 (As of May 20, 2004) *(Continued)*

| Company Name | Symbol | % OF Index (Adjusted) |
|---|---|---|
| Juniper Networks, Inc. | JNPR | 0.6497 |
| KLA-Tencor Corporation | KLAC | 0.9438 |
| Lam Research Corporation | LRCX | 0.2692 |
| Lamar Advertising Company | LAMR | 0.2917 |
| Level 3 Communications, Inc. | LVLT | 0.1936 |
| Lincare Holdings Inc. | LNCR | 0.2847 |
| Linear Technology Corporation | LLTC | 1.3846 |
| Marvell Technology Group, Ltd. | MRVL | 0.4255 |
| Maxim Integrated Products, Inc. | MXIM | 1.9148 |
| MedImmune, Inc. | MEDI | 0.5413 |
| Mercury Interactive Corporation | MERQ | 0.3849 |
| Microchip Technology Incorporated | MCHP | 0.4526 |
| Microsoft Corporation | MSFT | 8.4406 |
| Millennium Pharmaceuticals, Inc. | MLNM | 0.4058 |
| Molex Incorporated | MOLX | 0.2435 |
| Network Appliance, Inc. | NTAP | 0.5910 |
| Nextel Communications, Inc. | NXTL | 2.8564 |
| Novellus Systems, Inc. | NVLS | 0.3958 |
| NVIDIA Corporation | NVDA | 0.3296 |
| Oracle Corporation | ORCL | 2.1557 |
| PACCAR Inc. | PCAR | 0.8972 |
| PanAmSat Corporation | SPOT | 0.4114 |
| Patterson Dental Company | PDCO | 0.4082 |
| Patterson-UTI Energy, Inc. | PTEN | 0.2197 |
| Paychex, Inc. | PAYX | 1.1436 |
| PeopleSoft, Inc. | PSFT | 0.7705 |
| PETsMART, Inc. | PETM | 0.3553 |
| Pixar | PIXR | 0.3183 |
| QLogic Corporation | QLGC | 0.2248 |

**TABLE 1**   Corporations Making up the Nasdaq 100 (As of May 20, 2004) *(Continued)*

| COMPANY NAME | SYMBOL | % OF INDEX (ADJUSTED) |
|---|---|---|
| QUALCOMM Incorporated | QCOM | 5.5030 |
| Research in Motion Limited | RIMM | 0.8416 |
| Ross Stores, Inc. | ROST | 0.3226 |
| Ryanair Holdings PLC | RYAAY | 0.1700 |
| SanDisk Corporation | SNDK | 0.2952 |
| Sanmina-SCI Corporation | SANM | 0.4968 |
| Siebel Systems, Inc. | SEBL | 0.5056 |
| Sigma-Aldrich Corporation | SIAL | 0.3243 |
| Smurfit-Stone Container Corporation | SSCC | 0.3778 |
| Staples, Inc. | SPLS | 0.7783 |
| Starbucks Corporation | SBUX | 1.8686 |
| Sun Microsystems, Inc. | SUNW | 0.4835 |
| Symantec Corporation | SYMC | 1.3391 |
| Synopsys, Inc. | SNPS | 0.3583 |
| Tellabs, Inc. | TLAB | 0.1671 |
| Teva Pharmaceutical Industries Limited | TEVA | 1.2944 |
| VeriSign, Inc. | VRSN | 0.3557 |
| VERITAS Software Corporation | VRTSE | 0.9620 |
| Whole Foods Market, Inc. | WFMI | 0.4151 |
| Xilinx, Inc. | XLNX | 1.3856 |
| Yahoo! Inc. | YHOO | 1.5846 |

TABLE 2    Corporations Included in the S&P 500 (Components of S&P 500 on April 26, 2004)

| Symbol | Name | Last Trade | Volume |
|--------|------|-----------:|-------:|
| A | AGILENT TECH | 29.09 | 1,914,700 |
| AA | ALCOA INC | 32.34 | 3,780,800 |
| AAPL | APPLE COMP INC | 27.13 | 4,127,301 |
| ABC | AMERISOURCEBERGN | 58.45 | 1,735,700 |
| ABI | APPLIED BIOSYS | 18.83 | 1,686,400 |
| ABK | AMBAC FINL | 71.55 | 646,700 |
| ABS | ALBERTSONS INC | 23.15 | 1,445,400 |
| ABT | ABBOTT LABS | 44.00 | 3,505,800 |
| ACE | ACE LTD | 42.65 | 1,153,100 |
| ACS | AFFL COMPUTER | 47.93 | 2,595,300 |
| ACV | ALBERTO CULVER | 46.91 | 333,000 |
| ADBE | ADOBE SYS | 42.81 | 2,327,547 |
| ADCT | ADC TELECOMM | 2.81 | 5,564,909 |
| ADI | ANALOG DEVICES | 46.06 | 3,778,300 |
| ADM | ARCHER-DANIELS | 17.17 | 1,680,600 |
| ADP | AUTOMATIC DATA | 44.54 | 1,871,600 |
| ADSK | AUTODESK INC | 35.06 | 1,538,605 |
| AEE | AMEREN CORP | 43.73 | 418,700 |
| AEP | AMER ELEC PWR | 30.82 | 909,500 |
| AES | THE AES CORP | 7.76 | 1,908,300 |
| AET | AETNA INC | 89.26 | 1,053,200 |
| AFL | AFLAC INC | 41.30 | 1,038,900 |
| AGN | ALLERGAN INC | 91.44 | 669,000 |
| AHC | AMERADA HESS | 67.49 | 700,900 |
| AIG | AMER INTL GROUP | 73.03 | 4,548,500 |
| AIV | APT INV MANAGE | 28.19 | 533,200 |
| ALL | ALLSTATE CP | 46.41 | 1,902,700 |
| ALTR | ALTERA CORP | 21.26 | 4,382,531 |
| AM | AMER GREET A | 21.25 | 239,600 |
| AMAT | APPLIED MATL | 19.44 | 30,935,720 |

**Table 2**    Corporations Included in the S&P 500 (Components of
S&P 500 on April 26, 2004) *(Continued)*

| Symbol | Name | Last Trade | Volume |
|---|---|---|---|
| AMCC | APPLD MICRO | 5.45 | 6,363,103 |
| AMD | ADV MICRO DEVICE | 16.00 | 5,855,700 |
| AMGN | AMGEN | 59.06 | 9,197,988 |
| AN | AUTONATION INC | 16.76 | 1,174,900 |
| ANDW | ANDREW CORP | 20.57 | 1,643,563 |
| AOC | AON CORP | 26.49 | 4,331,600 |
| APA | APACHE CORP | 42.94 | 2,234,000 |
| APC | ANADARKO PETE | 54.77 | 1,706,700 |
| APCC | AMER POWR CONV | 24.02 | 1,328,831 |
| APD | AIR PRODS & CHEM | 52.19 | 718,300 |
| APOL | APOLLO GROUP | 95.10 | 1,420,390 |
| ASD | AMER STANDARD | 107.74 | 284,200 |
| ASH | ASHLAND INC | 48.91 | 1,928,000 |
| ASO | AMSOUTH BANCORP | 22.36 | 712,800 |
| AT | ALLTEL CORP | 51.90 | 1,036,000 |
| ATH | ANTHEM INC | 92.85 | 1,119,500 |
| ATI | ALLEGHENY TECH | 11.55 | 483,700 |
| AV | AVAYA INC | 16.34 | 3,101,900 |
| AVP | AVON PRODS INC | 78.91 | 588,500 |
| AVY | AVERY DENNISON | 64.86 | 398,600 |
| AW | ALLIED WASTE | 13.03 | 796,100 |
| AWE | AT&T WIRELS SVCS | 13.97 | 12,248,900 |
| AXP | AMER EXPRESS CO | 50.28 | 3,718,100 |
| AYE | ALLEGHENY ENERGY | 14.47 | 1,223,200 |
| AZO | AUTOZONE INC | 85.85 | 618,000 |
| BA | BOEING CO | 43.24 | 3,311,100 |
| BAC | BANK OF AMERICA | 81.19 | 4,515,200 |
| BAX | BAXTER INTL INC | 32.45 | 3,012,500 |
| BBBY | BED BATH BEYOND | 37.65 | 2,138,583 |
| BBT | BB&T CORP | 34.70 | 1,096,100 |

**TABLE 2**    Corporations Included in the S&P 500 (Components of S&P 500 on April 26, 2004) *(Continued)*

| SYMBOL | NAME | LAST TRADE | VOLUME |
|---|---|---|---|
| BBY | BEST BUY CO INC | 54.79 | 2,612,500 |
| BC | BRUNSWICK CORP | 42.94 | 461,500 |
| BCC | BOISE CASCADE | 36.14 | 498,900 |
| BCR | C R BARD INC | 111.00 | 529,100 |
| BDK | BLACK & DECKER | 59.85 | 531,900 |
| BDX | BECTON DICKINSN | 51.34 | 1,251,500 |
| BEN | FRANKLIN RES | 55.35 | 1,088,000 |
| BFb | BROWN FORMAN B | 48.05 | 69,500 |
| BHI | BAKER HUGHES INC | 36.55 | 1,401,500 |
| BIIB | BIOGEN IDEC | 60.09 | 3,253,275 |
| BJS | BJ SERVICES CO | 45.02 | 1,715,500 |
| BK | BANK OF NEW YORK | 30.26 | 1,804,900 |
| BLI | BIG LOTS INC | 14.62 | 242,300 |
| BLL | BALL CORP | 69.79 | 220,100 |
| BLS | BELLSOUTH CORP | 26.36 | 3,805,000 |
| BMC | BMC SOFTWARE | 19.97 | 1,307,800 |
| BMET | BIOMET INC | 40.25 | 1,213,745 |
| BMS | BEMIS COMPANY | 28.01 | 404,400 |
| BMY | BRISTOL MYERS SQ | 24.76 | 6,456,800 |
| BNI | BURL NTHN SANTA | 32.81 | 1,393,100 |
| BOL | BAUSCH & LOMB | 65.15 | 370,000 |
| BR | BURLINGTON RES | 67.95 | 1,189,300 |
| BRCM | BROADCOM CORP | 41.10 | 6,837,505 |
| BSC | BEAR STEARNS COS | 82.88 | 613,200 |
| BSX | BOSTON SCIEN CP | 41.20 | 3,915,200 |
| BUD | ANHEUSER BUSCH | 51.95 | 2,050,400 |
| C | CITIGROUP | 48.99 | 9,713,700 |
| CA | COMPUTER ASSOC | 29.17 | 8,876,900 |
| CAG | CONAGRA FOODS | 29.31 | 1,351,700 |
| CAH | CARDINAL HLTH | 71.80 | 1,613,100 |

**TABLE 2**   Corporations Included in the S&P 500 (Components of S&P 500 on April 26, 2004) (*Continued*)

| SYMBOL | NAME | LAST TRADE | VOLUME |
|--------|------|-----------:|-------:|
| CAT | CATERPILLAR INC | 80.90 | 2,473,800 |
| CB | CHUBB CORP | 69.31 | 1,581,600 |
| CBE | COOPER INDS LTD | 56.91 | 515,000 |
| CC | CIRCUIT CITY | 12.50 | 2,703,300 |
| CCE | COCA COLA ENT | 24.47 | 566,300 |
| CCL | CARNIVAL CORP | 43.77 | 2,457,000 |
| CCU | CLEAR CHANNEL | 43.74 | 1,173,400 |
| CD | CENDANT CP | 24.52 | 3,959,500 |
| CEG | CONSTELL ENERGY | 39.93 | 454,000 |
| CF | CHARTER ONE FINL | 34.00 | 548,300 |
| CFC | COUNTRYWIDE FNCL | 59.22 | 2,268,600 |
| CHIR | CHIRON CORP | 47.30 | 2,621,059 |
| CI | CIGNA CORP | 65.82 | 990,600 |
| CIEN | CIENA CORP | 4.84 | 6,537,526 |
| CIN | CINERGY CORP | 39.00 | 898,800 |
| CINF | CINCINNATI FIN | 43.04 | 238,614 |
| CL | COLGATE PALMOLIV | 57.06 | 3,309,300 |
| CLX | CLOROX CO | 51.58 | 946,900 |
| CMA | COMERICA INC | 52.43 | 367,000 |
| CMCSA | COMCAST CORP A | 29.85 | 9,481,150 |
| CMI | CUMMINS INC | 62.73 | 970,000 |
| CMS | CMS ENERGY CORP | 8.79 | 597,100 |
| CMVT | COMVERSE TECH | 18.30 | 1,788,851 |
| CMX | CAREMARK RX | 34.70 | 2,310,900 |
| CNP | CENTERPOINT | 11.25 | 2,141,300 |
| COF | CAP ONE FINAN | 67.68 | 3,481,600 |
| COL | ROCKWELL COLL | 32.96 | 450,500 |
| COP | CONOCOPHILLIPS | 72.70 | 1,950,600 |
| COST | COSTCO WHOLESAL | 38.09 | 2,339,018 |
| CPB | CAMPBELL SOUP CO | 27.50 | 698,100 |

TABLE 2    Corporations Included in the S&P 500 (Components of S&P 500 on April 26, 2004) (*Continued*)

| Symbol | Name | Last Trade | Volume |
|--------|------|-----------:|-------:|
| CPN | CALPINE CORP | 4.67 | 2,659,900 |
| CPWR | COMPUWARE CORP | 8.39 | 1,466,555 |
| CR | CRANE CO | 32.50 | 311,000 |
| CSC | COMPUTER SCIENCE | 41.75 | 1,160,100 |
| CSCO | CISCO SYSTEMS | 23.14 | 40,779,736 |
| CSX | CSX CORP | 30.54 | 1,074,000 |
| CTAS | CINTAS CORP | 45.69 | 1,297,439 |
| CTB | COOPER TIRE & RB | 22.42 | 817,400 |
| CTL | CENTURYTEL INC | 26.90 | 822,100 |
| CTX | CENTEX CORP | 49.82 | 1,539,700 |
| CTXS | CITRIX SYSTEMS | 20.78 | 3,125,172 |
| CVG | CONVERGYS CP | 15.27 | 573,300 |
| CVS | CVS CORPORATION | 39.10 | 2,012,300 |
| CVX | CHEVRONTEXACO | 92.25 | 2,831,100 |
| CZN | CITIZENS COMMS | 13.44 | 3,267,800 |
| D | DOMINION RES INC | 63.72 | 765,600 |
| DAL | DELTA AIR LINES | 6.73 | 2,438,600 |
| DCN | DANA CORP | 21.37 | 706,700 |
| DD | DU PONT CO | 44.99 | 3,667,100 |
| DDS | DILLARD CL A | 17.60 | 550,400 |
| DE | DEERE & CO | 70.26 | 1,923,500 |
| DELL | DELL INC | 35.71 | 11,376,332 |
| DG | DOLLAR GEN | 18.71 | 2,135,500 |
| DGX | QUEST DIAG | 85.02 | 748,200 |
| DHR | DANAHER CORP | 93.57 | 768,000 |
| DIS | WALT DISNEY CO | 24.38 | 9,034,600 |
| DJ | DOW JONES & CO | 47.04 | 409,200 |
| DLX | DELUXE CORP | 41.17 | 232,200 |
| DOV | DOVER CORP | 41.84 | 649,600 |
| DOW | DOW CHEMICAL CO | 41.71 | 2,304,600 |

Table 2    Corporations Included in the S&P 500 (Components of S&P 500 on April 26, 2004) *(Continued)*

| Symbol | Name | Last Trade | Volume |
|---|---|---|---|
| DPH | DELPHI CORP | 10.58 | 1,381,600 |
| DRI | DARDEN REST | 23.18 | 1,834,300 |
| DTE | DTE ENERGY | 39.43 | 633,500 |
| DUK | DUKE ENERGY | 21.60 | 3,188,100 |
| DVN | DEVON ENERGY | 62.24 | 873,800 |
| DYN | DYNEGY INC | 3.82 | 1,187,600 |
| EBAY | EBAY INC | 81.58 | 9,805,580 |
| EC | ENGELHARD CORP | 30.25 | 255,400 |
| ECL | ECOLAB INC | 29.04 | 1,005,600 |
| ED | CONSOL EDISON | 41.93 | 1,329,100 |
| EDS | ELECTR DATA | 19.23 | 3,793,700 |
| EFX | EQUIFAX INC | 25.48 | 539,200 |
| EIX | EDISON INTL | 23.13 | 883,100 |
| EK | EASTMAN KODAK | 26.43 | 2,308,100 |
| EMC | EMC CORP | 11.99 | 19,134,700 |
| EMN | EASTMAN CHEM | 43.80 | 297,500 |
| EMR | EMERSON ELECTRIC | 62.11 | 937,200 |
| EOG | EOG RESOURCES | 49.19 | 681,500 |
| EOP | EQUITY OFFICE | 25.40 | 1,218,700 |
| EP | EL PASO CORP | 7.09 | 1,456,500 |
| EQR | EQ RESIDENT | 27.96 | 888,100 |
| ERTS | ELECTRONIC ART | 51.60 | 2,610,895 |
| ESRX | EXPRESS SCRIPTS | 79.28 | 757,657 |
| ET | E*TRADE FINCL CP | 12.15 | 4,664,300 |
| ETN | EATON | 60.11 | 671,000 |
| ETR | ENTERGY CP | 56.24 | 958,700 |
| EXC | EXELON CORP | 66.75 | 1,256,300 |
| F | FORD MOTOR CO | 15.69 | 14,003,000 |
| FCX | FRPRT-MCM GD | 32.79 | 2,587,900 |
| FD | FED DEPT STRS | 50.73 | 1,707,100 |

TABLE 2    Corporations Included in the S&P 500 (Components of S&P 500 on April 26, 2004) *(Continued)*

| Symbol | Name | Last Trade | Volume |
|--------|------|-----------:|-------:|
| FDC | FIRST DATA CORP | 46.18 | 4,688,100 |
| FDO | FAMILY DLR STRS | 32.28 | 1,345,600 |
| FDX | FEDEX CORP | 72.98 | 1,653,200 |
| FE | FIRSTENERGY | 39.30 | 836,300 |
| FHN | FIRST HORIZN NTL | 43.60 | 473,600 |
| FII | FED INVESTORS | 29.78 | 628,100 |
| FISV | FISERV INC | 37.85 | 2,630,058 |
| FITB | FIFTH THR BNCP | 55.00 | 1,240,263 |
| FLR | FLUOR CORP | 38.83 | 293,400 |
| FNM | FANNIE MAE | 69.62 | 5,201,200 |
| FO | FORTUNE BRANDS | 77.74 | 778,400 |
| FON | SPRINT FON GP | 18.65 | 8,811,800 |
| FPL | FPL GROUP INC | 65.10 | 1,307,500 |
| FRE | FREDDIE MAC | 57.45 | 2,178,600 |
| FRX | FOREST LABS | 65.50 | 2,024,400 |
| G | GILLETTE CO | 39.47 | 2,963,200 |
| GAS | NICOR INC | 34.15 | 243,100 |
| GCI | GANNETT CO INC | 89.00 | 745,400 |
| GD | GENERAL DYNAMICS | 94.30 | 406,300 |
| GDT | GUIDANT CORP | 66.16 | 2,420,600 |
| GDW | GOLDEN WEST FIN | 103.47 | 390,600 |
| GE | GENERAL ELEC CO | 30.75 | 16,629,800 |
| GENZ | GENZYME GEN | 46.12 | 2,626,431 |
| GIS | GENERAL MILLS | 48.33 | 1,327,300 |
| GLK | GREAT LAKES CHEM | 26.35 | 592,600 |
| GLW | CORNING INC | 12.05 | 13,312,300 |
| GM | GENERAL MOTORS | 49.33 | 6,206,700 |
| GP | GEORGIA-PACIFIC | 37.71 | 2,862,700 |
| GPC | GENUINE PARTS CO | 36.68 | 346,900 |
| GPS | GAP INC | 22.80 | 4,638,700 |

TABLE 2    Corporations Included in the S&P 500 (Components of
S&P 500 on April 26, 2004) (*Continued*)

| SYMBOL | NAME | LAST TRADE | VOLUME |
|---|---|---|---|
| GR | GOODRICH CORP | 30.01 | 340,800 |
| GS | GOLDM SACHS GRP | 101.15 | 2,243,900 |
| GT | GOODYEAR TIRE | 9.19 | 1,428,500 |
| GTW | GATEWAY INC | 5.66 | 2,584,700 |
| GWW | W W GRAINGER INC | 53.82 | 620,500 |
| HAL | HALLIBURTON CO | 30.90 | 2,992,400 |
| HAS | HASBRO INC | 20.21 | 1,728,500 |
| HBAN | HUNTGTN BKSHR | 21.67 | 467,823 |
| HCA | HCA INC | 41.16 | 1,768,900 |
| HCR | MANOR CARE INC | 33.73 | 597,100 |
| HD | HOME DEPOT INC | 36.35 | 4,549,700 |
| HDI | HARLEY-DAVIDSON | 57.46 | 1,387,200 |
| HET | HARRAHS ENTER | 55.87 | 641,500 |
| HIG | HARTFORD FINL | 63.20 | 795,800 |
| HLT | HILTON HOTELS CP | 17.52 | 1,898,300 |
| HMA | HEALTH MGMT | 23.14 | 1,589,600 |
| HNZ | H J HEINZ CO | 38.81 | 1,556,400 |
| HON | HONEYWELL INTL | 34.76 | 1,670,400 |
| HOT | STARWOOD HOTELS | 41.11 | 1,334,800 |
| HPC | HERCULES INC | 12.01 | 554,300 |
| HPQ | HEWLETT-PACKARD | 21.67 | 8,824,800 |
| HRB | H & R BLOCK INC | 46.74 | 577,400 |
| HSY | HERSHEY FOODS CP | 90.09 | 1,067,900 |
| HUM | HUMANA INC | 17.71 | 3,514,300 |
| IBM | INTL BUS MACHINE | 90.43 | 4,533,600 |
| IFF | INTL FLAV & FRAG | 36.37 | 213,500 |
| IGT | INTL GAME TECH | 39.27 | 16,248,100 |
| INTC | INTEL CORP | 27.15 | 56,768,944 |
| INTU | INTUIT INC | 43.35 | 1,371,394 |
| IP | INTL PAPER CO | 42.44 | 2,270,900 |

TABLE 2    Corporations Included in the S&P 500 (Components of S&P 500 on April 26, 2004) (*Continued*)

| SYMBOL | NAME | LAST TRADE | VOLUME |
|--------|------|-----------|--------|
| IPG | INTERPUBLIC GRP | 16.15 | 1,141,600 |
| IR | INGERSOLL-RAND | 68.60 | 1,463,900 |
| ITT | ITT INDS INC | 79.65 | 598,700 |
| ITW | ILLINOIS TOOL WK | 87.04 | 1,349,100 |
| JBL | JABIL CIRCUIT | 28.94 | 995,600 |
| JCI | JOHNSON CONTROLS | 57.84 | 512,400 |
| JCP | J C PENNEY CO | 35.87 | 1,402,500 |
| JDSU | JDS UNIPHASE | 4.18 | 21,098,288 |
| JHF | J HANCOCK FINL | 46.71 | 1,832,000 |
| JNJ | JOHNSON&JOHNSON | 53.90 | 5,911,500 |
| JNS | JANUS CAPITAL GP | 15.55 | 1,541,400 |
| JNY | JONES APPAREL | 37.21 | 797,300 |
| JP | JEFFERSON PILOT | 53.30 | 369,100 |
| JPM | JP MORGAN CHASE | 38.55 | 7,071,300 |
| JWN | NORDSTROM INC | 38.05 | 568,100 |
| K | KELLOGG CO | 42.93 | 1,817,400 |
| KBH | KB HOME | 71.74 | 666,300 |
| KEY | KEYCORP NEW | 29.98 | 1,164,100 |
| KG | KING PHARM | 17.00 | 1,234,500 |
| KLAC | KLA TENCOR | 45.16 | 6,896,219 |
| KMB | KIMBERLY-CLARK | 65.15 | 1,499,600 |
| KMG | KERR MCGEE CORP | 49.86 | 772,500 |
| KMI | KINDER MORGAN | 61.93 | 558,100 |
| KO | COCA COLA CO | 50.69 | 5,580,800 |
| KR | KROGER CO | 17.50 | 2,921,700 |
| KRB | MBNA CORP | 25.35 | 4,370,800 |
| KRI | KNIGHT RIDDER | 79.22 | 397,700 |
| KSE | KEYSPAN CORP | 36.30 | 309,200 |
| KSS | KOHL'S CORP | 41.70 | 3,705,000 |
| LEG | LEGGET & PLATT | 23.74 | 671,500 |

**TABLE 2**    Corporations Included in the S&P 500 (Components of S&P 500 on April 26, 2004) *(Continued)*

| SYMBOL | NAME | LAST TRADE | VOLUME |
|--------|------|-----------:|-------:|
| LEH | LEHMAN BROS | 77.27 | 1,776,500 |
| LIZ | LIZ CLAIBORNE | 37.19 | 690,500 |
| LLTC | LINEAR TECH | 37.49 | 5,644,369 |
| LLY | ELI LILLY | 73.35 | 2,140,300 |
| LMT | LOCKHEED MARTIN | 46.50 | 2,099,800 |
| LNC | LINCOLN NATL | 46.94 | 674,900 |
| LOW | LOWES COMPANIES | 53.29 | 3,076,500 |
| LPX | LOUISIANA PACIF | 25.83 | 1,238,900 |
| LSI | LSI LOGIC | 9.04 | 4,038,000 |
| LTD | LIMITED BRANDS | 21.07 | 2,045,100 |
| LTR | LOEWS CORP | 59.10 | 468,100 |
| LU | LUCENT TECH | 3.81 | 44,654,400 |
| LUV | SW AIRLINES | 14.73 | 3,060,600 |
| LXK | LEXMARK INTL | 93.58 | 732,400 |
| MAR | MARRIOTT INTL | 47.50 | 848,600 |
| MAS | MASCO CORP | 28.31 | 1,713,300 |
| MAT | MATTEL INC | 17.55 | 5,090,000 |
| MAY | MAY DEPT STORES | 31.34 | 1,330,900 |
| MBI | MBIA INC | 60.70 | 494,500 |
| MCD | MCDONALDS CORP | 27.32 | 3,956,900 |
| MCK | MCKESSON CORP | 31.32 | 1,231,600 |
| MCO | MOODY'S CORP | 66.33 | 375,600 |
| MDP | MEREDITH CORP | 50.19 | 75,200 |
| MDT | MEDTRONIC INC | 49.85 | 4,438,100 |
| MEDI | MEDIMMUNE INC | 24.07 | 3,284,706 |
| MEL | MELLON FINL CORP | 30.96 | 1,098,000 |
| MER | MERRILL LYNCH | 57.19 | 3,415,500 |
| MERQ | MERCURY INTRACT | 46.45 | 1,799,879 |
| MET | METLIFE INC | 34.35 | 1,314,800 |
| MHP | MCGRAW HILL COS | 77.40 | 531,000 |

**TABLE 2**    Corporations Included in the S&P 500 (Components of
S&P 500 on April 26, 2004) (*Continued*)

| SYMBOL | NAME | LAST TRADE | VOLUME |
|--------|------|-----------|--------|
| MHS | MEDCO HLTH SOLN | 35.25 | 2,204,100 |
| MI | MARSHALL ILSLEY | 36.97 | 403,200 |
| MIL | MILLIPORE CP | 54.44 | 502,300 |
| MKC | MCCORMICK & CO | 33.83 | 265,800 |
| MMC | MARSH & MCLENNAN | 45.61 | 3,312,100 |
| MMM | 3M COMPANY | 87.75 | 2,909,000 |
| MNST | MONSTER WRLDWIDE | 28.81 | 2,100,795 |
| MO | ALTRIA GROUP | 55.58 | 4,333,900 |
| MOLX | MOLEX INC | 32.63 | 939,439 |
| MON | MONSANTO CO | 36.25 | 975,900 |
| MOT | MOTOROLA INC | 20.27 | 17,980,900 |
| MRK | MERCK & CO | 46.64 | 3,642,200 |
| MRO | MARATHON OIL | 34.34 | 2,240,100 |
| MSFT | MICROSOFT CP | 27.24 | 89,393,168 |
| MTB | M&T BANK CORP | 84.78 | 264,600 |
| MTG | MGIC INV CP | 74.44 | 1,032,800 |
| MU | MICRON TECH | 15.47 | 7,016,700 |
| MWD | MORGAN STANLEY | 53.66 | 4,893,800 |
| MWV | MEADWESTVACO CP | 27.81 | 812,900 |
| MXIM | MAXIM INTEGRTD | 46.03 | 5,948,851 |
| MYG | MAYTAG CORP | 28.58 | 512,000 |
| MYL | MYLAN LABS | 24.42 | 1,866,000 |
| NAV | NAVISTAR INTL | 48.17 | 602,900 |
| NBR | NABORS INDS LTD | 45.92 | 1,049,400 |
| NCC | NATIONAL CITY | 35.00 | 1,809,100 |
| NCR | NCR CORP | 46.85 | 269,800 |
| NE | NOBLE CORP | 37.99 | 1,165,700 |
| NEM | NEWMONT MINING | 40.55 | 3,694,900 |
| NFB | N FORK BANCP | 37.16 | 1,472,200 |
| NI | NISOURCE INC | 21.25 | 974,400 |

TABLE 2    Corporations Included in the S&P 500 (Components of S&P 500 on April 26, 2004) (*Continued*)

| SYMBOL | NAME | LAST TRADE | VOLUME |
|---|---|---|---|
| NKE | NIKE INC CL B | 73.54 | 1,822,300 |
| NOC | NORTHROP GRUMMAN | 99.13 | 627,800 |
| NOVL | NOVELL INC | 11.54 | 4,809,033 |
| NSC | NORFOLK SOUTHERN | 23.93 | 2,185,900 |
| NSM | NATL SEMICONDUCT | 43.89 | 5,021,200 |
| NTAP | NETWK APPLIANCE | 21.23 | 3,241,824 |
| NTRS | NORTHERN TRUST | 43.95 | 2,019,350 |
| NUE | NUCOR CORP | 65.31 | 771,600 |
| NVDA | NVIDIA CORP | 22.33 | 3,583,364 |
| NVLS | NOVELLUS SYS | 29.79 | 5,425,762 |
| NWL | NEWELL RUBBERMD | 24.89 | 1,053,300 |
| NXTL | NEXTEL COMMS | 25.37 | 7,406,209 |
| NYT | NY TIMES | 46.60 | 518,100 |
| ODP | OFFICE DEPOT | 17.90 | 1,705,600 |
| OMC | OMNICOM GP INC | 78.29 | 1,483,000 |
| ONE | BANK ONE CORP | 50.80 | 5,170,100 |
| ORCL | ORACLE CORP | 12.31 | 35,321,264 |
| OXY | OCCIDENTAL PETE | 48.69 | 1,717,400 |
| PAYX | PAYCHEX INC | 36.86 | 3,073,004 |
| PBG | PEPSI BOTTLING | 29.74 | 859,000 |
| PBI | PITNEY BOWES INC | 43.73 | 632,400 |
| PCAR | PACCAR INC | 58.72 | 821,018 |
| PCG | PG&E CORP | 28.38 | 1,820,500 |
| PCL | PLUM CREEK TIMB | 30.12 | 535,000 |
| PD | PHELPS DODGE CP | 69.41 | 2,262,800 |
| PEG | PUBL SVC ENTER | 44.81 | 925,700 |
| PEP | PEPSICO INC | 54.69 | 3,030,200 |
| PFE | PFIZER INC | 36.29 | 12,463,300 |
| PFG | PRINCIPAL FINL | 35.20 | 831,300 |
| PG | PROCTER & GAMBLE | 105.24 | 2,032,700 |

TABLE 2    Corporations Included in the S&P 500 (Components of S&P 500 on April 26, 2004) (*Continued*)

| SYMBOL | NAME | LAST TRADE | VOLUME |
|---|---|---|---|
| PGL | PEOPLES ENERGY | 42.58 | 110,200 |
| PGN | PROGRESS ENERGY | 44.29 | 695,100 |
| PGR | PROGRESS CORP OH | 88.06 | 580,500 |
| PH | PARKER-HANNIFIN | 57.33 | 926,300 |
| PHM | PULTE HOMES INC | 49.03 | 1,604,500 |
| PKI | PERKINELMER | 20.10 | 830,900 |
| PLD | PROLOGIS | 29.86 | 612,300 |
| PLL | PALL CORP | 24.74 | 584,400 |
| PMCS | PMC-SIERRA INC | 14.64 | 3,735,572 |
| PMTC | PARAMETRIC | 5.12 | 1,451,651 |
| PNC | PNC FINL SVC | 52.40 | 1,018,200 |
| PNW | PINNACL WEST CAP | 39.62 | 729,100 |
| PPG | PPG IND | 61.92 | 413,200 |
| PPL | PPL CORP | 44.02 | 489,300 |
| PRU | PRUDENTIAL FINL | 44.54 | 1,387,900 |
| PSFT | PEOPLESOFT INC | 18.41 | 3,955,551 |
| PTV | PACTIV CORP | 23.48 | 752,800 |
| PVN | PROV FIN | 13.00 | 1,916,600 |
| PWER | POWER ONE | 10.69 | 571,987 |
| PX | PRAXAIR INC | 37.75 | 708,700 |
| Q | QWEST COMMS INTL | 4.29 | 3,287,700 |
| QCOM | QUALCOMM INC | 66.15 | 5,955,571 |
| QLGC | QLOGIC CORP | 26.48 | 6,196,111 |
| R | RYDER SYSTEM INC | 39.34 | 923,500 |
| RBK | REEBOK INTL LTD | 37.97 | 1,085,600 |
| RDC | ROWAN CO INC | 22.36 | 1,640,500 |
| RF | REGIONS FINANCL | 35.51 | 628,700 |
| RHI | ROB HALF INTL | 28.52 | 1,001,200 |
| RIG | TRANSOCEAN INC | 27.51 | 1,999,600 |
| RJR | RJR TOBACCO HLDS | 59.18 | 799,400 |

**TABLE 2**   Corporations Included in the S&P 500 (Components of S&P 500 on April 26, 2004) *(Continued)*

| Symbol | Name | Last Trade | Volume |
|---|---|---|---|
| RKY | ADOLPH COORS | 66.90 | 409,000 |
| ROH | ROHM & HAAS CO | 39.11 | 809,400 |
| ROK | ROCKWELL AUTOMAT | 34.41 | 1,004,600 |
| RRD | DONNELLEY & SONS | 30.02 | 524,300 |
| RSH | RADIOSHACK | 32.75 | 926,200 |
| RTN | RAYTHEON CO | 31.96 | 1,291,200 |
| RX | IMS HEALTH | 26.06 | 1,007,100 |
| S | SEARS ROEBUCK | 41.56 | 1,149,300 |
| SAFC | SAFECO CORP | 44.06 | 732,543 |
| SANM | SANMINA-SCI CP | 11.34 | 6,801,894 |
| SBC | SBC COMMS | 25.50 | 9,942,900 |
| SBL | SYMBOL TECH | 13.26 | 1,194,200 |
| SBUX | STARBUCKS CORP | 39.00 | 2,118,616 |
| SCH | CHARLES SCHWAB | 10.88 | 2,787,500 |
| SDS | SUNGARD DATA SYS | 26.62 | 1,020,300 |
| SEBL | SIEBEL SYSTEMS | 11.15 | 4,149,715 |
| SEE | SEALED AIR CP | 49.93 | 335,800 |
| SFA | SCIENTIFIC ATL | 36.06 | 1,643,700 |
| SGP | SCHERING-PLOUGH | 17.13 | 3,002,600 |
| SHW | SHERWIN-WILLIAMS | 38.86 | 726,300 |
| SIAL | SIGMA ALDRICH | 57.40 | 440,952 |
| SLB | SCHLUMBERGER LTD | 61.98 | 2,326,300 |
| SLE | SARA LEE CORP | 23.41 | 3,068,200 |
| SLM | SLM CORPORATION | 38.11 | 1,787,000 |
| SLR | SOLECTRON CORP | 5.38 | 3,631,200 |
| SNA | SNAP-ON INC | 33.63 | 213,700 |
| SNV | SYNOVUS FINAN | 24.27 | 427,600 |
| SO | SOUTHERN CO | 29.49 | 1,851,600 |
| SOTR | SOUTHTRUST CP | 31.36 | 1,011,005 |
| SPG | SIMON PROP GRP | 49.75 | 629,600 |

**TABLE 2**    Corporations Included in the S&P 500 (Components of S&P 500 on April 26, 2004) *(Continued)*

| SYMBOL | NAME | LAST TRADE | VOLUME |
|--------|------|-----------:|-------:|
| SPLS | STAPLES INC | 26.84 | 3,399,681 |
| SRE | SEMPRA ENERGY | 31.75 | 523,800 |
| STA | ST PAUL TRAVLRS | 42.55 | 2,339,300 |
| STI | SUNTRUST BKS | 69.72 | 527,300 |
| STJ | ST JUDE MEDICAL | 77.55 | 1,013,400 |
| STT | ST STREET CP | 49.65 | 1,556,000 |
| SUN | SUNOCO INC | 64.00 | 798,300 |
| SUNW | SUN MICROSYS | 4.30 | 31,551,484 |
| SVU | SUPERVALU INC | 31.11 | 430,100 |
| SWK | STANLEY WORKS | 44.53 | 917,600 |
| SWY | SAFEWAY INC | 23.15 | 3,190,700 |
| SYK | STRYKER CORP | 100.10 | 675,600 |
| SYMC | SYMANTEC CORP | 48.91 | 4,046,440 |
| SYY | SYSCO CORP | 39.05 | 2,043,400 |
| T | AT&T CORP | 18.31 | 4,863,200 |
| TE | TECO ENERGY | 13.52 | 895,800 |
| TEK | TEKTRONIX | 31.25 | 782,300 |
| TER | TERADYNE INC | 22.53 | 3,725,900 |
| TGT | TARGET CORP | 44.22 | 2,207,900 |
| THC | TENET HEALTHCR | 11.40 | 2,372,000 |
| TIF | TIFFANY & CO | 40.72 | 708,200 |
| TIN | TEMPLE INLAND | 63.15 | 311,600 |
| TJX | TJX CO INC | 25.64 | 1,893,200 |
| TLAB | TELLABS INC | 9.73 | 2,615,607 |
| TMK | TORCHMARK CORP | 50.89 | 489,400 |
| TMO | THERMO ELECTRON | 30.86 | 1,059,600 |
| TNB | THOMAS & BETTS | 24.80 | 1,642,400 |
| TOY | TOYS R US CORP | 16.38 | 922,900 |
| TRB | TRIBUNE CO | 48.50 | 842,300 |
| TROW | T ROWE PRICE GP | 50.25 | 686,819 |

**TABLE 2**   Corporations Included in the S&P 500 (Components of S&P 500 on April 26, 2004) *(Continued)*

| SYMBOL | NAME | LAST TRADE | VOLUME |
|---|---|---|---|
| TSG | SABRE HLDGS | 25.53 | 817,100 |
| TWX | TIME WARNER INC | 16.76 | 10,144,700 |
| TXN | TEXAS INSTRUMENT | 27.46 | 10,087,100 |
| TXT | TEXTRON INC | 58.05 | 960,500 |
| TXU | TXU CORP | 33.53 | 13,178,900 |
| TYC | TYCO INTL | 28.97 | 8,106,900 |
| UCL | UNOCAL CORP DEL | 38.41 | 1,238,100 |
| UIS | UNISYS CORP | 13.53 | 894,800 |
| UNH | UNITEDHEALTH GP | 63.90 | 4,801,200 |
| UNM | UNUMPROVIDENT | 15.08 | 825,300 |
| UNP | UNION PACIFIC CP | 59.47 | 1,630,400 |
| UPC | UNION PLANTERS | 28.83 | 675,500 |
| UPS | UNITED PARCEL B | 71.54 | 2,248,200 |
| USB | US BANCORP | 26.00 | 3,779,400 |
| UST | UST INC | 37.12 | 473,400 |
| UTX | UNITED TECH CP | 89.50 | 2,159,600 |
| UVN | UNIVISION COMM | 35.56 | 1,391,600 |
| VC | VISTEON CORP | 11.38 | 1,278,800 |
| VFC | VF CORP | 47.32 | 207,200 |
| VIAb | VIACOM CL B | 40.19 | 4,519,200 |
| VMC | VULCAN MATRLS | 47.69 | 269,200 |
| VRTSE | VERITAS SOFTWARE | 28.93 | 4,966,207 |
| VZ | VERIZON COMMS | 37.74 | 3,657,300 |
| WAG | WALGREEN CO | 33.40 | 2,809,200 |
| WAT | WATERS CORP | 39.48 | 496,500 |
| WB | WACHOVIA CORP | 45.90 | 4,209,300 |
| WEN | WENDYS INTL | 39.85 | 978,800 |
| WFC | WELLS FARGO & CO | 56.39 | 3,628,300 |
| WHR | WHIRL POOL CORP | 68.25 | 515,100 |
| WIN | WINN-DIXIE STRS | 7.78 | 1,253,500 |

**TABLE 2**    Corporations Included in the S&P 500 (Components of S&P 500 on April 26, 2004) *(Continued)*

| SYMBOL | NAME | LAST TRADE | VOLUME |
|--------|------|-----------:|-------:|
| WLP | WELLPNT HLTH NET | 116.20 | 518,000 |
| WM | WASHINGTN MUTUAL | 40.20 | 3,816,000 |
| WMB | WILLIAMS COS INC | 10.39 | 2,544,000 |
| WMI | WASTE MANAGEMNT | 29.78 | 864,500 |
| WMT | WAL-MART STORES | 58.14 | 7,017,100 |
| WOR | WORTHINGTON INDS | 19.04 | 424,500 |
| WPI | WATSON PHARM | 42.90 | 525,600 |
| WWY | WM WRIGLEY JR | 61.31 | 420,900 |
| WY | WEYERHAEUSER CO | 64.14 | 3,262,400 |
| WYE | WYETH | 39.50 | 3,594,300 |
| X | US STEEL CORP | 35.50 | 2,284,000 |
| XEL | XCEL ENERGY | 16.85 | 2,184,100 |
| XL | XL CAPITAL LTD | 77.30 | 343,300 |
| XLNX | XILINX INC | 35.32 | 7,728,781 |
| XOM | EXXON MOBIL | 43.01 | 9,476,400 |
| XRX | XEROX CORP | 13.56 | 12,400,000 |
| YHOO | YAHOO INC | 57.00 | 10,982,837 |
| YUM | YUM! BRANDS INC | 38.94 | 1,124,900 |
| ZION | ZIONS BANCORP | 57.51 | 401,613 |
| ZMH | ZIMMER HLDGS | 84.00 | 1,117,600 |
| | | | 1,501,224,672 |

# REFERENCES

A 6th-Sense-Plus Attack On Motorola Costs; 'Mr. Fixit' Follows Six Sigma Plan. *Chicago Tribune* (July 27, 2003, 1).

Adams, J. L. (1980). *Conceptual blockbusting: A guide to better ideas.* Norton.

Agrawal, V., & Farrell, D. Who wins in offshoring? *Mckinsey Quarterly*, 2003, Special Edition.

Akao, Y. (1990). *Quality function deployment: Integrating customer requirements into product design.* Cambridge, MA: Productivity Press.

Albrecht, K. *At America's service.* (1992). New York: Warner Books.

Altshuller, G. (H. Altov) Translated by Lev Shulyak (1996). *And suddenly the inventor appeared: TRIZ, the theory of inventive problem solving.* Worcester, MA: Technical Innovation Center, Inc.

Altshuller, H. *Creativity as an exact science.* (1988). New York: Gordon and Beach Science Publishers.

Andersen, B., & Fagerhaug, T. (2000). *Root cause analysis: Simplified tools and techniques.* Milwaukee WI: ASQ Quality Press.

Anupindi, R., Chopra, S., Deshmukh, S. D., Van Miegham J. A., & Zemel, E. (1997). *Managing business process flow.* Upper Saddle River, NJ: Prenctice Hall.

AT&T (1988), *Process Quality Management and Improvement Guidelines*, Issue 1.1.

Baldrige National Quality Program (2004), *Criteria for Performance Excellence*, US NIST, Gaithersburg, MD, http://www.quality.nist.gov/.

Barassard, M., & Ritter, D. (Ed.) (1994). *The Memory Jogger.* Methuen, MA: GOAL/QPC.

Beatty, R. W., Huselid, M. A., & Schneier, C. E. (2003), "New HR Metrics: Scoring on the Business Scorecard," Organizational Dynamics, Vol. 32, No. 2, pp. 107–121.

Berger, C., Blauth, R., & Boger, D. (1993). Kano's Methods for Understanding Customer-defined Quality. *Center for Quality of Management Journal*, 2, 4.

Berry, L. L. *On great service: A framework for action.* (1995). New York: The Free Press.

Bitner, M. J., Faranda, W. T., Hubbert, A. R., & Zeithaml, V. A. (1997). Customer contributions and roles in service delivery. *International Journal of Service Industry*, 8, 3, 193–205.

Bodek, N. (2003), "Quality Conversation with Gary Convis", Quality Digest, Vol. 23, No. 10. October.

Bounds, G. M. (1996), *Cases in Quality*, Irwin Publishing, Chicago, IL.

Bowen, D. E., & Lawler, E. E. III (1992, Spring), The empowerment of service workers: what, why, how, and when. *Sloan Management Review*, 33, 3, 31–9.

Bowen, D. E., & Lawler, E. E. III (1995, Summer), Empowering service employees, *Sloan Management Review*, 36, 4, 73–84.

Byron, E. Speaking of Success. *The Wall Street Journal*, July 12, 2004; R10.

Buckingham, M. & Clifton, D. (2001), *Now, Discover Your Strengths*, The Free Press, New York, NY.

Buckingham, M. & Coffman, C. (1999), *First, Break All the Rules: What the World's Greatest Managers Do Differently*, Simon & Schuster, New York, NY.

Buzan, T. *The mind map book.* (1994). Penguin Group.

Carlzon, J. (1987), *Moments of Truth*, Harper Collins, New York, NY.

Carr, C. (1990), *Front-Line Customer Service—15 Keys to Customer Satisfaction*, John Wiley & Sons, New York, NY.

Chawan, A., Chopra, S., & Tyagi, R. K. Service chain design. (2003). A white paper, Kellogg School of Management, Northwestern University.

Charnes, A., Cooper, W., Lewin, A., & Seiford, L. (1994). *Data envelopment analysis: Theory, methodology, and applications.* Boston: Kluwer Academic Publishers.

Chase, R. B., & Stewart, D. M. (1994). Make your service fail safe. *Sloan Management Review*, 35, 3, 35–45.

Christian, H., & Baeyer, V. (1993). *The fermi solution.* Random House Inc.

Christine, H., & Muhlemann, A. (1997). *Service Operations Management: Strategy, design and delivery.* Hertfordshire, UK: Prentice Hall Europe.

Chowdhury, S., *Design for Six Sigma*. (2002). Chicago: Dearborn Trade Publishing.

Citibank Increases Customer Loyalty With Defect-Free Processes. *Journal for Quality & Participation*; Fall 2000, Vol. 23 Issue 4, 32.

Clifton, D. O. & Nelson, P. (1992), *Soar with Your Strengths*, Delacorte Press, New York, NY.

Collier, C. D. (1991). The service quality process map for credit card processing. *Decision Science*, 22, 406–420.

Covey, S. R. (2004), "Win-Win Partnerships", http://www.franklin-covey.com/ez/library/winwin.html, June.

Covey, S. R. (1989), *The 7 Habits of Highly Effective People*, Simon & Schuster, New York, NY.

Creveling, C. M., Slutsky, J. L., & Antis D., Jr. (2003). *Design for Six Sigma: In technology and product development*. Englewood, NJ: Prentice Hall.

Davenport, T. H., & Glaser, J. (2002, July). Just-in-time delivery comes to knowledge management. *Harvard Business Review*, 5–9.

de Bono, E. (1970). *Lateral thinking-creativity step by step*. Perennial Library.

Delong, T., & Ashish, N. (2003). *Professional Services: Text & Cases*, New York: McGraw-Hill Irwin.

Deming, W. E. (1992). *Out of the crisis*. Cambridge, MA: MIT, 183–247.

DePree, M. (1989), *Leadership Is An Art*, Doubleday, New York, NY.

Dunn, K. (1991), "McGuffey's Restaurants Corporate Culture", Presentation at AT&T 4th Annual Quality Conference, Murray Hill, NJ.

Ehrlich, B. H. (2002). *Six Sigma and Lean Servicing*. Boca Raton, FL: St. Lucie Press.

European Foundation for Quality Management (2004), *EFQM Excellence Model*, The Netherlands, http://www.efqm.org/.

Fitzgerald, L., Johnston, R., Brignall, S., Silvestro, R., & Voss, C. (1991). *Performance measurements in service businesses*. London: CIMA.

Fitzsimmons, J. A., & Fitzsimmons, M. J. (2004). *Service management: Operations, strategy, and information technology*, (4th ed.) New York: McGraw-Hill Irwin.

Flamholtz, E. C. (1999), *Human Resource Accounting: Advances in Concepts, Methods, and Applications*, Third Edition, Kluwer Academic Publishers, Boston, MA.

Flamholtz, E. C. (1990), *Growing Pains: how to make the Transition from an Entrepreneurship to a Professionally Managed Firm*, Revised Edition, Jossey–Bass Publishers, San Francisco, CA.

Freiberg, K., & Freiberg J. (1997), *Nuts!* Broadway Books, New York, NY.

Gardner, H. *Creating Minds*. (1993). Basic Books.

George, M. L. (2003). *Lean six sigma for service, how to use lean speed and six sigma quality to improve services and transactions.* New York: McGraw-Hill.

Ghobadian, A., Speller, S., & Jones, M. (1994). Service Quality—concepts and models. *International Journal of Quality and Reliability Management*, 11, 9, 43–66.

Gilbert, P., Slavney, D., & Tong, D. (2003), "10 Best Practices for Employee Surveys," *Workforce Online*, February. http://www.workforce.com/section/09/feature/23/39/90/.

Gordon, W. J. J. (1961). *Synectics*. Harper & Row.

Grapentine, T. (Winter/Spring 1999). The history and future of service quality assessment. *Marketing Research*, 5–20.

Gross Domestic Product by Industry—U.S. Department of Commerce, Bureau of Economic Analysis—www.bea.gov.

Hart, C. W. L. (1990). The power of unconditional service guarantees. *Harvard Business Review* (July-August), 54–62.

Harvey, J. D. (1998). Service quality: A tutorial. *Journal of Operations Management*. 16, 5, 583–598.

Hauser, J. R., & Clausing, D. (1988, May–June). The house of quality. *Harvard Business Review*, 66, 3:63–73.

Haywood-Farmer's model of service quality (1988). *International Journal of Operations and Production Management*, 8, 6, 19–29.

Headley, D. E., & Miller, S. J. (1993). Measuring service quality and its relationship to future consumer behavior. *Journal of Health Care Marketing*, 13, 4, 32–41.

Hernon, P., & Altman, E. (1998). *Assessing service quality: satisfying the expectations of library customers*. Chicago: American Library Association.

Heskett, J. L., Sasser, W. E., Jr., & Schlesinger, L. S. (1997). *The service profit chain: how leading companies link profit and growth to loyalty, satisfaction, and values.* New York: The Free Press.

Heskett, J. L. (1986). *Service economy.* Boston.

Hope, C., & Muhlemann, A. (1997). *Service operations management: strategy, design and delivery.* Hertfordshire, UK: Prentice Hall Europe.

Hostage, G. M. (1975, July-August). Quality control in a servic business. *Harvard Business Review*, 98–106.

How Technology Is Redefining the Role of the Firm: Lessons For Corporate Strategists. Knowledge@wharton.com.

How Warburg Pincus Views Prospects And Perils In Outsourcing Deals. Knowledge@wharton.com.

In a Global Economy, Competition Among BPO Rivals Heats Up— Knowledge@wharton.com.

In Pursuit of Perfection, Like Several Major Manufacturers, Navistar International Has Embraced The Six Sigma System To Reduce Defects; The Company And Proponents On The Line Hail Its Successes, But Some Workers See More Subtle Flaws. *Chicago Tribune* (April 4, 1999, 1).

International Organization for Standardization (2004), "Quality Management Principles," http://www.iso.org/iso/en/iso9000-14000/iso9000/qmp.html.

Johnston, R. (1995). The zone of tolerance: exploring the relationship between service transactions and satisfaction with the overall service, *International Journal of Service Industry Management*, 6, 2, 46–61.

Kano, N., Seraku, N., Takahashi, F., & Tsuji, S. (1990). Attractive quality and must be quality. *Quality*, 14, 2, 39.

Kaplan, R. S., & Norton, D. P. (1996), *The Balanced Scorecard*, Harvard Business School Press, Boston, MA.

Keefe, M., & Fanning, E. How the 100 Best Places to Work in IT Were Chosen. *ComputerWorld*, June 14, 2004.

Khoo, H. H. & Tan, K. C. (2003), "Managing for quality in the USA and Japan: differences between the MBNQA, DP and JQA," *The TQM Magazine*, Vol. 15, No. 1, 14–24.

Kimes, S. E., & Thompson, G. M. (Summer 2004). Restaurant revenue management at Chevys: Determining the best table mix. *Decision Sciences*, 35, 3, 371–392.

Kimes, S., Chase, R. B., Choi, S., Lee, P. Y., & Ngonzi, E. N. (1988 June). Restaurant revenue management: Applying yield management to the restaurant industry. *Cornell Hotel and Restaurant Administration Quarterly.* 32–39.

Kirkpatrick, D. L. (1998), *Evaluating Training Program: The Four Levels*, Berrett Koehler Publishers, San Francisco, CA.

Kolbe, K. (1990), *The Conative Connection: Uncovering the Link Between Who You Are and How You Perform*, Addison-Wesley Publishing Company, Inc., Reading, MA.

Lovelock, C., & Wirtz, J. (2004). *Service marketing: people, technology, strategy*, 5 th ed. Upper Saddle River, NJ: Prentice Hall.

Lowenthal, J. *Defining and analyzing a business process. A Six Sigma pocket guide.* (2003). Milwaukee, WI: ASQ Quality Press.

Magrab, E. B. *Integrated product and process design and development.* (1997). New York: CRD Press.

Mazur, G. (1993). *QFD for service industries: From voice of customer to task deployment. The fifth symposium on quality function deployment.* Novi, MI.

McAdam, R., & Canning, N. (2001). ISO in the service sector: Perceptions of small professional firms. *Managing Service Quality,* 11, 2, 80–92.

McCowen, R. A., Bowen, U., Huselid, M. A., & Becker, B. E. (1999), "Strategic Human Resource Management at Herman Miller," *Human Resource Management,* Vol. 38, No. 4, 303–308.

McKim, R. H. (1972). *Experiences in visual thinking.* Brooks-Cole.

Meister, D. H. (1993). *Managing the professional service firm.* New York: Free Press Associates, Published by Simon & Schuster.

Metters, R., King-Metters, K., & Pullman, M. (2003). *Successful Service Operations Management.* Thomson South-Western.

Mudge, A. E. (1971). *Value engineering: A systematic approach.* New York: McGraw-Hill.

Mundy, R. M., Passarella R., & Morse, J. (1986). Applying SPC in service industries. *Survey of Business,* 21, 3, 24–9.

Nelson, B. (1997), *1001 Ways to Energize Employees,* Workman Publishing, New York, NY.

Nelson, B. (1994), *1001 Ways to Reward Employees,* Workman Publishing, New York, NY.

Normann, R. (2000). *Service management: Strategy and leadership in service business.* London: John Wiley & Sons.

Osborn, A. F. (1953). *Applied Imagination.* New York: Scribners.

Parasuraman, A., Berry, L. L., & Zeithaml, V. A. (1991). Refinement and reassessment of the SERVQUAL scale. *Journal of Retailing,* 67, 4, 420–50.

Parasuraman, A., Zeithaml, V. A., & Berry, L. L. (1985). A conceptual model of service quality and its implications for future research. *Journal of Marketing,* 49, 4, 41–50.

Parasuraman, A., Zeithaml, V. A., & Berry, L. L. (1994). Reassessment of expectations as a comparison standard in measuring service quality: implications for further research. *Journal of Marketing,* 58, 1, 111–24.

Phadke, M. S. (1989). *Quality engineering using robust design.* Englewood Cliffs, NJ: Prentice Hall.

Prince, J. B. (1994), "Performance Appraisals and Reward Practices for Total Quality Organizations," *Quality Management Journal,* Vol. 1, Issue 2, 36–46.

Pugh, S. (1991). *Total design: Integrated methods for successful product engineering,* Reading, MA: Addison-Wesley.

Reichheld & Sasser, *Harvard Business Review,* Sept–Oct 1990.

Retail Payments Research Project, *A Snapshot of the U.S. Payment Landscape*—Research sponsored by the Federal Reserve System 2002.

Richard, N. (2000). *Service management: Strategy and leadership in service business* (3rd ed.) New York: John Wiley & Sons.

Rioux, S. M., Bernthal, P. R., & Wellins, R. S. (2000), "The Globalization of Human Resource Practices", HR Benchmark Group, DDI, Bridgeville, PA. http://www.ddiworld.com/pdf/CPGN56.pdf

Rosen, R., Digh, P., Singer, M., & Phillips, C. (2000), *Global Literacies: Lessons on Business Leadership and National Culture,* Simon & Schuster, New York, NY.

Rosen, L. D., & Karwan, K. R. (1994) Prioritizing the dimensions of service quality, *International Journal of Service Industry Management,* 5, 4, 39–52.

Rotondaro, R. G., & de Oliveira, C. L. (2001). *Using failure mode effects analysis (FMEA) to improve service quality.* Proceedings of the Twelfth Annual Conference of the Production and Operations Management Society, POM-2001, March 30-April 2.

Rucci, A. J., Kirn, S. P., & Quinn, R. T. (1998), "The Employee-Customer-Profit Chain at Sears," *Harvard Business Review*, Vol. 76, No. 1, 82–97, January–February.

Rust, R. T., Zahorik, A. J., & Keningham, T. L. (1995, April). Return on quality (ROQ): Making service quality financially accountable. *Journal of Marketing*, 59, 58–70.

Rust, R. T, Zahoric, A. J., & Keiningham, T. L. (1994). *Return on quality: measuring the financial impact of your company's quest for quality*. Chicago: Probus Publishing Company.

Sasser, W. E., & Reichheld, F. F. "Zero Defections: Quality Comes to Services" *Harvard Business Review*, September–October 1990.

Sauerwein, E., Bailom, F., Matzler, K., & Hinterhuber, H. H. (1996). The Kano model: How to delight your customer. *International Working Seminar on Production Economics*. Innsbruck, Igls, Austria, 313–327.

Savransky, S. D., & Stephan, C. (1993). *The methodology of inventive problem solving*. The Industrial Physicist, American Physics Society.

Schlesinger, L. A., & Heskett, J. L. (1991, Spring). Breaking the cycle of failure in services. *Sloan Management Review*, 17–28.

Schmenner, R. W. (Summer 2004). Service businesses and productivity. *Decision Sciences*, 35, 3, 333–347.

Schmenner, R. W. (1998). *Plant and service tours in operations management* (5th ed.) Upper Saddle River, NJ: Prentice Hall.

Shannon, C.E., & Weaver, W. (1949). *The mathematical theory of communication*. Urbana, IL: University of Illinois Press.

Shostack, L. (1982). Designing a service. *European Journal of Marketing*, 16, 1, 49–63.

Shostack, L. (1984). Designing services that deliver. *Harvard Business Review*, 62, 1, 133–9.

Shostack, L. (1987). Service positioning through structural changes. *Journal of Marketing*, 51, 1, 34–43.

Skinner, W. (1974). The focused factory. *Harvard Business Review*, 35, 3, 113-21.

Smith, L. R., & Sudjianto, A. (1997, May). *Principle based approach to product development, 6th Industrial Engineering Research Conference Proceedings*. Institute of Industrial Engineers.

Spencer, J. (2002, May 8). In search of a real, live operator: Firm spends billions to hide them. *The Wall Street Journal*.

Sterman, J. D. (2000). *Business dynamics: System thinking and modeling for a complex world*. New York: McGraw-Hill.

Suh, N. P. *The principles of design: Oxford series on advanced manufacturing*. (1990). New York: Oxford University Press.

Suh, N. P. *Axiomatic design: Advances and applications*. (2001). New York: Oxford University Press.

Taguchi, G. (1986). *Introduction to quality engineering: designing quality into products and processes*. Asian Productivity Organization.Tax, S. S., & Brown, S. W. (1998). Recovering and learning from service failure. *Sloan Management Review*, 49 1, 75–88.

Ulrich, K. T., & Eppinger, S. D. (1976). *Product design and development*. New York: McGraw-Hill.

U.S. Department of Labor, Bureau of Labor Statistics, Industry at a Glance—www.bls.gov.

U.S. General Accounting Office Study (1991), *Management Practices: U.S. Companies Improve Performance Through Quality Efforts* (GAO NSIAD 91-190). May.

Van Dierdonck, R., & Brandt, G. (1988). The focused factory in service industries. *International Journal of Operations and Production Management*, 8, 3, 31–8.

Vora, M. K. (2004), "Creating Employee Value in a Global Economy through Participation, Motivation, and Development," *Total Quality Management & Business Excellence*, Vol. 15, No. 5–6, 793–806, July–August.

Vora, M. K. (2003), "Global Quality Management without Boundaries," Invited Guest Editorial, Zairi, M. (Ed.), *The TQM Magazine*, Vol. 13, No. 2, 69–70, March.

Vora, M. K. (2002a), "Business Excellence through Quality Management," *Total Quality Management*, Vol. 13, No. 8, 1151–1159, December.

Vora, M. K. (2002b), "Creating Customer Value through Voice of the Customer Management," 7th World Congress for TQM, Proceedings, Vol. 2, 73–81, Verona, Italy, June 27.

Vora, M. K. (2001), Business Excellence Model. *Workforce Management Online* (2004), http://www.worksforce.com/, Vol. 2, No. 12, June 24.

Wheelwright, S. C., & Clark, K. B. (1992). *Revolutionizing product development*. New York: The Free Press.

Why Some Companies Succeed at CRM (and Many Fail). Knowledge@wharton.com.

Womack, J. P., & Jones, D. T. (1996). *Lean thinking*. New York: Simon & Schuster.

*Workforce Management* Online (2003), http://www.workforce.com/section/09/ article/23/41/94, April.

*Workforce Management* Online (2003a), http://www.workforce.com/section/09/article/23/55/58, November.

*Workforce Management* Online (2003b), http://www.workforce.com/section/09/article/23/53/40, October.

Wysocki, B., Jr. (2002, May 30). Doctors lead a crusade to replace office visits as standard procedure. WSJ Online.

Yes, Six Sigma Can Work For Financial Institutions. *ABA Banking Journal*, Sept. 2003, Vol. 95 Issue 9, 93.

Zeithmal, V. A., Berry, L. L., & Parasuraman, A. (1993). The nature and determination of customer expectations of service. *Journal of the Academy of Marketing Science*, 21, 1, 1–12.

# INDEX

NOTE: Boldface numbers indicate illustrations.

# ABOUT THE AUTHORS

**Parveen S. Goel** is a PE, Six Sigma Black Belt, and Global Chief Engineer of Six Sigma & Reliability Engineering for Steering and Suspension Systems Engineering at TRW Automotive. His Global Team's responsibilities include Six Sigma, Design for Six Sigma (DFSS), Engineering Policies, and Roadmaps. Parveen has a Ph.D. in Industrial Engineering from Wayne State University in Detroit, Michigan. He lives in Windsor, Ontario Canada.

**Praveen Gupta** is the author of the best-selling Six Sigma book *Six Sigma Business Scorecard: Creating a Comprehensive Corporate Performance Measurement System* and *The Six Sigma Performance Handbook*. Praveen participated in the development of Six Sigma at Motorola in the mid-eighties. He has taught Six Sigma at Motorola University for more than a decade. Praveen is a Six Sigma Master Black Belt and an ASQ Fellow. Praveen consults in the area of corporate performance improvement and innovation. He lives in Lisle, Illinois.

**Rajeev Jain** provides leadership in the operational effectiveness and shareholder value improvement utilizing Six Sigma methodologies. He directs efforts in corporate development and international strategy for Hewitt Associates, a global human resources outsourcing and consulting firm. Instrumental in developing the organizational design and providing intellectual leadership for many performance enhancement initiatives at Hewitt, Mr. Jain is also a CPA (international), an MBA with a specialization in strategy, and a Fellow of the Chartered Institute of Management Accountants and the Chartered Association of Certified Accountants. He lives in Hinsdale, Illinois.

**Rajesh K. Tyagi** is a visiting assistant professor at the College of Commerce, DePaul University and adjunct assistant professor at Kellogg School of Management, Northwestern University. He teaches service operations management, operations management, and quality management and decision making for managers. Rajesh received his Ph.D. in engineering from the University of Ottawa, Canada, and his MBA form the Kellogg School of Management, Northwestern University. He lives in Evanston, Illinois.

## CONTRIBUTORS

**Dr. Manu Vora** is Chairman and President of Business Excellence, Inc., a global quality management services firm located in Naperville, Illinois. Manu has over 30 years of leadership experience and has successfully guided Fortune 500 companies (AT&T Bell Laboratories and Lucent Technologies) with Malcolm Baldrige assessments in the areas of customer satisfaction, employee well-being and satisfaction, and continuous process improvement. He has authored 30 publications and has made over 120 presentations around the world in Asia Pacific, the Caribbean and Central America, Europe, Latin America, the Middle East, and the United States. Manu holds an MBA with Marketing Management, and a B.S., M.S., and Ph.D. in chemical engineering. As an Adjunct Professor, he teaches "Quality Management" in the MBA Program at the Stuart Graduate School of Business at Illinois Institute of Technology in Chicago. Dr. Vora has received numerous awards such as American Society for Quality (ASQ) "2001 Grant Medal", Testimonial Awards, and ASQ Chicago Section's Joe Lisy, and First Founder's Award.

**Arvin Srivastava** has about 15 years of business process improvement experience in several industries. He holds a BSME degree from IIT Roorkee, and an MBA degree from Benedictine University, Illinois. He has worked with suppliers in Malaysia, Singapore, and India. Arvin has helped QTC clients in Six Sigma projects, Green Belt training, and project management. Arvin has previously contributed chapters in *The Six Sigma Performance Handbook.*

**Om Prakash Yadav** is an assistant professor at North Dakota State University, Fargo. He has a Ph.D. in Industrial Engineering from Wayne State University, Detroit. He has over 15 years of academic and industrial experience. Om Prakash has published and presented more than 15 research papers in international journals and conferences.